# JAPANESE TECHNOLOGY ASSESSMENT

## Computer Science, Opto- and Microelectronics, Mechatronics, Biotechnology

by

J. Albus, R. Bauer, J. Bentley, T. Binford, M. Brady,
D. Brandin, F. Capasso, N. Caplan, D. Collins, C. Cooney,
T. Gannon, M. Harrison, K. Hess, D. Jackson, H. Kroger,
M. Kutcher, P.J. MacVicar-Whelan, G.L. Miller, J. Nevins,
D. Oxender, J. Riganati, F. Ris, L. Rossol, G. Sato, R. Scace,
K. Schultz, N. Sondheimer, W. Spicer, W.-T. Tsang,
R. Wickner, H. Wieder, J. Wilson, J. Woodall

for

Science Applications International Corporation
La Jolla, California

NOYES DATA CORPORATION
Park Ridge, New Jersey, U.S.A.
1986

Library of Congress Catalog Card Number 86-17995
ISBN: 0-8155-1096-9
Printed in the United States

Published in the United States of America by
Noyes Data Corporation
Mill Road, Park Ridge, New Jersey 07656

10 9 8 7 6 5 4 3 2 1

Library of Congress Cataloging-in-Publication Data

Japanese technology assessment.

Studies originally prepared by Science Applications International Corporation for the U.S. Dept. of Commerce, 1984–1985.
Includes bibliographies and index.
1. High technology--Japan. I. Albus, James Sacra, II. Science Applications International Corporation.
T27.J3J39 1986 338.4'76'0952 86-17995
ISBN 0-8155-1096-9

# JAPANESE TECHNOLOGY ASSESSMENT

# Foreword

This series of studies on Japanese technology assesses Japanese research and development in four of today's most visible high-technology areas: computer science, opto- and microelectronics, mechatronics (a term created by the Japanese to describe the union of mechanical and electronic engineering necessary to produce the next generation of machines, robots, and the like), and biotechnology.

The evaluations were conducted by panels of leading U.S. scientists—chosen from academia, government, and industry—who are actively involved in research in their areas of expertise. The studies were prepared for the purpose of aiding the U.S. response to Japan's technological challenge.

The main focus of the assessments is on the current status and long-term direction and emphasis of Japanese research and development. Other important aspects covered include evaluation of the state of the art; identification of key Japanese researchers, R&D organizations, and resources; and comparative U.S. efforts. The general time frame of the studies corresponds to future industrial applications and potential commercial impacts spanning approximately the next two decades. This comprehensive document should help to encourage new programs and incentives to meet the Japanese challenge.

The information in the book is from:

> *JTECH Panel Report on Computer Science in Japan,* prepared by D. Brandin, J. Bentley, T. Gannon, M. Harrison, J. Riganati, F. Ris, and N. Sondheimer. The study was prepared for Science Applications International Corporation under contract to the U.S. Department of Commerce, December 1984.

*JTECH Panel Report on Opto- and Microelectronics,* prepared by H. Wieder, W. Spicer, R. Bauer, F. Capasso, D. Collins, K. Hess, H. Kroger, R. Scace, W.-T. Tsang, and J. Woodall. The study was prepared for Science Applications International Corporation under contract to the U.S. Department of Commerce, May 1985.

*JTECH Panel Report on Mechatronics in Japan,* prepared by J. Nevins, J. Albus, T. Binford, M. Brady, N. Caplan, M. Kutcher, P.J. MacVicar-Whelan, G.L. Miller, L. Rossol, and K. Schultz. The study was prepared for Science Applications International Corporation under contract to the U.S. Department of Commerce, March 1985.

*JTECH Panel Report on Biotechnology in Japan,* prepared by D. Oxender, C. Cooney, D. Jackson, G. Sato, R. Wickner, and J. Wilson. The study was prepared for Science Applications International Corporation under contract to the U.S. Department of Commerce, June 1985.

The table of contents is organized in such a way as to serve as a subject index and provides easy access to the information contained in the book.

Advanced composition and production methods developed by Noyes Data Corporation are employed to bring this durably bound book to you in a minimum of time. Special techniques are used to close the gap between "manuscript" and "completed book." In order to keep the price of the book to a reasonable level, it has been partially reproduced by photo-offset directly from the original reports and the cost saving passed on to the reader. Due to this method of publishing, certain portions of the book may be less legible than desired.

## NOTICE

# Japanese Technology Evaluation Program

### Purpose

The Japanese Technology Evaluation Program (JTECH) is operated for the Federal Government by Science Applications International Corporation (SAIC) to provide definitive technical assessments of emerging Japanese thrusts in selected high-technology areas. JTECH is a pilot program, initiated in September 1983 by the Department of Commerce with support also provided by other government agencies. The objectives of this pilot program are to develop and refine the appropriate assessment methodology, to demonstrate the effectiveness of the approach by evaluating Japanese advanced technology in several selected technical areas, and to evaluate the assessment experience and provide recommendations regarding the Japanese science and technology literature sources and their utility. The evaluations are intended to serve as a sound technical basis for allowing proper responses to the Japanese challenge, particularly with respect to providing technical input for potential users charged with making technology forecasts and competitive assessments, determining viable U.S. industry responses, and establishing directions for U.S. research and trade policies.

### Approach

The assessments are performed by panels of U.S. technical experts, with about six panel members for each subject area. Panel members are selected by the following criteria: leading authority in the field, technically active with recent "hands-on" experience, and knowledge of Japanese and U.S. research programs. Each panel member devotes approximately one month to literature review, assessments, and report writing on a part-time basis over a period of nominally six months. To balance perspectives, members for each panel are selected from industry, academia, and government.

### Assessments

The main forcus of the assessments is on the current status and long-term direction and emphasis of Japanese research and development efforts. Other important aspects include the background evolution of the state-of-the-art; key Japanese researchers, R&D organizations, and resources; and comparative U.S. efforts; The general time frame of the R&D forecasts is up to approximately ten years, corresponding to future industrial applications and potential commercial impacts somewhat farther in time, say 5 to 20 years. Subtopics for consideration within a technical area are selected primarily on the basis of expected commercial impact of the Japanese R&D on U.S. industry.

**Literature**

SAIC provides Japanese S&T literature and translation services to the panelists using a selective and iterative approach. Initially, general source material, mostly in English, is provided together with Japanese R&D literature sources of information, for panel members' review. Panel members identify from translated titles of papers and reports information of interest in their area, selected abstracts are then translated (often by telephone), and then full papers are translated as required.

In addition, special efforts are made under the JTECH approach to provide panelists with timely semi-open source material, such as informal proceedings from *ad hoc* seminars and conferences in the Japanese high-tech community, results from recent technical committee meetings on national Japanese R&D projects, and from contacts at R&D centers in Japanese high-tech industries. Literature requests and dissemination take place on a continued, interactive basis throughout the course of the panel assessments.

**Staff**

The SAIC JTECH program staff members help select the topics to be assessed, identify and enlist qualified experts as panelists, organize and coordinate panel activities, provide literature/translation support, guide and assist in the preparation of panel reports, and provide general administrative support.

Dr. Tony W. Armstrong
Director of JTECH
Science Applications International Corporation
1200 Prospect Street
La Jolla, CA 92037

Debra J. Allen
Administrative Assistant to the Director
Science Applications International Corporation
1200 Prospect Street
La Jolla, CA 92037

Dr. George Gamoto
(SAIC Sr. Consultant)
Science Coordinator of JTECH
The Institute of Science and Technology
University of Michigan
Ann Arbor, MI 48109

Dr. Young B. Kim
Assoc. Dir. of JTECH for Literature
Advanced Materials Technology, and
University of Southern California
1449 Carmelina Ave.
Los Angeles, CA 90025

# Contents and Subject Index

## PART I
## COMPUTER SCIENCE

# PART III
# MECHATRONICS

## PART IV
## BIOTECHNOLOGY

# Part I

# Computer Science

The information in Part I is from *JTECH Panel Report on Computer Science in Japan,* prepared by D. Brandin, J. Bentley, T. Gannon, M. Harrison, J. Riganati, F. Ris, and N. Sondheimer. The study was prepared for Science Applications International Corporation under contract to the U.S. Department of Commerce, December 1984.

# Acknowledgements

The Japanese Technology Evaluation Program (JTECH) is indebted to the panel members for their efforts in completing the assessment in the short time allotted while continuing to meet other commitments and pursuing ongoing research interests.

We appreciate the contributions made by Mr. George Mu from the Department of Commerce, International Trade Administration, Office of Japan. His help and guidance as the DoC Contracting Officer was invaluable. We express our gratitude to the National Science Foundation for their financial support for this study.

The panel acknowledges the formal and informal support from the U.S. Embassy staff in Tokyo, the Office of Naval Research, and the National Academy of Science/National Research Council Panel on International Developments in Microelectronics and Computer Science.

# About the Authors

## David H. Brandin

Mr. Brandin holds a B.S. in Mathematics from the Illinois Institute of Technology and is presently the Vice President and Director of the Information Services and Systems Division at SRI International. Since joining SRI in 1972 he has also served as Vice President of the Computer Science Division. His technical background is in computational analysis, simulation and communications. He is the immediate Past President of the Association for Computing Machinery, a past Director of the American Federation of Information Processing Societies, and a member of the National Academy of Sciences - National Research Council Panel on International Developments in Microelectronics and Computer Science.

## Jon L. Bentley

Jon Louis Bentley received his B.S. degree in Mathematical Sciences from Stanford University in 1974 and his M.S. and Ph.D. degrees in Computer Science from the University of North Carolina at Chapel Hill in 1976. Dr. Bentley has served as a consultant to Bell Laboratories, IBM Corporation, Stanford Linear Accelerator Center, Xerox Palo Alto Research Center, and various other companies. He is currently an Associate Professor of Computer Science and Mathematics at Carnegie-Mellon University. He has been a California State Scholar, University of North Carolina Graduate School Fellow, and National Science Foundation Fellow. His primary research interests are the design, analysis, and application of computer algorithms and the mathematical foundations of computation. Other research areas in which he has worked include software engineering tools and novel computer architectures.

## Thomas F. Gannon

Thomas F. Gannon received his Ph.D. from Stevens Institute of Technology in Electrical Engineering and Computer Science. He is currently the Technical Director of Technology Development Programs for Digital Equipment Corporation, focussing on artificial intelligence, database modeling and architecture, human interface systems, parallel processing architectures, and software engineering methodologies. He recently managed a multi-corporate team on the development of an advanced computer architecture program for the Microelectronics and Computer Technology Corporation (MCC). Dr. Gannon is a reviewer for several professional journals and is presently writing a text book on fault tolerant software systems design.

## Michael A. Harrison

Michael A. Harrison received his B.S. and M.S. degrees in Electrical Engineering from Case Institute of Technology in 1958 and 1959, respectively, and the Ph.D. degree from the University of Michigan in Ann Arbor in 1963. He has been with the University of California at Berkeley from 1963 and a Professor of Computer Science since 1971. He has been a Guggenheim Fellow and spent sabbaticals at MIT, Stanford University, University of Frankfurt, and Hebrew University in Jerusalem. He is an editor of Discrete Mathematics, Discrete Applied Mathematics, Future Generation Computer Systems, Information

Processing Letters, Theoretical Computer Science, Journal of Computer and System Science. He has served for several years on the Computer Science and Technology Board of the National Academy of Science, and is the Chairman of a Panel of the National Research Council on International Developments in Computer Science. He has written five books and approximately one hundred technical publications. His current interest is in software systems.

**John P. Riganati**

Since 1979 Dr. Riganati has been Division Chief at the Institute for Computer Sciences and Technology (ICST) of the National Bureau of Standards (NBS). Dr. Riganati's Computer Systems Components Division addresses computer data communication, advanced computer architectures, data interchange media and mechanisms, and standards for large and small scale computer interfaces. Dr. Riganati has written over 37 papers, reports and book chapters and holds seven patents. He currently represents the Computer Society on the IEEE Standards Board, serves on NBS's Research Advisory and Post-Doctoral Qualifications Committees, and serves on IEEE's Committees for Communication and Information Policy and for Supercomputers. Since October 1981, when he was a panelist at the Fifth Generation Computer Systems Conference in Tokyo, he has focused on technical and strategic developments in Japan and has been active in formulating appropriate U.S. responses.

**Frederic N. Ris**

Dr. Ris received a B.A. from Harvard College in Chemistry and Physics and a D.Phil. in Mathematics from Oxford University, which he attended as a Rhodes Scholar and a National Science Foundation graduate fellow. He is a consultant to the manager of Symbolic and Numeric Computation in the Computer Sciences Department at the IBM Thomas J. Watson Research Center in Yorktown Heights, N.Y. His responsibilities include following the Japanese Fifth Generation and Supercomputer national projects and collateral efforts around the world. He is currently recording secretary to the IEEE standards committee for format-free, floating-point arithmetic.

**Norman K. Sondheimer**

Norman K. Sondheimer received his Ph.D. in Computer Sciences from the University of Wisconsin-Madison in 1975. He spent 1975-1978 teaching at the Ohio State University. From 1978 to 1982, he was a research scientist at Sperry Univac. Since that time, he has been on the staff of the University of Southern California's Information Sciences Institute. Dr. Sondheimer's research has been on natural language processing and man-machine interaction. In 1981, he served as president of the Association for Computational Linguistics.

# Executive Summary

The purpose of the JTECH Panel on Computer Science was to examine a narrow domain of computer science in Japan and draw comparisons with similar technologies in the United States.

Table 1 lists the topics selected by the Panel. Note that each topic has not received equal attention. The rationale for topic selection and the attention given to each topic is covered in the text.

General Conclusions

The Panel chose to compare American and Japanese technology in three areas: basic research, advanced development, and product engineering. For this report, basic research means theoretical or laboratory-based inquiries; advanced development means the development of prototypes (hardware or software); and product engineering is the ability to manufacture and/or ship products.

Our overall conclusions are:

- Research - Japan is far behind the United States and slipping further behind.
- Advanced Development - Japan is behind the United States and holding that relative position
- Product Engineering - Japan is comparable with the United States and beginning to pull away.

Table 2 summarizes the panel's assessments in each major category. These assessments vary according to the individual components as discussed in detail in the text. While no technological break-throughs were identified, the panel was surprised by the emphasis the Japanese placed on machine translation and very high bandwidth transmission/optical computing.

Table 1 Topics Selected by the Panel

- Software:
  - Software Engineering
  - Operating Systems
  - Applications Software (Packages)
  - Languages
  - Data Base

- Artificial Intelligence and Man-Machine Interface:
  - Japanese Character Processing
  - Speech Recognition
  - Machine Translation (of natural languages)
  - Expert Systems and MITI's Fifth Generation Computer Systems (FGCS)
  - Language, Processors, and Tools (for AI R&D)
  - Natural Language Understanding

- Processor Architecture and Computer Organization:
  - Parallel Processors
  - Supercomputers (hardware and software)
  - Workstations
  - Clones (i.e., IBM compatible)

- Communications:
  - Local Area/Value Added Networks
  - Hardware
  - Protocols/Software
  - Facsimile/Office Automation

TABLE 2 ASSESSMENT BY CATEGORY

| CATEGORY | BASIC RESEARCH | ADVANCED DEVELOPMENT | PRODUCT ENGINEERING |
|---|---|---|---|
| SOFTWARE | Far behind the U.S.<br>Falling further back | Behind in development of prototype software<br>Falling further back | Ahead of the U.S. in the development of applications code<br>Holding |
| ARTIFICIAL INTELLIGENCE & MAN-MACHINE INTERFACE | Far behind the U.S. and slipping | Behind and holding | Behind and holding |
| PROCESSOR ARCHITECTURE & COMPUTER ORGANIZATION | Far behind the U.S. and slipping | On par with the U.S. but catching up | Ahead of the U.S. and improving |
| COMMUNICATIONS | Far behind the U.S. and slipping | Hardware — even and holding<br>Software — behind and holding | Ahead and pulling away<br>Behind but improving |

The U.S. is expected to dominate basic computer science research over the next decade because of its present base and momentum. The Japanese are aware of their deficiencies. While not presently well-structured to perform basic research, they have begun to address this issue. Also, the market share of U.S. companies impacts the profits available for maintaining the present U.S. research base. The U.S. cannot take the position that its edge in basic research can be maintained indefinitely.

Commercial Assertions

The panel's emphasis was technical rather than commercial. However, some commercial implications are covered. U.S. research investments are sufficiently greater than Japan's in most categories indicating that immediate commercial upsets are not expected. An exception could be in supercomputers where Japan exceeds the U.S. in both government funding and the amount of commercial interest.

Recommendations

New incentives that encourage basic researchers to stay with their ideas through advanced development and product engineering are needed to reduce the technology-transfer time within the U.S. Adapting the Japanese practice of requiring a prototype for government-funded research would force movement in the direction of advanced development. Greater incentives for universities and industry to cooperate would be useful along with programs that spur cooperation between engineers and scientists. We would also recommend placing greater emphasis in the training of scientists and engineers in the U.S. universities. More cooperative commercial ventures and joint research efforts such as the Microelectronics and Computer Corporation (MCC), along with the necessary relaxation of government regulations, are recommended.

Literature Aspects

The panel was asked to document their experience related to the utility of the Japanese literature and information sources provided during the study. The purpose was to identify important sources in the computer science area and to aid in improving the methodology and effectiveness for future panel assessments. These comments and recommendations are contained in this report. A discussion of the JTECH literature support approach and a compilation of the materials, sources, and translations provided to the panel is contained in the appendix.

An example of useful sources uncovered was the existence of semi-public Japanese Working Groups in computer science and related disciplines (similar to the DARPA Principal Investigator meetings). It would be useful to obtain information from such meetings in a timely manner.

# 1. Introduction

This is the final report of the JTECH Panel on Computer Science prepared for the Department of Commerce under contract to Science Applications International Corporation.

The panel, under the chairmanship of David H. Brandin, SRI International, was composed of seven computer technologists drawn from government, industry, and academia. Each member of the panel accepted responsibility for a subspecialty of computer science. The panel was chartered to examine those portions of Japanese computer science which the panel felt were important R&D areas and were within the panel's available resources.

## 1.1 Methodology

After selecting subtopics, the panel selected sources of information, reference material, and identified documents for translation. Dr Young Kim, JTECH staff, provided initial source materials, containing translations of a variety of Japanese documents. Dr. Kim acquired literature and provided translation services throughout the study. The panel worked closely with the National Academy of Science/National Research Council Panel on International Developments in Microelectronics and Computer Science (through a panel overlap of the corresponding Chairmen). This NAS/NRC panel made two visits to Japan during the JTECH panel's deliberations which provided important data.

Documents were circulated to all panelists along with important references and translations of the Japanese table of contents of key journals. Panelists identified those titles, usually isolated chapters, that they felt would be appropriate to translate. After translation, these documents were circulated as well. A compilation of all source material examined is contained in the appendix.

The panel then prepared draft reports for their given areas and identified missing data. An additional NAS visit to Japan was utilized to obtain some of these data. Working group meetings were held in January, March, May, and July 1984.

Individual draft reports were reviewed at a working group meeting in May 1984, and a draft consensus was prepared. After report integration, another working group meeting was held in July 1984 at the National Computer Conference in Las Vegas.

## 1.2 Subtopic Selection

The panel considered subtopics within computer science which emphasized long-term Japanese R&D efforts up to 10 years. The primary criterion for selection being the expected commercial impact of Japanese R&D on U.S. industry. To some extent, this choice of subtopics was dictated by the panel's expertise and the limited time and resources available.

The depth to which each subtopic was explored was a function of the panelist's experience and the availability of relevant Japanese data. We have described some U.S. efforts in the report only when we felt it was important to the understanding of our assessments (e.g., artificial intelligence and architecture).

Subtopics which the panel considered important but which were not covered in the study are given in Table 3. The panel ignored microelectronics because both a separate JTECH panel and a National Academy of Science panel are studying this topic. Also, a separate JTECH panel will address robotics. The remainder of the subtopics were not considered due to the factors mentioned above (lack of particular expertise on panel, limited time and resources, etc.).

Table 3. Topics Not Covered By This Study

CAD/CAM (Mechanical or Electrical)
Microelectrics (Hardware, VLSI)
Graphics
Vision, Image Understanding
Robotics
Personal Computers

1.3 Drawing Comparisons

There are significant differences between the political and social systems of the United States and Japan. These differences cause some uncertainty in the conclusions. Fortunately, we are concerned here with comparative positions and trends rather than absolute findings.

It is useful, however, to discuss the differences to help understand why their emphasis may or may not be placed in a particular technology.

1.3.1 National Differences

Generally, a technology evolves from basic research through advanced development, into product design and development, and ultimately into the marketplace. Figure 1* illustrates the distribution of cost across technology development phases. In Japan, the Ministry of International Trade and

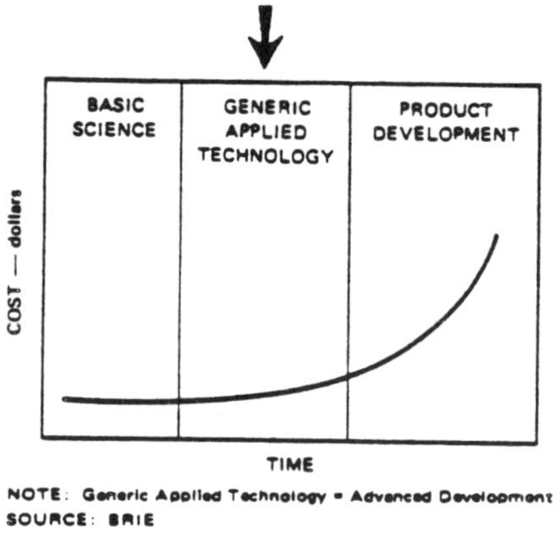


FIGURE 1 MITI RESEARCH EMPHASIS

---

* (These representations are drawn from Advanced Computing: The Commercial Impact of Japanese Programs on U.S. Competitiveness, a Berkeley Roundtable on the International Economy Seminar, University of California at Berkeley, June 27, 1984).

Industry (MITI) places its research support on advanced development (which leads to the development of considerable proprietary innovations). This is shown in Figure 1, where the arrow indicates the research emphasis.

The U.S. government emphasis is illustrated in Figure 2. The U.S. government gives little support to the development of corporate proprietary interests. The major consideration is public domain development of basic and applied technology and defense.

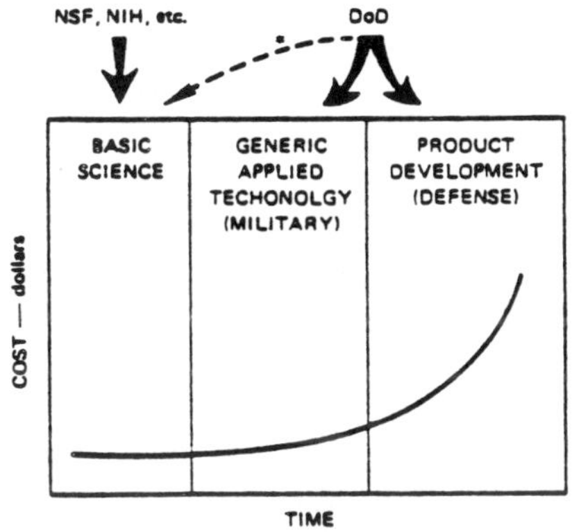


*NAS/NRC PANEL ON MICROELECTRONICS: DoD SUPPORT IN COMPUTER SCIENCE BASIC RESEARCH IS ACTUALLY *GREATER* THAN NSF SUPPORT.

SOURCE: BRIE

FIGURE 2 U.S. GOVERNMENT RESEARCH POLICY EMPHASIS

One result of this difference in emphasis is a Japanese weakness (and U.S. strength) in basic research, but a Japanese strength (and U.S. weakness) in advanced development.

### 1.3.2 Financial Differences

Japanese financing of research and development is different from U.S. practice, and this leads to some uncertainties in comparing program size, cost, return on equity, return on capital, and other company positions. Japanese saving rates are dramatically higher than the U.S. rate, Japan's interest rates are lower, Japanese businesses are willing to trade profits for market share and higher employment, and professional labor rates are lower in Japan. Cost of capital is considerably lower* and, therefore, in the capital-intense research domain of computer technology, Japanese investments will buy more than an equivalent U.S. investment.

Japanese accounting for overhead expense is at least as variable as in the United States. Confusing numbers, language problems, the natural propensity to minimize ones costs and maximize their output, and so forth, led to uncertainty in the panel's determination of research overhead costs. For example, we were told that the MITI Institute for New Computer Technology (ICOT) overhead was paid by the participating companies, but we were unable to determine whether that comprised part of the company's annual budget. Similarly, variations and inconsistencies in such data as percent of sales committed to R&D, head count, and annual research budget made it difficult to assess the overhead expense. (Hitachi Central Research Labs and NEC were studied.)

There are other financial differences as well. Japanese companies that participate in the national research programs (funded by MITI or other agencies) reportedly** invest 2-to-3 dollars for every dollar supplied by the government. This is not the case in the United States. Therefore, a $25 million dollar MITI program in Japan may actually be equivalent to a $100 million U.S. program.

Finally, salary differentials exist across the board. Entry level computer scientists in Japan draw the same salary as entry level secretaries, about $700/month with benefits. At age 30 their salary is about $30,000/year. Tax rates between United States and Japan are not comparable until the salary range reaches about $40,000/year. These salaries are substantially lower than the corresponding U.S. salaries.

---

* There was some controversy at the BRIE seminar on June 27, 1984 as to "how much lower," but there was unanimous agreement that it was lower.

** This was a uniformly consistent comment from all of the organizations we visited.

### 1.3.3 Education and Training Differences

There are fundamental differences between Japanese and American emphasis on education and training. For example, there is little basic research in Japanese universities compared to that in the United States. Japanese firms provide perhaps three years of intense industrial training to entry level engineers before they are even asked to pick a specialization, whereas U.S. entry level engineers are expected to plunge into the fray. The panel's assessment was that this also contributed to a U.S. edge in basic research. Section 3.5.1 contains a more thorough analysis of university programs in Japan.

### 1.3.4 Some Social Differences

Other system differences between the United States and Japan also complicate the comparisons in technology. The Japanese social system rewards and encourages people to be technicians. They operate in less creative production modes with more concern for reliability. This leads to much lower error rates in production software than found in the United States. The work force is more loyal which makes training investments more rational and worthwhile.

In the precompetitive phase, Japan is further along the production cycle than in most U.S. cooperative efforts. For example, it is common Japanese practice to require a prototype in cooperative national research projects.

There are other differences in such areas as industrial policy, anti-trust laws, the syndicated Zaibatsu (system of cross-owned companies), trading companies, and basic cultural differences. The findings of the panel should be viewed in the light of these comments.

## 1.4 Organization of this Report

Section 2 contains the overall findings and assessments including a summary of the problems associated with collecting data.

Section 3 contains more detailed assessments by topic. There are also descriptions and analyses of related subjects such as the training of computer science personnel.

Section 4 contains comments related to commercial observations and general recommendations.

Section 5 contains references which are indexed to specific parts of section 3. Unnumbered references pertain to the subsection in general.

Recommendations related to the acquisition of Japanese scientific and technical information are contained in the Appendix.

# 2. Summary Findings and Assessments

This section presents a summary of technical assessments along with comments related to data collection.

## 2.1 General Findings

The panel made comparisons in basic research, advanced development, and product engineering. These areas are defined as follows:

Basic Research: Theoretical or laboratory-based inquiry. In software it refers to exploring new concepts and the creation of innovative software.

Advanced Development: The development of hardware or software prototypes.

Product Engineering: The ability to manufacture and/or ship products (hardware) and to develop production codes (software).

Figure 3 is an explanation of our coding system. The comparisons represent the state, or rate of change, of Japan's computer science as compared to the United States.

| ABSOLUTE COMPARISON | RATE OF CHANGE IN ABSOLUTE POSITION |
|---|---|
| < FAR BEHIND | ↑ PULLING AWAY OR CATCHING UP QUICKLY |
| — BEHIND | ↗ GAINING GROUND OR CATCHING UP |
| O EVEN | → HOLDING POSITION |
| + AHEAD | ↘ LOSING GROUND |
| > FAR AHEAD | ↓ SLIPPING QUICKLY |

FIGURE 3 CODING SYSTEM FOR COMPARING THE STATE OF JAPANESE COMPUTER TECHNOLOGY WITH U.S. COMPUTER TECHNOLOGY

Figure 4 illustrates the panel's overall assessment. This assessment is carried to more detailed levels later in the report. The reasons for each assessment in Figure 5 are discussed in Detailed Findings (2.2) and Assessments by Topic (3).

| | BASIC RESEARCH | ADVANCED DEVELOPMENT | PRODUCT ENGINEERING |
|---|---|---|---|
| OVERALL ASSESSMENT | < ↘ FAR BEHIND AND LOSING GROUND | − → BEHIND AND HOLDING | ○ ↗ EVEN BUT GAINING GROUND |

FIGURE 4 OVERALL ASSESSMENT OF JAPANESE COMPUTER TECHNOLOGY

| | BASIC RESEARCH | ADVANCED DEVELOPMENT | PRODUCT ENGINEERING |
|---|---|---|---|
| SOFTWARE | < ↓ | − ↘ NOTE 1 | + ↑ NOTE 2 |
| ARTIFICIAL INTELLIGENCE | < ↘ | − → | − → |
| PROCESSOR ARCHITECTURE & COMPUTER ORGANIZATION | < ↘ | ○ ↗ | + ↗ |
| COMMUNICATIONS | < ↘ | NOTE 3 ○ → | + ↗ |
| | | NOTE 4 − → | < ↘ |

NOTES.

1. Refers to development of software using state-of-the-art concepts.
2. Development of production code; tailored, low-error-rate applications code.
3. Hardware
4. Software, in particular, protocols and systems software.

FIGURE 5 SUMMARY ASSESSMENT BY TOPIC

## 2.2 Detailed Findings

The next level of discussion was divided into the following four topics: software, artificial intelligence, processor architecture and computer organization, and communications. A summary assessment for each of these topics appears in Figure 5. Each topic is discussed in Section 3.

There is little fundamental research in software, but there is growing interest. New investments suggest that Japan will probably hold their relative position. The lack of emphasis in fundamental research leads to software problems in advanced development. However, the panel felt that Japan's ability to deliver older forms of IBM-compatible, highly reliable software was far superior to the U.S. and increasing. Further, there is data that suggests the Japanese are writing superior compilers and utilities.

The Japanese are placing considerable emphasis on certain aspects of Japanese language processing. They are particularly interested in the man-machine interface, Kanji character recognition, and speech recognition. We could not discern a significant level of basic research in the artificial intelligence spectrum. The Institute for New Generation Computer Technology (ICOT) is the fledgling medium for this research. The Japanese have accomplished much in a short period in their Fifth Generation Computer System effort.

Japan's efforts in processor architecture, despite serious attention, cannot match the magnitude of basic research activities in the U.S. However, the Japanese emphasis on prototypes is an advantage in advanced development. Similarly, their ability to ship IBM-clone hardware and new devices gives them an advantage over the U.S. With respect to supercomputers, we concluded that there was considerably more prototyping in the U.S. than Japan despite some impressive benchmark performance. However, the Japanese are committed in this arena, and they are closing the prototyping gap. U.S. vendors currently dominate the supercomputer market, but superior Japanese software for supercomputers, such as vectorizing FORTRAN compilers, is increasing the competition.

Japan has only recently taken interest in protocal and communication software research which is the reason for the low ranking in communications basic research. Japan's lead in areas such as fiber optics caused us to rank advanced development and product engineering in communications hardware as superior to the U.S.

## 2.3 Findings Related to Japanese S&T Literature

### 2.3.1 Conclusions and Recommendations

The Panel was requested to record its experience related to the acquisition of Japanese literature and data. Conclusions and recommendations are given below.

- Japanese Working Group meetings are used to disseminate recent research information. This type of information is important in assessing Japanese R&D. A means of monitoring and disseminating such semipublic Japanese R&D information on a continuing basis is needed.

- The language imbalance should be addressed. The study of the Japanese language in high schools and universities especially for students studying computer science and related disciplines should be encouraged.

- Translation services should be expanded, but with high selectivity by technically knowledgeable people.

- Expand U.S.-Japanese scientific interchange by small, delegations with appropriate credentials.

This first JTECH Panel experienced some of the common difficulties in obtaining Japanese technical information. It is hoped that efforts in this area will assist future studies.

### 2.3.2 Recommendations related to the Acquisition of Japanese S&T Data

## The Language

There is a need to encourage the study of Japanese in our high schools and universities especially with students in computer science and related disciplines. Programs such as those at MIT, NC, and NYU are steps in this direction. Financial incentives for these programs are needed.

## Working Groups

Most members of the Panel were surprised to learn the existence of semipublic Japanese Working Groups similar to DARPA Principal Investigator meetings. The meetings are occasionally advertised, and it would be useful to obtain information from such meetings in a timely manner. This requires Japanese-speaking computer scientists if any improvement is going to be made in the rate at which the U.S. learns of Japanese accomplishments. English-speaking conferences occur infrequently and there are delays in publishing which render the English publications less timely and less valuable.

## Translation

Translation services are essential, and we must accept the costs.

## U.S. Government in Japan

Personnel knowledgeable in computer technology are needed on U.S. government agency staffs in Japan. These people are required for timely, on-scene assessments and to establish liaison between Japanese and American scientists.

## Delegations

Detailed planning is particularly important for professional visits to Japan. Planning requires time and effort, but pays dividends in attaining predetermined goals.

Considerable attention must be given to the scope of the visit which is a function of the duration of the stay and the size of the delegation. Generally, time is limited which means developing realistic itineraries and specific agendas. Large delegations should be divided into small working committees which are more effective. This allows particular areas of interest to be investigated in a less formal atmosphere, and it helps prevent the dilution of expertise often experienced in the "guided tour" treatment of large groups.

It is recommended that information concerning Japanese culture and the current political climate be included in the planning stage.

# 3. Assessments By Topic

## 3.1 Software

The panel identified the following software categories for assessment:

- Software engineering
- Operating systems
- Applications software/packages
- Programming languages
- Data base software

The overall software conclusion is shown in Figure 6, and conclusions for each of the above software categories are summarized in Figure 7.

The following summary comments apply to the conclusions presented in Figure 7:

- Software Engineering
  - Basic research: No basic research is being done at all.
  - Advanced development: Western ideas are being evaluated and tested. Constant improvements are being made.
  - Product engineering: Attention to detail and customized guaranteed products with lower error rates than seen in the West.

- Operating Systems
  - Basic research: None.
  - Advanced development: None.
  - Product engineering: Clones of Western systems often outperform the originals. The Japanese are highly competent at performance evaluation and system tuning.

- Applications Software Packages
  - Basic research: Little basic research

| | BASIC RESEARCH | ADVANCED DEVELOPMENT | PRODUCT ENGINEERING |
|---|---|---|---|
| SOFTWARE | < ↓ | — ↘ NOTE 1 | + → NOTE 2 |

NOTES:
1. Refers to development of prototype software systems.
2. High-reliability, IBM-compatible, tailored applications code.

FIGURE 6 SOFTWARE CONCLUSION (SUMMARY OF FIGURE 7)

| | BASIC RESEARCH | ADVANCED DEVELOPMENT | PRODUCT ENGINEERING |
|---|---|---|---|
| SOFTWARE ENGINEERING | < ↘ | — ↗ | > ↑ |
| OPERATING SYSTEMS | < ↓ | < ↓ | + → |
| APPLICATIONS SOFTWARE PACKAGES | < ↓ | — ↘ | UNKNOWN |
| LANGUAGES | < ↓ | — ↘ NOTE 1 | > → |
| DATA BASE SYSTEMS | < ↓ | < ↘ NOTE 2 | < ↓ |

FIGURE 7 SOFTWARE DETAIL

NOTES:
1. Refers to work in vectorizing compilers.
2. Work on the ICOT DELTA machine was considered in this evaluation.

- Advanced development: This area is mixed. Some customized software is excellent. At the same time, the Japanese are purchasing large quantities of foreign software. (See MITI White Paper, 1983/1984 edition).

- Languages
  - Basic research: None.
  - Advanced development: Mainly FGCS
  - Product development: The Japanese are engineering first-rate compilers.

- Database Systems
  - Basic research: None.
  - Advanced development: Only the Delta machine at ICOT, and early reports are inconclusive.
  - Product engineering: There are some IBM clones, but we have been unable to evaluate them.

### 3.1.1 Software Engineering

Software engineering is concerned with the construction of multi-person, multiversion, multiyear programs. We could discern no areas of great Japanese strength in software engineering research. On the other hand, many Japanese companies seem to have done an excellent job of translating research results to practice in their "software factories," and most of the large computer companies have a software factory of one form or another. One of the more visible efforts is the SWB (Software WorkBench) factory of Toshiba Fuchu works under the direction of Yoshihiro Matsumoto.

This factory employs 2000 technical people, many of whom work at individual workstations consisting of a terminal and a small printer in one of two large, single-room buildings. The factory delivers 4,000,000 lines, or 2000 lines per programmer, of equivalent assembly language code per month. This output per programmer is considered exceptionally high in the U.S.

The factory specializes in process-control software for systems such as nuclear power plants, steel mills, and flight guidance. Specialization allows them to achieve a reuse rate of roughly 65%. That is, of 3000 lines of delivered code, 1000 lines are new and 2000 lines are reused. The tasks they are solving are quite sophisticated. A typical process-control system might be built around two large mainframe processors, six minicomputers, and a dozen microcomputers. Their codes are of extremely high quality. They report an error rate of 0.3 bugs per thousand lines of code (typical U.S. rates are a factor of ten greater). This explains the principal reason we gave them such high marks in software product engineering. Much of their software comes with a ten-year warranty. They will debug any software product in that period at no charge to the customer.

Producing software in a specialized factory-like environment seems to contribute to the high quality output and to the high levels of code reuse. However, such regimented working conditions would probably not be as acceptable to American workers.

The idea of software reusability has been an important topic in software engineering research for some five years, and a workshop was held in Newport, Rhode Island in late 1983 to give the effort some direction. Unfortunately, most of the U.S. and European papers in the proceedings and presentations were rather vague and philosophical. However, the Toshiba speaker described the SWB system and related some of the quantities described above. Many of the workshop participants were amazed at the progress represented by the SWB factory. We would not be at all surprised to see this phenomenon repeated in the next few years. Applying old research to new problems, Japanese applied software engineers may find themselves ahead of their Western colleagues—a familiar pattern.

The success of the Toshiba factory is due to careful application of methods developed by the software engineering research community in the mid-1970's. The details of the methods used can be found in two excellent articles by Y. Matsumoto.[1] Critical methods include the SADT specification technique, HIPO coding methods, and careful performance estimation. The SWB factory uses simple but effective software tools such as a source code control system and extensive simulators. The success of this approach comes from the fact that the techniques are scrupulously applied. There is also great emphasis placed on software metrics, performance, and quality control.

Most other Japanese firms have adopted, or are adopting, the software factory concept. Outputs vary but high-quality, low-error-rate production code is consistent.

### 3.1.2 Operating Systems

There appears to be no coordinated and organized plan of research in operating systems in Japan. The major manufacturers, Hitachi and Fujitsu, have embarked on a strategy of marketing IBM compatible computers. This eliminates a major part of their problem with operating systems as the machines can run any IBM compatible software whether produced by IBM or by a third party.

Many of the smaller software producers have become interested in UNIX. There are a number of important users of UNIX in Japan. In June 1984, the Software Industry Association (SIA) sponsored a tutorial week on UNIX. It is anticipated that more work will be done in UNIX as the Japanese consensus is that UNIX is a powerful and useful system. Recent visitors to Japan have seen UNIX systems in operation. The Berkeley 4.2 version is widely available, and Japanese UNIX-based systems are now on sale in the U.S.

One area of application related to operating systems is the interface between computers in local area networks. There has been significant research on local area networks in Japan particularly in fiber optics. Japanese manufacturers are heavily involved in network protocols and are expert in this area. It is safe to predict that this is an area where important Japanese products will be developed.

### 3.1.3 Applications Software

We detected no basic research in this category. Most efforts in advanced development seem to be comparable to efforts developed 10 years ago in this country (e.g., menu-driven office systems). However, the low error rate of their developments leads to higher quality software.

The Japanese practice of tailoring applications to each customer's specification is a fundamental difference between the Japanese and the U.S. market (the U.S. abandoned this some time ago). This practice consumes quantities of programmer time and consequently limits the resources available for more innovative work. On the other hand, it is recognized that software maintenance can consume up to 50% or more per year of the original software investment. Under these conditions, the Japanese may gain more in software maintenance than they lose in tailoring their codes. We have received reports of excellent Japanese commercial accounting packages.

### 3.1.4 Programming Languages

Except for the development of production compilers, e.g., COBOL, we could not discern any meaningful research or advanced development in programming languages. (LISP/PROLOG-like systems are discussed under Artificial Intelligence Section 3.2.)

### 3.1.5 Database Technology

Current work on database systems in Japan can be divided into three parts. The major effort involves the large manufacturers watching U.S. firms closely and producing copies of IBM compatible products and other commercial systems. Following the U.S. lead, they have recently decided that relational database systems and distributed databases are important and work is beginning on those areas as well. So far, the developments are derivative. Ultimately, the Japanese will produce good systems of these types, but this approach is bound to leave them behind innovative U.S. companies.

A second area of activity is in Japanese universities and research laboratories. An excellent summary of this work appears in a special issue of Database Engineering.[2] The general impression is that the work is not original and is again following research directions laid out in the West.The third effort is the database effort centered around the Delta machine which is part of the ICOT Project. The preliminary reports make it clear that an ordinary Hitachi system will provide the management of the database itself. Details of the relational engine being used as an interface between the PSIMs and the standard database system are not available at this time. It will be worthwhile to observe how database access is integrated into the user interface.

To summarize, Japan is far behind in all aspects of the database field. Their efforts to copy standard systems will be successful, and the United States can hold its lead only by continuing further R&D in this area. In this regard, the decline in database research in the United States is quite noticeable.

### 3.2 Artificial Intelligence & Man-Machine Interface

In the areas of artificial intelligence (AI) and man-machine interface, the panel emphasized six topics:

- Typed input of Japanese characters
- Speech recognition
- Machine translation
- Expert systems
- Natural language understanding
- Support languages, processors and tools

In these areas, the United States generally has a substantial lead in basic research.

The United States has long been the world leader in AI. Currently, AI R&D is receiving unprecedented support. In spite of the much heralded Japanese Fifth Generation Computer System (FGCS) research effort, the U.S. has far larger commitments and a substantial research tradition. We expect that the U.S. research lead in FGCS will widen.

The majority of U.S. support in AI is going towards advanced development in testing research conjectures. However, the Japanese have long standing efforts in Japanese character processing and speech as well as a new initiative in machine translation. The Japanese effort on FGCS has also shown rapid progress in a number of advanced development areas. However, assuming a sustained level of U.S. commitment, the summary shows the U.S. with a lead that is likely to hold.

In spite of the recent major Japanese effort in product engineering, the United States has the only commercially available machine translation systems. The U.S. markets AI support products, natural language understanding systems, and expert system tools. Only in speech recognition is there competition. The U.S. is ahead with the dollar amounts invested in overall advanced development and is likely to maintain that lead; however, each area needs to be considered individually.

A summary is shown in Figure 8, and an itemized analysis is presented in Figure 9. This is a sampling of artificial intelligence and man-machine interface. The panel excluded such topics as vision/image understanding, graphics, and robotics. Further, workstations are considered under the architecture subtopic and appear in a later section.

Japanese character processing is an area of great national concern. The number of research papers on character processing indicates the high interest in this area.

Japan is one of the significant participants in the speech recognition market. Japan's basic research in speech recognition is on a par with the United States. Over the last several years they have been developing an impressive variety of prototypes and products. Based on the number of large companies active in the area, their long-term commitment appears stronger than in the United States.

America has the only commercial machine translation products available. Japan has a long-term interest here which has recently blossomed into a major government-sponsored R&D initiative that will eventually change the competitive situation.

| | BASIC RESEARCH | ADVANCED DEVELOPMENT | PRODUCT ENGINEERING |
|---|---|---|---|
| ARTIFICIAL INTELLIGENCE & MAN-MACHINE INTERFACE | < <br> ↘ | – <br> → | – NOTE <br> → |

NOTE:
Evaluation is difficult because of incomparable product offerings

FIGURE 8 ASSESSMENT IN ARTIFICIAL INTELLIGENCE (SUMMARY OF FIGURE 9)

| | BASIC RESEARCH | ADVANCED DEVELOPMENT | PRODUCT ENGINEERING |
|---|---|---|---|
| JAPANSESE CHARACTER PROCESSING | O NOTE 1 <br> → | > <br> ↑ | > <br> ↑ |
| SPEECH RECOGNITION | O <br> → | + <br> ↗ | + <br> ↗ |
| MACHINE TRANSLATION | O NOTE 2 <br> → | O <br> ↑ | – <br> ↗ |
| EXPERT SYSTEMS | < <br> ↓ | < <br> ↘ | < <br> ↘ |
| LANGUAGE, PROCESSORS, & TOOLS | < <br> ↘ | – <br> → | – <br> → |
| NATURAL LANGUAGE UNDERSTANDING | < <br> ↓ | < <br> ↓ | < <br> ↓ |

NOTES:
1. Affected by work in character processing research.
2. Affected by work in basic natural language processing research.

FIGURE 9 DETAILED ASSESSMENTS IN ARTIFICIAL INTELLIGENCE AND MAN-MACHINE INTERFACE

Before the Japanese FGCS effort, Japan had little experience with expert systems. While dwarfed by the U.S. achievement and level of effort, there is evidence that Japan now has the basis for significant work in this area. However, assuming the current plans for U.S. R&D investment in expert systems, our lead should continue to grow.

The Japanese FGCS efforts have been primarily in support systems. Japan's ability to focus on a specific technology and engineer high quality results is evident in their work on LISP processors and PROLOG systems. However, American R&D efforts have produced a growing list of products in this area. Although, the Japanese emphasize this area, there will be continuing improvement coming from American efforts.

While the United States has emphasized all phases of work in natural language understanding, Japan has not had major efforts outside of the machine translation area. As with expert systems, American investments will continue to push us further ahead.

The remainder of this section will look at each of these topics in detail. The last subsection discusses the potential risks in projecting the future of the AI marketplace.

### 3.2.1 Japanese Character Processing

Written Japanese involves four character sets:

(1) Kanji (or Kanzi): Chinese ideographs, with essentially one or two pictures per word. Kanji is used for the main meaning bearing nouns, verbs, adjectives, etc. The average educated adult knows about 5,000 characters and typical dictionaries contain about 15,000 characters.

(2) Hiragana (or Hira-kana): A small phonetic character set used for conjunctions, prepositions, auxiliary verbs, the inflection of verbs and adjectives, the pronunciation of words, and for words with nonexistent Kanji representation.

(3) Katakana: Another small phonetic character set used for telegrams, emphatic words, and Japanese cognates of foreign words and phrases including many technical terms.

(4) Roman (or Romaji): The Roman alphabet.

Hiragana and Katakana are collectively called the Kana characters. Although the Kanas have been used for some time, the older, more traditional Kanji is strongly preferred. Sources compare the expectation of a general conversion of the Japanese to Kana-only to predictions that Americans will convert to using the International Phonetic Alphabet for spelling. This situation has had enormous effect on the Japanese commercial world and has limited the adoption of computer technology. This is especially true with respect to Western technology. The problems to be overcome include the following:

- Storage and Manipulation: More than eight bits per character is required.

- Output: Devices require significantly larger character sets and finer resolution than is necessary in the West.

- Input: Keyboard entry devices based on the limited Western character sets are inappropriate for Kanji. Character recognition of Kanji (especially handwritten characters) is difficult.

The Japanese computing industry has attempted to respond to the Kanji problem. Standard Japanese Intermediate Structure (JIS) character sets were established in 1978. Level 1 covers 2,965 characters, and Level 2 set raises the total to 6,349. With respect to output, the Japanese have an active market for the highest resolution output devices, e.g., 24 x 24 dot matrix printers, laser printers, and bit-mapped CRTs. Finally, they have attacked the problem of Kanji character input in a variety of ways.

The Kanji input problem is easy to state and difficult to solve. It has motivated a wide spectrum of research and development efforts including speech recognition, optical character recognition, and innovative keyboard entry methods. Each has resulted in commercial products. This section will show the Japanese commitment to Kanji processing, by keyboard entry methods; speech recognition will be covered in the next subsection.

Dozens of systems on the market today support typed Kanji input.[1] The most common methods of typed Kanji are as follows:

- Large Keyboards: hundreds of keys with characters from standard sets, often manipulated with a pen.

- Multishift or Chord Keyboard: multiple shift keys interacting with a main keyboard, requiring two handed typing.

- Multistroke Keyboard: two or more strokes on a simple keyboard for each Kanji character, using a variety of codes and heuristic techniques such as Kana combination cognates.

- Kana-to-Kanji Translation: a typist enters a phonetic spelling and the computer translates the input into Kanji.

The two most significant techniques appear to be large keyboards and Kana-to-Kanji translation. Between March 1980 and October 1982, papers were presented at meetings of the Working Group on Natural Language Processing of the Information Processing Society of Japan (IPSJ) showing Kana-to-Kanji translation efforts at NTT, JICST (Japan Information Center for Science and Technology), IBM Japan, Fujitsu, and KDD.

The Toshiba JW-10 word processor,[2] introduced in about 1978, appears to have been the first commercial Kana-to-Kanji translating system. It is noteworthy for its dependence on linguistic information to a degree rarely, if ever, evidenced in American mass-market products. Figure 10 shows the information sources and data flow. Input arrives as Kana characters with spaces to indicate pause groups. Pause groups contain a stem and a variety of prefixes and suffixes. An attempt is made to uncover the stem in each group using morphological, essentially grammatical information from a dictionary. Proper

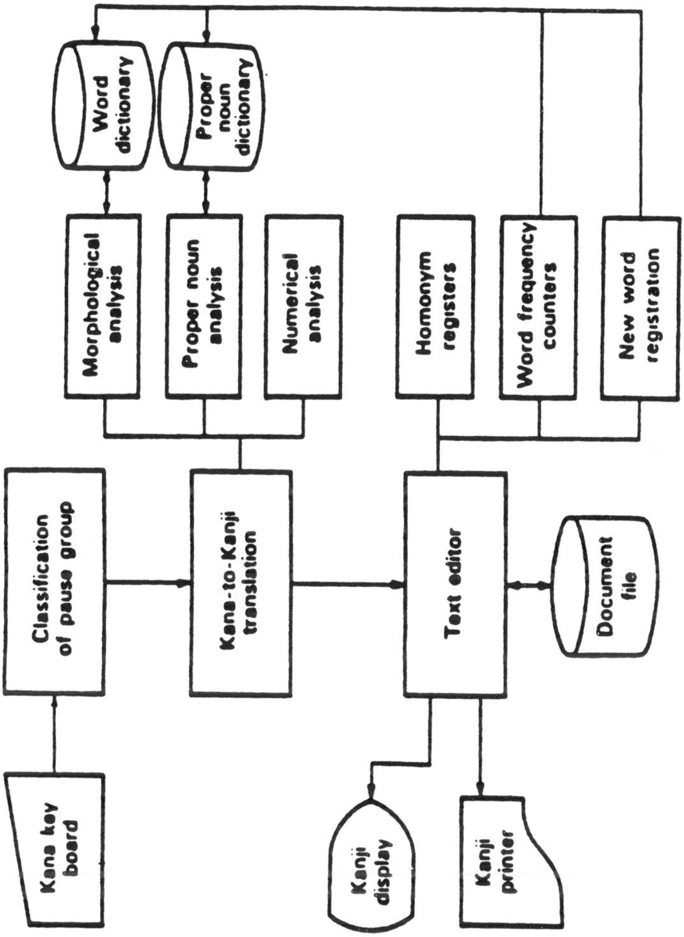


FIGURE 10 BLOCK DIAGRAM OF JAPANESE WORD PROCESSOR

names and numbers are given special treatment. Translation is complicated because of the high rate of homonyms. When no distinct Kanji character can be isolated, the typist is given a choice of the Kanjis found. These are presented in order of word frequency until the typist sees the correct one. This choice updates the word frequency count, and new words can be added at this time.

With the Kanji problem as crucial to the Japanese as it is, innovative keyboard entry methods will not disappear. In fact, more ideas will certainly be forthcoming. No obviously superior input method has arisen or appears imminent. Costs of hardware, economy of scale, and growing familiarity are likely to decide the winners. What is clear is that Kanji processing will be increasingly important. In terms of commercial market, the Japan Business Machine Makers Association reports a 71 billion yen market for Japanese language word processors in 1983. They predict a 500 billion yen market for 1984.

However, growth of Kanji processing could be slowed in several ways. For example, the cost/performance of Western computer systems could increase fast enough to offset the lack of Kanji processing. It is also possible that a lack of standardization will discourage the commitment of funds by potential customers. But, we project that the Kanji cultural imperative will not be denied.

The Japanese market is the second largest in the world. Currently, U.S.-related companies have a large share of the computing market in Japan. Looking at the overall problem, Japanese character processing will be necessary to maintain U.S. market shares in Japan. For example, manufacturers of printers and CRTs must support higher resolution than is necessary in Western markets in order to compete in the Japanese market. It is important to achieve compatibility with Kanji input devices. Standards of performance can be excellent barriers to market entry.

### 3.2.2 Speech Recognition

Speech recognition, a very complex problem, has for years been a natural and major interest in Japan partly because of the difficulty in inputting the written Japanese language.[1] They have a major presence in basic research, although the U.S. is still the leader and should remain so. The Japanese have been releasing innovative products for several years, and their major corporations appear to be quickening their efforts. We feel that they should be rated ahead of the U.S. in advanced development and product engineering, with a greater rate of growth in both areas.

There are several dimensions to automated speech recognition, which include speaker dependency and presentation.[2] In speaker independent systems, the computer system need not be made familiar with the user's voice before the system is employed. This technique is currently limited to small vocabularies. In speaker dependent systems, the computer system is trained to the user's voice. "Presentation" refers to the ability to recognize continuous or isolated words or phrases, and such "connected speech" is the most difficult.

In Japan, there has also been long-term interest in speech recognition, and they associate speech recognition with the Kanji entry problem. It has been estimated that Japan is second only to the U.S. in the number of speech researchers. Kyoto University produced a working demonstration of automatic speech recognition in 1963. NEC demonstrated a device for speaker independent recognition of spoken digits in 1964.

MITI sponsored a ten year effort, from 1971 through 1981, on Pattern Information Processing Systems (PIPS) which included a major speech recognition effort. The overall effort was funded for approximately 22 billion yen. ETL carried out the fundamental work, and NEC was able to demonstrate a isolated word, speaker independent, 100 word vocabulary speech-recognition system. The current state-of-the-art system employs spectral-template matching using linear, non-linear, or dynamic time warping. Costs ranged from $1,000 to $20,000.

Experimentation and product release continues. For example, a recent issue of the Journal of the Robotics Society of Japan[4] showed activity at the following sites:

- Toshiba: speaker independence, voice word processing, integrated circuit realization.
- Hitachi: speaker independence, sequential words.
- OKI Electric Industry Co.: speaker independence.
- Matsushita Electric Industrial Co.: integrated circuit realization.
- NEC: sequential words, speaker independence, units for personal computers, voice word processing.
- Fujitsu Laboratories: large vocabulary, voice word processing.

In addition, MITI's 1981 Fifth Generation Computer Systems plan includes speech recognition research.

In the United States, about two dozen companies are marketing speech recognition devices and many university and industry sites are performing research. One of the most significant recent developments is DARPA's inclusion of funding of speech research in their Strategic Computing Program.[5]

In the near term we expect (a) limited connected word recognition, speaker dependent, with vocabularies on the order of 50-250 words; (b) speaker independent maximum vocabularies of about 20 selected words; and (c), within four or five years, substantial improvements in the large vocabulary, continuous speech and/or speaker independent systems. The cost of isolated word recognizers should drop. The possibility of vertical integration in Japan will assist here. However, it is unlikely that a large profit, mass market will appear in the next few years. Judging by the size of the Japanese companies involved, their association with the Kanji input problem, and their long-term involvements, the Japanese will not abandon these projects.

Some potential surprises postulated by the panel were:

- Applications: Just as Texas Instruments "Speak & Spell" gave a major impetus to speech generation, a single major commercial success could radically change attitudes toward speech recognition. One technology that would be likely to support such a success would be a speaker-independent system. In fact, Japan currently supports the largest field application of speech recognition anywhere. This is a set of small vocabulary, speaker independent, isolated word systems serving the banking and financial industry in 40 cities. Another success would be a phonetic transcriber for use with Kanji dictation.

- Backlash: Cancellation of major speech recognition projects, by IBM or DARPA could substantially cool interest.

- Technical Surprises: It could be that an innovative combination of existing recognition methods would achieve significantly higher performance. Another possibility would be the use of specialized parallel architectures for speech recognition that could be realized in VLSI. Currently, there is a close analogy between computer chess and computer speech recognition because the most successful algorithms in each area have used primarily the brute force approaches (i.e., computationally intensive). There is, however, one school of research proposing alternate methods such as a combination of coarticulation and acoustic-phonetic processing and some optimism exists for success.

### 3.2.3 Machine Translation

The translation of information into and out of Japanese is a major problem. Of late, computerized translation efforts have been receiving major support from the Japanese government and industry. The major basic research in natural language processing has come from the U.S., and this is expected to continue. The majority of the money being spent on the development of machine translation is by the Japanese. While the only products being marketed today are American, this is expected to change given the magnitude of the Japanese efforts.

Note: brute-force machine translation was abandoned in the U.S. in the late 1960's.

The goal of machine translation is to replace all or part of the human effort in the translation of text by computational processing. Products range from on-line terminology databanks (dictionaries), to partial translation systems integrated with word-processing systems, to fully automatic systems. The human translation market worldwide is informally estimated at 10 to 20 billion dollars per year. A 1982 survey by the Japanese Electronics Industries Association showed that more than 2.5 billion dollars was spent in Japan on business and industrial translation alone. Technical translations are reviewed by senior translators and familiarity with a technical field is required. In Europe, university degrees in translation are offered. Automation could encourage significant growth because skilled translators are hard to find, many translation jobs are tedious, and high costs and delays discourage the use of existing translation services.

The Japanese feel a special attraction to machine translation. As a major trading nation, they need to have efficient means of getting information to and from their markets. Of the 79 papers presented to the Working Group on Natural Language Processing of the Information Processing Society of Japan between March 1980 and October 1982, the titles of 21 directly mention machine translation.[1] (In contrast, of the 67 papers presented at the meetings of the Association for Computational Linguistics in the U.S. during those three years, only 1 addressed the topic.) At least 18 different machine translation projects are active in Japan. These include two at Kyoto University, one each at Kyushu and Osaka Universities, and efforts at Fujitsu, Hitachi, Toshiba, and NEC, as well as NTT and ETL. In addition, the Japanese interests have invested in two American efforts, Weidner and LATSEC. The most popular pairs for translation are Japanese-to-English and English-to-Japanese.

The single most striking effort in this area in Japan is the Mu-project,[*2,3] funded by the Science and Technology Agency, a subdivision of the Prime Minister's office. The Mu-project began in 1982 as a three year, national machine translation effort. Eleven different companies are producing demonstration systems showing Japanese-to-English translation of scientific abstracts, each using the same basic technology. An English-to-Japanese

---

* Its full name is "Research on a Machine Translation System (Japanese-English) for Scientific and Technological Documents."

system is planned. Funding was at 1.32 billion yen for 1982, 1.63 billion yen for 1983 and 1.62 billion yen for 1984. In comparison, the much better publicized MITI-sponsored Fifth Generation effort received less money in 1982, its first full year of effort. In 1983, the Fifth Generation effort received 2.73 billion yen, while MITI's superspeed computer effort received 1.57 billion yen. As is often the case, the companies that were invited to participate were not supported to the level necessary to meet the technical goals. They are very likely contributing significant internal funds.

One basic technology is being employed. It emanates from the Kyoto University laboratory of Professor Makoto Nagao. The laboratory is acting as a subcontractor to the Ministry of Communication's Industrial Technology Committee, Electronic Technology General Research Center (DENSOKEN). Several other universities are also involved. The scientific terminology dictionaries are being produced at JICST, with systems integration being handled at Tsukuba RIPS.

The Mu-project's translation algorithm is shown in Figure 11. Individual Japanese sentences are analyzed according to a Japanese dictionary and knowledge of the grammar of Japanese (not shown). The resultant Japanese

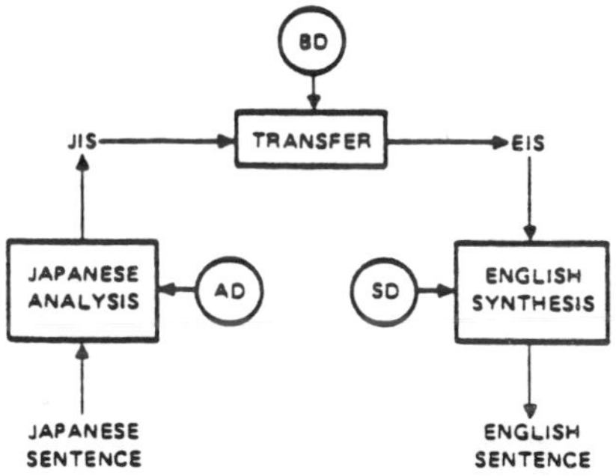


JIS : JAPANESE INTERMEDIATE STRUCTURE (Dependency Structure Based on Cases)
EIS : ENGLISH INTERMEDIATE STRUCTURE (Phrase Structure Tree)
AD : ANALYSIS DICTIONARY FOR JAPANESE
BD : BILINGUAL DICTIONARY
SD : SYNTHESIS DICTIONARY FOR ENGLISH

FIGURE 11 OVERALL CONSTRUCTION OF THE MT SYSTEM

Intermediate Structure (JIS) reflects the modificational structure present in the sentence through types of modifications identified by cases. To achieve an analysis that can be related to an English sentence, a translation procedure identified by the box labelled "Transfer" is applied to the JIS. The resultant English Intermediate Structure (EIS) is shown in an analysis form indicating the clauses, noun phrases, prepositional phrases, and other phrases that will make up the target English sentence. The EIS is transformed into an English sentence using an English language dictionary and a set of transformations that are intended to remove remaining artifacts of the Japanese original from the EIS.

The United States has significant efforts at only four small companies and one university center as summarized below. Nearly all have partial or controlling foreign interest and support, and the majority of their customers are foreign.

- LATSEC produces Systran, the earliest commercial system. It is generally driven by complex dictionary entries supplied by the customer. It has a variety of customers around the world including the U.S. Air Force Foreign Technology Division. It appears to have Japanese and German partial ownership.

- ALPS produces translation tools integrated with word processing equipment.

- LOGOS produces a German-to-English translation system integrated into WANG word processing systems.

- Weidner Communications Corp. offers a variety of language pair translation systems all integrated with word processing. Weidner has over 20 installations worldwide. Controlling interest in the company is in the hands of the Japanese who are said to be reducing its size and funneling attention to Japanese translation.

- The University of Texas Linguistic Research Center has produced a system for German-to-English translation under funding from Siemens of West Germany. It has passed production-style testing. Currently, it is being expanded in Germany for internal use and possible product release. In Texas, work on other language pairs has begun.

Although their performance is short of the Japanese goals, the LOGOS and Weidner systems represent the state-of-the-art in commercial systems. Other than the University of Texas effort, little research into machine translation is underway within the U.S.

Looking at the Mu-project, the use of the transfer stage places the system in the main stream of current thinking worldwide. Many of the details, such as the emphasis on lexically driven processes, are current in American research circles. On the negative side, many of the hard issues that are part of the machine translation problem still remain: massive lexical ambiguity, large variance between languages, the dependence on context, and world knowledge for interpretation of sentences.[4] Further, the system is still undergoing initial laboratory testing.

However, it can be expected that the work will result in increased interest in machine translation in Japan and will eventually produce useful systems. These systems will likely be used in-house first and, as they become increasingly sophisticated, later released for sale.

The time scale for the availability of useful systems is not clear as unsolved and potentially unsolvable problems remain. The real rewards will come from incremental gains in efficiency and quality. These are the sort of gains that arise from an engineering perspective and hands-on experience; the Japanese have the first and will gain the second.

One possible negative surprise for the Japanese efforts is that the heavy investment in a single translation algorithm, which certainly will not solve all problems, may lead to disenchantment. On the other hand, the volume of industrial effort and the pressures of international trade may produce commercial results far sooner than would otherwise be possible.

### 3.2.4 Expert Systems and MITI's Fifth Generation Computer Systems

(See Section 5 for references)

This section discusses two major AI topics: expert systems and the AI aspirations seen in MITI's plan for FGCS.*

---

* There has always been a major current in AI which uses the computer to uncover how humans process information; however, that is not of interest here. What is of interest is the growing movement within the field towards commercial application of AI. This is a movement which the Japanese have joined and would like to lead.

Expert systems are symbol manipulation systems that contain knowledge that a human expert could be said to have. Often they use identifiable rules in the form of "if that occurs then do this," with processing that is explainable in terms of human activities. The rules, and other knowledge, are realized in various knowledge representation systems; inference mechanisms apply the knowledge to solve problems using the rules. Expert systems have seen laboratory applications in medical diagnosis, geological analysis, computer fault diagnosis, and many other areas.

The field of expert systems has recently gained a great deal of attention partly because of technical progress. Much of the excitement, however, is in response to MITI's large commitment of funds and prestige in order to produce a FGCS targeted to be the innovative computer systems of the 1990's. This includes the knowledge-base and problem-solving technology of expert systems in a crucial way (see Figure 12). Since this is the largest Japanese effort related to expert systems, we will describe it here.

A Japanese group comprised primarily of university and corporate laboratory personnel began to capture substantial world attention following the "International Conference on Fifth Generation Computer Systems," held in Tokyo in October 1981. At that time a ten-year program of research and development was proposed to be funded by MITI to the level of approximately 100 billion yen ($430M). The first three years would be spent in basic investigation. The next four years were to be devoted to subsystem prototyping, and the final three years to construction of a total system prototype. The diverse characterizations of a FGCS as presented in the 1981 conference included:

Socioeconomic Goals

- Increase productivity in low-productivity areas
- Confront international competition
- Contribute to international cooperation
- Assist energy and resource conservation
- Cope with an aged society

End-User View

- Functions which do not require professional knowledge to use
- Behavior comparable to human judgment, decision making, and discourse

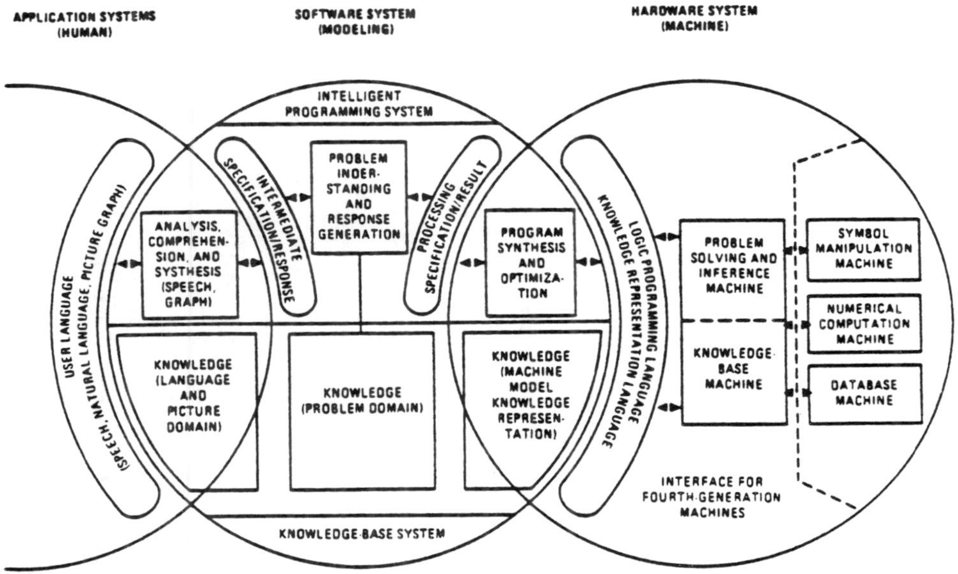


FIGURE 1-2 A CONCEPTUAL DIAGRAM OF A FIFTH—GENERATION COMPUTER SYSTEM FROM THE PROGRAMER'S STANDPOINT

- Flexible configurability to a wide range of jobs
- Functions to facilitate programming
- Reliability

Function

- Sensory (speech, handwriting, image)
- Cognitive (knowledge base, reasoning, problem solving)

Component Technologies

- VLSI Design Automation
- IC Technology
- Communications
- Parallel Processing
- Software Engineering
- Artificial Intelligence

To these ends, a set of 26 themes in research and development were identified and characterized in terms of objectives appropriate to the three phases of the program proposed. Numerous articles and books have been published on the details of the FGCS project.

Fifth Generation Computer Systems received a 426 million Yen (about $2M) allocation in the MITI budget for JFY 1982. The Institute for New Generation Computer Technology (ICOT) was established in April 1982, growing to a forty-odd man research staff in its first year. In JFY 1983, the program was allocated 2.72 billion Yen (about $12M), of which all but 20% was to be spent on ICOT's direct operations. The budget for JFY 1984 is 5.12 billion Yen bringing the first three years to a total of 8.3 billion Yen compared with the originally projected 10 billion. Contrary to past MITI practice in industrial development projects, 100% of the expenses are said to be paid by MITI rather than regarding development subsidies of less than 50% as advances against future invention royalties. ICOT estimates that 150 personnel in the eight Japanese electronics companies with whom it is affiliated were doing subcontracted tasks in JFY 1983, and that this number will grow to 350 in JFY 1984 while the ICOT staff itself will only grow to some 50 researchers. In 1985, the latter number is projected to double as the FGCS project enters its intermediate phase. However, there is evidence that the actual expenditure on ICOT-related activities is greater than these numbers indicate. As noted for other Japanese projects, the companies seem to be paying a portion of the overhead in running ICOT's physical operation in leased space in Tokyo. The companies also pay salary differentials for their ICOT assignee if the ICOT base scale falls below what the assignee would have been making at the company. Assignees from the MITI Electro-Technical Laboratory draw more than half their salaries from ETL while at ICOT.

The laboratory at ICOT is organized into three major internal groups supplemented by nine external advisory groups composed of university, industrial, and national laboratory representation. Table 4 lists these divisions.

Table 4. ICOT Divisions

Architecture Group

- Parallel inference and knowledge base machines
- Relational database
- VLSI CAD

Software Group

- Prolog extensions
- Application to inference and knowledge base subsystems
- Natural language interfaces
- Intelligent programming (application development) system

Prototyping group

- Sequential inference machine--HW, microcode, OS

External Advisory Groups

- Technology Trends
- Project Promotion Committee
    - Working Group on Parallel Processing
    - Working Group on Inference and Kernel Languages
    - Working Group on Natural Language Processing
    - Working Group on Consultation Systems
    - Working Group on Basic Theory
    - Working Group on Language and Software Dissemination

A principal accomplishment to date has been the design and development of a Personal Sequential Inference Machine (PSIM) intended for the single-user support of program development in PROLOG. This machine is comparable to the Symbolics 3600 LISP machine in size, scope, and performance. It is based on the characteristics of PROLOG and does not exploit any architectural innovation. Nevertheless, it is impressive that less than a year elapsed from conception to delivery of the first prototype. The machine executes a language called "Kernel Language Zero" (KLO), a PROLOG-like instruction set, by means of a microcoded implementation on a straightforward hardware organization. There is a large amount of memory attached (20 megabytes in the prototypes, 80 megabytes by architecture) to avoid having to deal with problems of garbage collection in a demand paging environment. An address translation mechanism is provided for ease in task switching. The machine is being fabricated for ICOT by Mitsubishi, and the unit price is expected to be in the neighborhood of 35 million Yen (about $150,000). A much faster engine is being talked about at NEC which would be a back-end processor with perhaps 4-5 times the performance of the PSIM. Work is reported to be underway on a database machine named "Delta" which will provide knowledge base services for one or more PSIMs connected through an Ethernet arrangement.

The Software Group has made a heavy effort in PROLOG language and language processor design. They also have considered extensions to facilitate concurrency, modularity, interfaces to a relational database, high-level control structures, and functional programming constructions. There is conjecture about adding set operations and pure logic programming in subsequent versions of the language.

The Parallel Architecture Group will have major responsibilities during the intermediate phase of FGCS.

The accomplishments to date are more narrow than the broad set of research topics listed above. Some, like machine translation and traditional supercomputers, are being covered elsewhere in Japan. Others, like the knowledge base management system, apparently are being subcontracted to the companies. Still others, like the relational algebra machine or systematization technology for computer architecture, seem not to be pursued at all. The major advances in expert systems and the other application areas of AI will not be forthcoming until the later stages of the project.

Numerous other smaller efforts exist in the universities, the major computer companies, and the smaller software companies. Personal LISP machines are under development at ETL, NTT, Fujitsu, NEC, and Kyoto University. Recently, expert systems constructed by the Science University of Tokyo and Fujitsu have been demonstrated. Speech recognition is a very competitive field in its own right as described earlier. Work on parallel inference engines is taking place at Tokyo University and almost certainly at several other places. Much of this work has been inspired by the FGCS program--either as attempts to support it or to provide technical opportunities for choices which may have not been made with full information.

U.S. Efforts in Expert Systems

In the United States, early support for expert systems and general AI research came from the U.S. government with DARPA, NIH, and NSF being especially important. The major computer science research centers all participated heavily in the effort (Stanford University, Massachusetts Institute of Technology, Carnegie-Mellon University, Bolt Beranek and Newman, SRI International, and USC/Information Sciences Institute). Most of the initial work came from these centers. These centers are still active and some have upwards of 100 active AI researchers.

Some examples of active start-up ventures in expert systems are:

| | |
|---|---|
| • Teknowledge | Developing prototypes for industry on contract and offering service, consulting, and customer education |
| • IntelliCorp (formerly IntelliGenetics) | AI software for genetic engineering and the KEE knowledge engineering framework |
| • Applied Expert Systems | Developing financial applications for large customers. |

The first two companies mentioned each have 50 to 100 employees. All commercially available expert system products have come from such companies.

In American industry, DEC has the most mature expert systems effort, employing at least as many people as the start up companies. Their XCON system is used to configure all VAX computers, and, they are developing XSEL, a system to advise salesmen. In addition, evidence from hiring ads indicate an interest in expert systems by other American companies.

Two planned American efforts are worth special attention. One is the Microelectronics and Computer Technology Corporation (MCC) recently established in Austin, Texas. MCC is a joint research venture of approximately 16 American companies. Among MCC's initial projects are several concerned with AI. These include expert systems work applications to database systems, and supportive work in architecture. MCC as a whole is starting off with a budget of 50 million dollars per year. AI and AI-related efforts may receive one-sixth to one-third of that amount.

The largest American AI effort is likely to be part of the "Strategic Computing" program sponsored by the Defense Advanced Research Projects Agency (DARPA). The program has an initial one-year funding level of 50 million dollars, and there are plans for it to grow to 150 million dollars per year for 10 years. All applications feature expert systems. Both efforts will be discussed further in the Computer Architecture section.

U.S. vs. Japan in Expert Systems

Despite worldwide publicity concerning the dramatic increase in Japanese activity, the United States has a far stronger and more diversified activity in expert systems. Similarly, announced plans for the near future show the American effort growing at a faster rate. The panel concludes that the Japanese are far behind in all areas. They are falling back in research and slipping further in advanced development and product engineering. This does not imply that Japan will not succeed in making significant advances.

### 3.2.5 Languages, Processors and Tools (See Section 5 for references)

AI programming languages are LISP (a U.S. preference) and PROLOG (a Japanese preference). Our opinions on LISP and PROLOG workstations are covered in more detail in a later section.

One of the key elements in the Fifth Generation Project is to produce KIPS (Knowledge Based Information Systems). The Japanese project is based on logic programming in the language PROLOG. This seems to be a curious choice since most artificial intelligence research is based on the use of LISP which has been in use for a quarter century. There are workstations and other sophisticated systems for LISP. Also, there are excellent textbooks for new LISP users.

PROLOG, on the other hand, is a somewhat obscure language with none of the support mentioned for LISP. There is no user community with much PROLOG experience, nor is there a network to assist in the development of such an ambitious project.

During a visit to ICOT, many of these concerns were raised with the ICOT Director. He explained the choice of PROLOG as follows. Choosing LISP would have led to a perception that the project was building on American ideas and might have been perceived as "nonoriginal." By choosing PROLOG, the work would appear more "Japanese."*

Finally, it is interesting to note that some of the other MITI-funded projects have chosen to work with LISP, although PROLOG is used as well. Fujitsu has just announced a backend LISP processor. It seems clear that both languages will be explored; and, if it is beneficial to switch languages to obtain the desired goals, then someone will do so (though not necessarily ICOT).**

Some of the most important tools for AI work are knowledge representations. Knowledge representations may be considered as very powerful database systems. Several American manufacturers of LISP machines supply these systems to augment R&D work done in their version of LISP. Some of the

---

* There is some irony in this because PROLOG was designed by a Frenchman, and its early use was in England.

** Many LISP related efforts are discussed in the Computer Architecture section of the report.

first commercial offering of the AI start-up companies have been knowledge representations. The design of a new system has been included in the ICOT initial work. However, no other work of equal significance has been noted in Japan. In fact, CSK, one of the largest independent software houses in Japan, has bought a minority interest in the American company IntelliCorp and they are initiating marketing of their knowledge representation system in Japan.

The overwhelming AI-research base gives a clear advantage to the U.S. in basic research. Similarly, the number of AI ventures gives a lead to the U.S. in advanced development. Since the only known production LISP systems are U.S. based, we scored the Japanese behind in this category as well. However, of all the areas in AI, this is the one in which they have shown the most rapid progress. We predict that the Japanese will not fall further behind, and they will add significantly to the state-of-the-art with their PROLOG work.

### 3.2.6 Natural Language Understanding (See Section 5 for references)

This section deals with the understanding of typed natural language such as Japanese or English. One of the major barriers to computer usage is the complexity of the formal computer languages. Typed natural language is one of the possible solutions. Speech recognition would work well with natural language understanding, but its development lags significantly. All of the interfaces mentioned in this section work from typed input.

There is a variety of natural language interfaces for data base systems commercially available in the United States. Artificial Intelligence Corporation produces the most established system. It is marketed by IBM and Cullinet, among others. Over 200 copies have been sold at the $70,000 level. Other systems are offered by Mathematica Products Group, a unit of Martin Marietta, and Microrim, Inc. Several startup companies offer custom made interfaces and natural language understanding tools. R&D efforts have been underway at many of major U.S. corporations, such as IBM, AT&T, Hewlett Packard, Burroughs, and GTE. Most of the major government-sponsored AI centers mentioned in the section on expert systems have done some work in natural language understanding.

Almost all American efforts have been for English language interfaces, but some interfaces have been produced for European languages. R&D work is being done in Europe with American-originated technology.

The Japanese have engaged in some natural language understanding R&D work. For example, IBM Japan was involved for a number of years with the California Institute of Technology producing technology that is still being explored by IBM in West Germany. ICOT has begun working on a natural language parsing system, and spinoffs from the machine translation efforts can be expected. It is likely that it will be 5-10 years before their commercial impact is felt. Finally, it is unlikely that the Japanese will be able to produce suitable English language interfaces and that Japanese character input methods will be sufficient to attract a wide audience for Japanese interfaces.

### 3.2.7 AI Market

Market forecasts for Artificial Intelligence such as the areas covered in the last three sections can be substantial:

- "The applications of intelligent software at home and office represents the largest growth segment of technology market in U.S. rising to 8.5 billion dollars in 1993," according to International Resource Development.

- DM Data predicts the combined knowledge systems, natural language software, computer-aided instruction, and voice recognition market to be 2.5 billion dollars by 1990.

The market figures are believable; the time scale is questionable. Current sales consist almost entirely of tools, educational services, consulting, and contract research. It is still too early to tell when a broad commercial market will emerge.

There can be serious problems for AI in a number of areas:

- The general public's expectations far exceed the realities of the technology.

- American industrial and venture capital may not be sufficient to cover the development costs.
- Trained manpower may be spread too thin to produce enough development teams.
- Major defense programs may dilute the commercial efforts.

## 3.3 Processor Architecture and Computer Organization

Computer architecture is the study of computer organization. This includes such diverse topics as sequential machines and distributed systems. Our overall general assessment in this area is illustrated in Figure 13.

Japan is clearly lagging the United States in creative applied research of advanced computer architectures and will not close this gap rapidly. We expect that Japan will rely upon the Western research community to provide creative applied research results and concepts which will drive Japan's future directions in prototype development and laboratory experimentation.

| | BASIC RESEARCH | ADVANCED DEVELOPMENT | PRODUCT ENGINEERING |
|---|---|---|---|
| COMPUTER ARCHITECTURE | < ↘ | ○ ↗ | - ↗ |

FIGURE 13 OVERALL ASSESSMENT OF COMPUTER ARCHITECTURE (SUMMARY OF FIGURE 14).

On the other hand, Japanese industry is rapidly closing with U.S. industry in advanced development of prototype architectures, successive refinement of those architectures through laboratory experimentation, and high-quality, low-cost product engineering/manufacturing. We expect that Japanese industry will continue to focus its resources in these areas to establish a favorable advantage over U.S. industry in the future.

The comparable United States and European R&D efforts described have been slow in gaining momentum despite Japan's progress over the past few years. The magnitude of the commercial significance of these efforts has yet to become apparent. The difference between Japanese and Western industry seems to be that Japan assumes that the world is technology driven, whereas the West assumes that the world is market driven.

Figure 14 illustrates our detailed findings in architecture in the subtopics: parallel processors, supercomputers, workstations, and (IBM) clones.

| | BASIC RESEARCH | ADVANCED DEVELOPMENT | PRODUCT ENGINEERING |
|---|---|---|---|
| PARALLEL PROCESSORS | < ↘ NOTE 1 | – ↑ NOTE 2 | N/A |
| SUPERCOMPUTERS/ HARDWARE | N/A | O ↗ NOTE 3 | – ↗ NOTE 4 |
| SUPERCOMPUTERS/ SOFTWARE | < ↓ NOTE 5 | O ↗ NOTE 5 | + ↗ NOTE 6 |
| PROFESSIONAL WORKSTATIONS | N/A | < ↓ NOTE 7 | < ↓ NOTE 8 |
| CLONES | N/A | + ↗ NOTE 9 | > ↑ NOTE 10 |

NOTES: See Text

FIGURE 14 DETAILED FINDINGS IN COMPUTER ARCHITECTURE

The following notes apply to Figure 14:

Although there is widespread interest and active study in these areas:

1. We find little evidence of basic research into interconnection issues, dataflow or reduction computational models, or shared memory/communicator protocols, all of which are mainstays of this kind of development. Because of the ICOT potential and because of the heavy prototyping activity, we give the benefit of the doubt in assessing the rate of change as ↘ instead of ↓.

2. While a larger number of pioneering architectural prototypes have been produced in the U.S., the rate of new prototyping in Japan is greater. At present, U.S. researchers are busy writing funding proposals while the Japanese are building more prototypes with fewer people.

3. Japanese supercomputers are produced by the large, vertically integrated mainframe companies using the most sophisticated technologies. These are prestige items which attract top engineering talent in a highly competitive development environment.

4. We believe the Cray X-MP to have a more robust processor organization than the Fujitsu VP-200 (which is marketed in the U.S. as the Amdahl 1200) and the Hitachi S-810/20. The Cray will handle a larger range of applications with more uniform performance because of its faster cycle time and scalar execution rates, its shorter vector startup times and pipeline structure, and the immediate capability to upgrade to 2-way and 4-way systems.

5. Fortran vectorization technology continues to be developed in the U.S., and it is being well exploited in Japan.

6. It already appears that the Fujitsu vector Fortran compiler is the "best-of-breed," and the gap can be expected to widen given Fujitsu's commitment to the technology.

7. There is an order-of-magnitude less effort on professional workstations than in the U.S.

8. Other than CAD/CAM and Graphics which we excluded from the study, the only professional workstations are Japanese language word processing systems.

9. In the second-source, IBM-mainframe-compatible area, the Japanese have a clear lead. They are now starting on microprocessors, and we anticipate vigorous competition in that area.

10. The conventional wisdom that the Japanese excell at low-cost, high-quality replication is particularly true here.

Substantial university, industry, and government research and development interests in advanced computer architecture have been renewed in the last several years as a result of Japan's FGCS and High-Speed Scientific Computer Programs. Additional impetus has resulted from the establishment of ICOT by MITI through which research and development programs are conducted by selected professionals from Japanese industry with funding provided by MITI.

In response to the challenges identified by the FGCS Program, several comparable research efforts have been established in the United States and Western Europe. These programs have established similar research goals and collaborative research relationships among industry, universities, and government agencies. The common goal of all these efforts is to develop advanced computer architectures that will serve as the basis for the next generation of computing and information processing systems.

To understand the potential evolution of advances in computer architecture, an account of current research and development work in Japan is discussed below.

The growing interests in advanced computer architecture research and development in Japan span a wide range of topics in the following areas:

- Sequential Architectures
- Parallel Architectures
- Distributed/Network Architectures
- Specialized Architectures

Within the area of sequential architectures, two major thrusts have been identified - microprocessor and mainframe realizations of existing sequential architectures (clones), and sequential inference architectures. NEC has recently announced a low-cost microprocessor which implements the instruction set of the Intel 8086 microprocessor as well as additional instructions for enhanced functionality. Fujitsu and Hitachi have announced implementations of the IBM S/370 which also includes additional instructions for enhanced functionality. Several efforts are underway in Japan to develop sequential inference architectures which will support expert system applications based upon PROLOG and LISP. ICOT is currently investigating a Sequential Inference Machine (SIM) to support a PROLOG environment, while Fujitsu is currently developing ALPHA (a LISP processor) which will support expert system applications based upon LISP.

R&D efforts are underway in Japan in the area of parallel architectures by universities and industry. Research projects within Japanese universities include the development of models for parallel execution of logic programs, general scientific applications, sparse matrix problems, linear equation solving, and distributed database applications. Several projects have progressed beyond the development of parallel execution models to the design, construction, and evaluation of prototype systems. Specific research projects that have been described in the literature include the following:

- Parallel Architectures for Logic Programming:
  - Parallel Inference Engine (PIE) - University of Tokyo
  - Multi-Microprocessor System for Concurrent LISP-Kyoto University
- General Parallel Scientific Computing Architectures:
  - Highly Parallel Processor Array for Scientific Applications (PAX-128) - University of Tsukuba
  - Sparse Matrix Solving Machine (SM**2) - Keio University

- Systolic Linear Equation Solver - Keio University
- Hierarchical Two-Level Microcode Architecture:
  - Hierarchical Micro-Architecture of a Two-Level Microprogramed Multiprocessor Computer (MUNAP) - Utsunomiya University
  - User Programmable Local Host Utilizing Low-Level Parallelism (QA-2) - Keio University
- Distributed Database Architecture:
  - Distributed Database Machine (GRACE) - University of Tokyo

In turn, the public laboratories and industrial sectors within Japan are focusing their efforts towards developing potential commercial products based upon the research results generated by Japanese and Western universities. Several commercial research and development projects that have been described in the literature include the following:

- Parallel Architectures for Logic Programming:
  - Parallel Inference Machine (PIM) - ICOT
  - Parallel Execution Model for Logic Programming - ETL
  - LISP-based Data Driven Machine (EM-3) - ETL
  - Dataflow Machine for List Processing - ECL/NTT
  - Reduction Machine for LISP - NEC
- General Parallel Scientific Computing Architectures:
  - Scientific Dataflow Processor Array System - ECL/NTT
  - Scientific Dataflow Machine (SIGMA-1) - ETL
  - High Speed Super Computer Program - MITI
- Distributed Database Processing Architectures
  - Distributed Data-Driven Processor (DDDP) - OKI

In addition to the distributed database architecture work at the Universities of Tokyo and OKI, research is underway at the University of Toyohashi in the area of regular tree expressions and their mapping to functional network architectures.

The approach taken by MITI to realize major advances in R&D for advanced computer architectures is to stimulate and support collaborative R&D relationships among university, industry, and government agency partners. The key partners from each of these sectors in Japan can be summarized as follows:

- Industry
  - Fujitsu
  - Hitachi
  - Matsushita
  - NEC
  - Toshiba
  - OKI Electric

- Universities
  - Keio University
  - Kyoto University
  - Tokyo University
  - University of Toyohashi
  - University of Tsukuba

- Government:
  - Electrotechnical Laboratory (ETL/MITI)
  - ICOT (MITI)
  - Nippon Telegraph and Telephone/Electro Communication Laboratory (NTT/ECL)

Given the observed progress at ICOT, this laboratory will have several prototypes of the Personal Sequential Inference Machine (PSIM) and associated data base machines for demonstration by the end of 1984. Given the experience gained during these research projects, ICOT should also have an initial (paper) model for the parallel execution of logic programs with supporting plans for the development of parallel architectures for inference and data base processors. The initial research work at ICOT in artificial intelligence and knowledge-based systems will use the PSIM and associated database machines. However, this work will migrate to the parallel architectures for inference and database processors once suitable prototypes can be constructed.

Advanced development work at NTT is focused on the development of dataflow architectures for LISP-based applications and general scientific computing applications. We expect that these projects will result in the introduction of commercial projects into the industry for knowledge-based system processors and scientific processors in the future. Research work underway at ETL is also focused on the development of parallel dataflow architectures for general applications. There appears to be some level of both competition and collaboration between ETL and NTT/ECL in these areas.

Most of the applied research and development work in the area of distributed architectures is being conducted at OKI Electric within the Distributed Data-Driven Processor Project (DDDP). The commercial application of this work is unclear at this time.

Several advanced development projects are underway at NEC to develop special purpose processors for LISP-based applications and signal processing applications. We expect that this work to continue and will result in the introduction of commercial semiconductor components into the industry for knowledge-based system processors and signal processing engines in the near future.

We expect ICOT NTT/ECL, ETL, NEC, OKI, and others to coordinate advanced R&D work and share results to accelerate efforts in developing commercial products in the areas of knowledge-based system processors, general scientific computing processors, distributed database processors, and special purpose signal processing components. In addition, we expect Japanese laboratories and industry to focus their efforts on their technical strengths in advanced development of prototype architectures and subsequent refinement to commercial products. Rather than developing a leadership position in applied research for advanced computer architectures, we expect that Japan will continue to utilize the wealth of creative applied research available from the Western research community to drive their advanced development work.

In response to the challenges identified by the FGCS Program, several comparable research efforts have been established in the United States and Western Europe. These programs have research goals which include the FGCS

Program and research which addresses the need for improved software development productivity, VLSI CAD/CAM system support and advanced system packaging, and interconnection technologies. In addition, these programs have established collaborative research relationships among industry members, universities, and government agencies.

The major comparable research programs include the following:

- Government-Sponsored Programs:
  - Esprit (European Economic Community)
  - Alvey (U.K.)
  - DARPA (U.S.)
- Industry-Sponsored Programs:
  - MCC (U.S.)
  - ECIRC (Europe)

The European Strategic Programme on Research in Information Technology (ESPRIT) is sponsored by the Commission of European Economic Communities (EEC) as part of their overall research program (total annual budget about $700 million). The purpose of this program is to stimulate multinational R&D projects among industry and universities within the EEC to improve standardization and portability of International Technology (IT) products and to develop strong, competitive, local IT industries among EEC nations. The commission has invited proposals for collaborative, "pre-competitive" R&D in five major areas: (1) Advanced (fast, submicron) Microelectronics, (2) Software Technology, (3) Advanced Information Processing, (4) Office Automation, and (5) Computer-Integrated Flexible Manufacturing. Advanced Information Processing includes all Japanese Fifth Generation themes. The program comprises projects proposed by industry under the review of a steering committee composed of twelve European companies. Each proposal must be in one of the five major areas, involve partners from at least two countries, and involve less than 50% EEC funding. In a twelve month pilot phase launched in mid-1983, approximately $10 million was committed to research projects. The main program has been projected at $300 million per year including both the EEC and industrial contributions.

The Alvey Programme for Advanced Information Technology is managed by a directorate in the Department of Trade and Industry within the U.K. to establish consortia of industry members (including British Telecom, Ferranti, and Plessey) and universities to implement a collaborative research program in Information Technology (IT). The purpose of this program is to mobilize the technical strengths of the U.K. in IT to improve the competitive position of the U.K. in the world market. Programs proposed by industry/university teams fall into four major research thrusts: (1) Software Engineering, (2) VLSI CAD/CAM, (3) Man-Machine Interfaces, and (4) Intelligent Knowledge-Based Systems. Of these, the third and fourth contain the Japanese Fifth Generation themes. Approximately $80 million has been projected to support research and training in U.K. academic institutions over the next five years, while $400 million has been projected to support industrial participation in advanced development of prototype architectures and software support over the same period. It appears that little of this funding is "new." It constitutes a consolidation and redirection of previous funding initiatives. The government provides 100% of the required funding for the university research and training programs. Industry is expected to pick up at least half of its expenses in the joint R&D projects in which it participates. While this program has substantial overlap with ESPRIT, it is intended to complement ESPRIT (in which the U.K. participates) rather than to compete with it.

The DARPA Program in Strategic Computing, introduced in the AI section, has stated its major goal is to "develop a broad base of machine intelligence technology to increase our national security and economic strength." Starting with a set of perceived requirements for military computing in 1990, the program has identified several targets of opportunity: (1) Expert Systems, (2) Speech and Image Recognition, (3) Natural Language Processing, (4) System Development and Prototyping Capabilities, (5) Advances in Computer Science Theory, (6) Parallel Computer Architectures, (7) Microsystem Design Methods and Tools, and (8) Microelectronic Fabrication Technology. Projects funded within the program are expected to be carried out by industry drawing on university research results. University/industry collaboration is expected in certain key areas of which computer architectures is one. The program has had $50 million appropriated in FY 1984, with a $95 million request for 1985, and a projection of $150 million or more per annum thereafter through the 1980s.

The Microelectronics and Computer Technology Corporation (MCC) has been established as a joint collaborative research venture by major U.S. computer system manufacturers, semiconductor manufacturers, and system development corporations to address mutual needs for long-term, high-risk research programs. The four major research programs established by MCC are Advanced Computer Architecture, Software Technology, VLSI CAD/CAM, and VLSI System Packaging/Interconnect Technology. The expenses of these research programs are fully supported by the industrial participants of MCC and will amount to over $50 million dollars over the first three years of MCC's ten year program.

The European Computer-Industry Research Center (ECIRC) is a multilateral research facility announced in late 1983 by its three industrial partners: Bull (France), ICL (U.K.), and Siemens (Germany). It is expected to operate at a budget of around $8 million per annum and will not initially be open to additional participation. The stated aim is to facilitate collaborative long-range research in artificial intelligence and expert systems, and has been viewed as insurance against the failure of ESPRIT due to political obstacles.

In addition to these major research initiatives, both the Department of Energy (DOE) and the National Science Foundation (NSF) have announced plans to sponsor joint university/industry/national laboratory applied research projects in advanced computer architecture within the United States.

The formation of these research projects has stimulated a number of research proposals for advanced computer architectures covering a wide range of problem domains. However, sufficient resources are not available from the university, industry, and government to pursue the scope of ideas that have been proposed. At present, no national focus or mechanism for the coordination of the diverse needs of these communities appears to exist within the United States.

The future of advanced computer architectures will depend upon unknown economic factors related to software support, migration of existing applications, new markets, and technological breakthroughs which could change existing industry trends for computing and information processing systems.

Potential surprises could result from a major breakthrough in parallel processing architecture for knowledge-based systems or general scientific computing applications.

### 3.3.1 Parallel Processors

Despite the in sequential, specialized, and parallel architectures, there is no effort in Japan that matches the number of parallel processing studies underway in the U.S. However, the Japanese have strong incentives to accelerate research in:

- Image Processing
- Kanji Character Recognition
- Graphics Processing
- Speech Recognition
- Facsimile Processing (Data Compression/Transmission)

This will lead to considerable efforts to reduce the gap. Because of their hardware interests in these technologies, the Japanese have made more progress in advanced development.

### 3.3.2 Supercomputers

The term "supercomputer" describes the most powerful numerical computers. Such computers execute more operations per second and make effective use of more storage than any other computer systems. It follows from this definition that supercomputers have always existed. In fact, they are the "bow wave" of the computational computer technology ship. Just as the spread of a bow wave from a ship begins at a point and becomes wider as the ship progresses, the technology associated with supercomputers begins as research and advanced development and enters the production phase and broad commercial use. From this point of view the market for supercomputers is necessarily limited because, once a design is in widespread use, new ideas and new structures have become the supercomputers of that moment.

Advances such as about 6 orders of magnitude in speed over the last 4 decades have occurred at constant cost. This dramatic increase and capability has made possible applications of supercomputers to a broad spectrum of areas and has given rise to the concept of "production versions" of supercomputers. Only the U.S. and Japan have a significant presence in this field.

An operational computer system consists of the dynamic interaction of the system hardware, system software, and the application software. Improvements in supercomputer performance have resulted from advances in each one of these areas. Hardware advances result from (1) improvement in the speed of the components, and (2) changes in the structure of the system that allow more operations to be performed at the same time which is referred to as "increased concurrency."

Improvements in the applications software take the form of (1) more efficient computational procedures, which employ fewer steps to achieve the same results, and (2) more appropriate computational procedures, which make use of the available hardware in a more efficient fashion. Improvements in systems software are necessarily related to both applications software and to the system architecture. Since much application software has been developed to achieve maximum performance on a specific computer, one task of improvements in systems software is to accept existing applications software and to automatically modify it for optimum performance on a new computer system. A second task of system software improvements is to provide extensions to existing widely known languages so that users may build upon their knowledge and tailor new developments to new system architectures. Still a third task of improvements to systems software is to address new language constructs which are optimumly matched to new system architectures.

One characteristic of American supercomputers during the last decade is a lack of sophistication in system software. Since a typical user of a supercomputer during this period of time has been a scientist or engineer with a FORTRAN programming ability, the user has been willing to assume the burden of creating both application and systems software.

Two factors have made the migration of users from one supercomputer to the next difficult. First, users have had to invest considerable resources and knowledge in creating software for existing machines; second, the available comparison techniques for projecting performance of new system architectures has been unsatisfactory. Workload-related kernels are a measure of the performance of a supercomputer system, but no one regards them as definitive. Their use though, does illustrate that non-American supercomputers exist and that these can be considered seriously from a performance point of view.

### 3.3.2.1 Japanese Efforts

By the end of 1983, Fujitsu and Hitachi machines were undergoing testing. The NEC machine is scheduled for completion in March of 1985. The National Super Speed Program was announced in January 1981 and is expected to result in a prototype machine in 1989. The commercial supercomputer development and the National Super Speed Program are distinct efforts. The National Super Speed Program has a number of goals which involves the development of technology which has applications beyond supercomputers. The National Super Speed Project is administered by MITI through its Agency of Industrial Science and Technology (AIST). Funding is by a research contract to the specially created Scientific Research Association. OKI, Toshiba, NEC, Hitachi, Fujitsu, and Mitsubishi are actively collaborating with the Electro-Technical Laboratory and Nippon Telephone and Telegraph. Interaction with a number of Universities is maintained through the Council for Industrial Technology and its Committee for National Development Programs which is composed of a number of government and university technical experts.

The supercomputer market has not been large (probably 0.1% of the world computer market during the last decade). One factor that could dramatically expand this market would be software compatibility and portability with IBM hardware.

Preliminary testing indicates that both the Hitachi and Fujitsu machines are capable of executing application software developed for high-end IBM machines with essentially no modifications. Even though some fifteen to twenty percent of IBM's commercial base involves scientific applications, IBM has not manufactured a machine in the supercomputer class in the past decade. Sales of supercomputers have been increasing. During 1983, CRAY Research announced sales approximately fifty percent larger than those forecast at the beginning of the year. The entry of the Japanese into the supercomputer market, with costs for application software and special system software potentially less than American supercomputers is a new factor in the supercomputer marketplace. The effects will become clear during the next two years as delivery of Japanese supercomputers occurs and as new American products appear.

Surprise must be measured relative to a projected expectation. The following base case expectations are assumed as a reference:

1. The Japanese supercomputers will prove to be more producible and considerably more user friendly than previous American made supercomputers.

2. IBM will introduce capabilities to augment its high end machines, which will diffuse the impact of Japanese supercomputers on the scientific portion of IBM's installed base.

3. American supercomputers from ETA (the GF10), CRAY (the CRAY 2 and the CRAY 3), and Denelcor (the HEP II), will retain over 90% of the follow-on business from their existing market, and sales will continue to grow at an increasing rate.

4. Japanese supercomputers will capture 25% of new supercomputer applications in the United States during the remainder of the 1980's.

5. No further sales of U.S. supercomputers into Japan will occur.

Outside of the United States and Japan, only France has a supercomputer development effort. However, R&D efforts associated with FGCS could also be considered as contributing to supercomputer development. In this regard the Alvey program in the United Kingdom, DRET and INRIA efforts in France, the European Centers for computer research in Munich, and the ESPRIT program of the European Economic Commission are all relevant. However, the strategic objective of the ESPRIT program straightforwardly and aptly summarizes the computational state-of-the-art in Europe: "The achievement of technological parity with, if not superiority over, world competitors within ten years." In the context of this overall strategy, the ESPRIT intent to examine "new economics of logic, storage cells and their interconnection brought about by VLSI" is interesting but not a significant factor in the evolution of the technology. Nevertheless, their exploration of very highly parallel architectures involving "few cell types with a high degree of replication, computational locality in groups of cells, short and regular control of data flow, minimal use of high fan out/wire or routing, highly contextual intercell/group/node communication, with high degrees of a synchronous concurrency among cells/groups/nodes" provides a European suggestion that the evolution of supercomputer technology during the next decade is likely to be considerably more rapid than it has been during the past decade.

Only CRAY Research and Control Data Corporation currently manufacture supercomputers in the United States. ETA and Denelcor have announced that they are currently working on supercomputers. Other potential commercial contributors include IBM, Texas Instruments, Burroughs, and an international cooperative link between ICL of the UK and Three Rivers of the United States. The efforts of the Microelectronics and Computer Technology Corporation (MCC), the Semiconductor Research Corporation (SRC), programs of DARPA (on strategic computing), the National Science Foundation (to provide computer access), the Department of Energy (to advance the state-of-the-art in parallel computing), and the National Bureau of Standards (to develop a computer research facility to explore characterization of concurrent processing concepts) should also be considered relevant.

Both Hitachi and Fujitsu have gone to considerable trouble to deliver the first versions of their machines on schedule. For example, the largest Fujitsu machine, the VP200 at Numazu, has been fabricated from 16K RAM rather than the 64K RAM parts expected to be used in the final version, and contains only 0.5 million words of memory in lieu of the 32 million words of memory announced for the final version. The largest Hitachi machine, the S-810/20 at the University of Tokyo, underwent modification just prior to final delivery changing its peak performance rate from 630 MFLOPS to 857 MFLOPS. The Hitachi machine employs water cooling. While neither these facts nor the observations of scientists and engineers indicates the machines cannot be mass produced, it does point out significant problems yet to be resolved. A second factor relates to the degree of success the Japanese will have with a larger base of supercomputer users, and the IBM response if that base should prove to be a significant demand of IBM machines. A third factor relates to the development of wafer scale integration and the related "computer on a chip" projects. Both of these promise to significantly improve the performance attainable in the mainstream computer marketplace, and the success of these efforts could influence the sales of supercomputers outside of the scientific base. A fourth factor relates to the use of Japanese supercomputers in Japan. The American supercomputer base is well established, and the performance of American supercomputers and their delivery is not a question of speculation.

Summarizing, basic research in supercomputers covers all topics and is therefore not assessed as a single line item. The Japanese lag in advanced development for historical reasons but are moving to close the gap. Similarly, the U.S. ships many more machines than Japan, but we can expect the Japanese to try to catch up. Finally, the Japanese software for supercomputers is superior, but U.S. firms are expected to catch up as the market increases the incentive.

### 3.3.3 Professional Workstations

The term "professional workstation" refers to a class of single user computers more powerful than personal computers and generally equipped with bit-mapped graphics and local area networks. We define the IBM PC/AT and the Apple Macintosh as marginally outside of our category of "professional workstation" and the Apple LISA as just inside it. Professional workstations systems have been used for different applications such as office automation and LISP programming. Some of the better known general purpose offerings come from Apollo Computer, Perq Systems Corporation, and Sun Microsystems. The primary vendors of LISP machines are Xerox, Symbolics, and LISP Machine. The Apple LISA, as well as the Xerox Star, are examples of office workstations. All of these examples are American.

The panel identified some similar work in Japan. The ICOT Personal Sequential Inference Machine, which we have described in the AI section, is a prototype system optimized for the PROLOG language. Research is underway on more advanced machines, including LISP machine work, at ITL, NTT, Hitachi, NEC, and Kyoto University.* The work on Japanese word processors has encouraged the use of more powerful hardware in word processing. In fact, FUJI Xerox offers a Japanese word processor on the same hardware as used by the Xerox Star.

### 3.3.4 Clones

NEC is manufacturing an INTEL 8086 microprocessor clone and both Fujitsu and Hitachi manufacture IBM S/370 clones. These clones include additional instructions and extensions for enhanced functions. These are software compatible systems and are highly price competitive.

Because clone systems require no basic research, we have not assessed this subject. However, in the tradition of Japan's interests in accelerating its entry into new markets, we expect them to continue to lead and, in fact pull away, in both advanced development and product engineering.

## 3.4 Communications

Communications is an integral part of computer science and R&D in communications parallel other computer science programs.

Figure 15 is our summary assessment of communications technology.

Digital communications is particularly complex for the Japanese because of the Kanji character set. Encoding Kanji characters creates input problems, bandwidth problems (because they take two bytes/character as opposed to one byte for Roman characters), and display and output problems. As a

---

* Professional CAD/CAM and graphics workstations were excluded from consideration by the panel.

result, there is considerable interest in Japan in the use of facsimile. These factors create variations in the assessment presented in Figure 16.

| | BASIC RESEARCH | ADVANCED DEVELOPMENT | PRODUCT ENGINEERING |
|---|---|---|---|
| COMMUNICATIONS | < ↘ | NOTE 1: ○ → | + ↗ |
| | | NOTE 2: — → | < ↘ |

NOTES:
1. Hardware.
2. Software.

FIGURE 15 ASSESSMENT OF COMMUNICATIONS TECHNOLOGY (SUMMARY OF FIGURE 16).

| | BASIC RESEARCH | ADVANCED DEVELOPMENT | PRODUCT ENGINEERING |
|---|---|---|---|
| LOCAL AREA NETWORKS | < ↗ | — → | ○ → |
| HARDWARE | ○ → | ○ → | ○ ↗ |
| PROTOCOLS AND SOFTWARE | < → | — → | < ↗ |
| FACSIMILE/OFFICE AUTOMATION | < → | + → | + ↗ |

FIGURE 16 DETAILED ASSESSMENT OF COMMUNICATIONS TECHNOLOGY

### 3.4.1 Communications Research

Japanese research in hardware for communications is not rated uniformly low as is their research related to other computer science topics. The problems mentioned above have driven the Japanese to increasingly fast digital communications systems, and they are investing considerable research in fibers, networks, optical computing and communications, materials, etc.

Research at the Electrical Communications Laboratories (ECL) of Nippon Telephone and Telegraph (NTT) is especially interesting. The charter of ECL is to conduct telecommunications R&D and to write draft specifications for equipment purchased by NTT. The four laboratories conduct related but different research.

Musashino (one of four ECL facilities) concentrates on integrated network, switching and communications processing systems, peripheral memory equipment (e.g., magnetic disks), and fundamental research. Yokusuka Labs work on data communications systems, transmission systems (optical, satellite), visual communication systems (TV teleconferencing, facsimile), and information terminal devices (telephone equipment). Ibaraki Labs studies component parts and materials (e.g., fiber optic cable and other devices) and develops outside plant systems. Atsugi ECL concentrates on IC research (LSI, optical communications, VLSI CAD, and Semiconductor Components (GaAs).

These laboratories spend about 70 billion Yen in direct research expenses not counting salaries for about 3100 researchers out of a total staff of 3500.

ECL has identified a variety of technologies for development in advancing communication networks such as satellite and optical transmission capabilities, switching, and information processing. Their emphasis is clearly on digital technology, and LSI is the medium. There is some attention to pattern recognition. NTT's emphasis is on electronic switching systems (ESS). Their brochure (June 1982) states:

> . . .ESS. . .will form the core of future communication networks in Japan . . .research and development activities are being conducted on new service facilities and switching systems in which the ESS technique is applied. These include a mobile communication system (maritime mobile telephone and automobile telephone), a home use facsimile communication system. . . The ECLs are taking advantage of the rapid progress in LSIs and digital technology and have been conducting research and development on digital telephone switching systems. . .

The emphasis at Musashino is in the development of an integrated network system (INS) — an overall system incorporating telephone, telex, facsimile, and video, to name a few.

Some of the technical aims of the INS prototype are:

(1) Digital Interface for Subscribers
(2) Service Integration by Digital Switching Systems
(3) Communications Processing
(4) Digital Terminals (Not surprising, the goal is to digitize the subscriber's lines.)
(5) Picture Service Switching (Video switching requires broad band channels and digital switching. ECL ESS systems are embodied in their D-10 and D-20 systems. Development research is currently on a common channel signaling network, a high level language for ESS and small size ESS's such as an electronic PABX or EPABX.)

The Research Division at Musashino was established in the mid-60's. About 6% of their personnel work in pure basic research or objective basic research while more than 80% of the staff work in applied research, fundamental development, and development.

### 3.4.2 Local Area Networks

The Japanese need to move high bandwidth data has driven the efforts in Local Area Networks (LAN). The panel was not exposed to considerable data in this arena, but work at Oki Electric, Mitsubishi Electric, NEC, ECL, ETL, and Toshiba was reviewed.

We concluded that the Japanese lag in this category of basic research but will attempt to catch up.

Value Added Networks (VAN), such as TELENET, TYMNET, and COMPMAIL, that embody the concept of LANs over a large geographic area are significantly more prominent in the U.S. than in Japan. The U.S. experience in ARPANET is now approaching 10 years and the proliferation of this technology is simply not matched in Japan. Nevertheless, the NTT efforts in Integrated Network Services, the fact that it is a coordinated national effort, and the heavy reliance on new technology, points to increasing Japanese competition.

### 3.4.3 Hardware

We have already commented on the extensive research effort in communications hardware in Japan. Every organization contacted or reviewed had programs in hardware technology, optical computing, fibers, electronic switching, and more. For these reasons we consider the Japanese even with the U.S. in all three assessments, and we predict their natural market forces will close the gap in shipped products.

### 3.4.4 Protocols and Software

Nippon Electric Corporation (now NEC Corporation) is particularly interested with software productivity and quality. They have established efforts in software product engineering, microcomputer software (embedded systems), and software management. Their main concern is in developing applications software, and they have a company-wide program for reusing software.

Musashimo ECL has research sections committed to artificial intelligence, architecture, and software. Their research topics include programming environments, natural language, and packet switching. They have developed protocol models similar to the ISO OSI (International Standards Organization's Open System Interconnection Standard). Their Data Communications Network Architecture carries two extra functions with parallel overlaps.

Mitsubishi Electric (Kamakura), the organization that built the ICOT SIM/PROLOG machine, has an Information Systems and Electronics Development Lab with interests in packet switching, LAN, and office automation. For example, their Optical Communications Systems Department has an operational 32M bit optical LAN. Growing from 150 to 1800 R&D people in 3 years, they are an engineering oriented developer of software. They use flowchart tools developed by NTT, Ada-like semiformal design languages, and they are familiar with integrated design software production systems and tools. They place some emphasis on software complexity meters.

The Panel concluded that Japan's position in basic research and advanced development is behind the United States, but the inevitable catchup in production engineering would occur.

### 3.4.5 Facsimile/Office Automation

Technology in office automation is concentrated on solving problems associated with the Japanese language.[1] The problems of processing Kana and Kanji is unlike any language problem in Western cultures. The Japanese place heavy reliance on facsimile transmission and they are emphasizing new communications technology (fiber optics, optical computing, integrated network services).

The Japanese devote considerable resources to the development of office automation hardware (LAN's, copiers, facsimile machines) and Kana-to-Kanji input devices. We could not discern research program of consequence in related technologies such as relational database systems, electronic mail, or wordprocessing.

Most word processors are 16-bit based. This creates some difficulty since it limits address space, and Kanji characters require twice the resident memory as Roman characters (2257 Byte). Thus, in both input and processing efficiency, Japanese wordprocessor systems have fundamental competitive deficiencies.

The input problem cannot be minimized. Most of the major venders place substantial marketing resources on their "Japanese Language" capabilities.

| | |
|---|---|
| • Fujitsu | Japanese Processing Extended Feature |
| • Hitachi | Kanji Information Processing System with a high speed laser output |
| • NEC | Japanese Language word processor including an OCR that reads handwritten characters |
| • OKI | Personal Computer Model 30 featuring Japanese language processing and Japanese language word processors |

Leading Japanese electronics firms were still demonstrating prototype "forms-driven" OA systems for their clerical staff. This system, without electronic mail resources was considered to be 1970 technology. We concluded that the Japanese were far behind in basic research except in the area of communications technology. Japanese firms bring new products to market faster than U.S. firms. This advantage is somewhat mitigated by their relative weakness in software and protocols. Because of the need to build high band width communications devices and their general production strengths, we granted Japan the edge in product engineering. Again software deficiencies mitigated the rate of change.

## 3.5 Miscellaneous Topics

Two additional subjects attracted the attention of the panel: university programs in computer science and software protection.

## 3.5.1 University Programs in Computer Science

### 3.5.1.1 The Educational System

The secondary schools are divided into lower and upper sections. The upper secondary schools introduces some technical work. A "technical college" might include the upper technical school as well the next level, the junior college. The highest level is the university. A discussion of the technical schools from the point of view of the computer field is found in "Computer Education in Technical Schools."[1]

### 3.5.1.2 The Universities

The most important public universities are: the Universities of Tokyo, Kyoto, Osaka, Tohoku, Hokaido, Nagoya, and Kyushu. Tokyo University (or Todai) is the most prestigious. In addition, Tsukuba University and the Tokyo Institute of Technology contribute to the computer science field.

Private universities are less important in Japan than in the United States. Two private universities that offer computer science are Keio University and Waseda University. There is also theoretical work at Kyoto Sangyo University.

### 3.5.1.3 Statistical Aspects of Japanese Technical Education

We do not have adequate information on the statistical aspects of the Japanese computer related education. The numbers quoted in the references do not match data quoted by NSF. According to data from "Computer Education in Technical Schools,"[1] the total number of undergraduate students in computer sciences related disciplines in 1981 was 1420; there were also 460 MS students and 110 PhD students. These figures seem unrealistically low, and it may be that they do not include computer engineering programs. "Science and Technology Education in Japan - Prospects & Future Tasks"[2] gives information on the number of researchers in the technologically advanced countries as well as a breakdown of the degrees by subject.

### 3.5.1.4 How Computer Science and Engineering Are Organized

There is a variety of academic organizations that house computer-related subjects in Japan. At the University of Tokyo, for example, there is both a Department of Electrical Engineering and a Department of Electronics Engineering. There is also a Department of Information Science, which appears to be within the College of Engineering, as well as another program in the College of Science. Graduate programs appear to span departmental groups. There is a Department of Mathematical Engineering and Industrial Engineering at the University of Tokyo, and some robotics work is done in the Department of Precision Engineering.

Most of the older Japanese universities are organized around the Chair model that is found in the European institutions. A Chair is a discipline which consists of a full professorship, an associate professorship, and perhaps several assistant professorships. Advancement through the ranks is slow and depends on a vacancy occurring. Universities founded after World War II tend to follow the American model.

### 3.5.1.5 Research in Universities

The single striking contrast between Japanese and American universities is the area of graduate level programs and research. In the United States, much of the basic research in computer science is conducted in universities. In Japan, basic research activity is predominant in industrial centers. The Japanese emphasis on research in industry is, in part, due to the unique relation of the Japanese employee to the corporation. Because the Japanese employee remains with a firm much longer than his American counterpart, there is more incentive to invest in training programs by Japanese corporations. This training extends to sophisticated areas including basic research.

### 3.5.1.6 Financial Support of Universities

The Ministry of Education funds some research. Research funding also comes from the Agency for Industrial Science and Technology (AIST), an agency of MITI, which appears to work something like the U.S. National Science Foundation.

Japanese central computing centers are better funded and equipped than their U.S. counterparts. The Tokyo University Computing Center has a Hitachi S810 model 20 which is one of the world's fastest computers. There is a variety of M280H dual-processor computers including array processors, and there are at least three such machines coupled through a local area network. The M280H unit processor is rated between 12 and 17 mips. One of these is equipped with several integrated array processors, whose peak throughput is 33 megaflops. (The manufacturer rates the S810/20 at 630 megaflops peak). This computing center also serves as a regional center networked to other universities which obtain some computing services at Tokyo. Other university and nonregional computing centers have an impressive amount of equipment.

Laboratories are more modern and well equipped than other university facilities which tends to show where the Japanese are placing emphasis.

There is little VLSI equipment and research being carried out at Japanese universities. However, large computer manufacturers provide prototypes of their new models to the universities to assist in debugging.

The computing environment in Japanese universities is devoted towards interactive use of large central facilities. FORTRAN appears to be the language of choice. The collection of individual or group VAXs and personal workstations seen in U.S. universities is just beginning to materialize in Japan. This is probably holding up software development in the universities.

#### 3.5.1.7 University Personnel

Many senior Japanese professors serve on national committees, the ICOT board, and other important positions. They are thoroughly integrated into the Japanese computing community. We were told that professors at public universities could not be consultants to private industry. On the other hand, we did discover industrial projects at universities, although these tend to be mostly development rather than research work. We also learned that, while a professor may not consult, he may advise a company. While it would be inappropriate for him to be paid for consulting, he can accept honoraria and equipment in his laboratory in exchange for his advice. The growth prospects for the universities may be hampered by the political difficulties of establishing new chairs. The way around this problem seems to be starting new departments. This may be why the number of computer-related departments proliferates. It is worthwhile exploring the structure of two universities, Tokyo and Tsukuba.[3]

The Department of Electrical Engineering at the University of Tokyo has eleven full professors. Two hold computer-related positions — one in artificial intelligence and the other in computer architecture. At Tokyo, there is also a Department of Electronic Engineering with four full professors — one each in digital communications, electronic materials, microwaves, and optics. There are three computer-related full professors at the University of Tokyo in the Department of Mathematical Engineering. The assignment of professors in the computer field among various technical departments tends to diffuse the overall information science effort in the undergraduate area. It is at the graduate level where collaboration takes place from different departments.

The University of Tsukuba was patterned after the American rather than the European system. The organization consists of a cluster of colleges or schools, a graduate school, and a research organization with specialized institutes and projects. There is a College of Information Sciences at Tsukuba, with a quota of eighty students. In the same cluster, there is a College of Engineering Sciences with 160 students. The Japanese prefer the term "information science" rather than "computer science." The Institute for Information Science at Tsukuba, headed by Professor Nishino, is well funded and includes professors from different departments. There are forty-eight associated faculty members, of whom fifteen are full professors and roughly the same number are either Associate Professors and Assistant Professors. There are research projects on a wide variety of subjects. This is one of the better balanced programs that we observed. It covers all branches of information science as well as research projects on a wide variety of subjects, although VLSI and hardware are missing. Presumably, electrical engineering covers these topics at Tsukuba, but this is not confirmed.

The number of university graduates is too low, and there is a shortage of trained people at every level.

3.5.1.8 Conclusions

Japanese universities are selecting highly qualified high school students and training them reasonably well. They lack the latest and most advanced computing systems and they rely on central, well-funded facilities which provide several cycles in an IBM-compatible mode. It seems likely that they will switch over to the VAX/UNIX combination which has been so productive in the U.S. VLSI equipment and research is absent in the universities.

The single distinction between the Japanese and American educational system is the concentration of Japanese research in industry. Although there are good entry-level educational opportunities in industry, continuing education is not always available and not always adequate. The lack of strong academic research centers and a large graduate student body hampers long-range prospects for innovation.

3.5.2 Software Protection

Despite our assessments on computer science software product engineering, the Japanese feel that they are behind in the development and use of computer software. This is considered by them to be a result of the decision to build IBM compatible machines. They feel (as we did) that the gap is most intense in such areas as object-oriented systems and distributed computing.

There is concern in Japan that access to new technology can be cut off at any time by the U.S. government. More specifically, they are concerned with things such as UNIX licenses, delayed access to new software systems, and export controls in general.

One idea mentioned by two different Japanese companies was the concept of creating software research laboratories in the United States. Such an approach would be an effective way of exporting technology from the United States to Japan.

The Japanese have also begun to think about ways of protecting their domestic software marketplace. In December 1983, a proposal was sent to the Japanese Diet for a new law concerning software. The full title of the document was "Proposal with Respect to Rearrangement of the Foundation for Software—Aiming at Securing Legal Protection for Software—an Interim Report." This bill was proposed by the Industrial Structure Council and the Information Industry Committee. The principal authors include a number of senior computer technologists who serve on the boards of other important committees and organizations such as ICOT.

The level of the people corresponds to presidents of small colleges and vice presidents of major corporations. The three stated goals of the document are:

1. The promotion of the software development
2. The prevention of duplicate investment
3. The promotion use of software

According to the bill, authors gain rights to the programs they have written through registration. The rights to the programs are for 15 years, which is a shorter period than most countries. The form of program registration proposed includes detailed documentation and source code listings. Section 12 is the most interesting: "(12) Arbitral (Compulsory) System—A system should be established whereby newly produced programs based on existing programs are patented inventions; programs in the public interest, programs which have not been worked, and other such programs are available for use and reproduction etc. in exchange for appropriate consideration upon certain conditions and subject to arbitration as necessary."

Other parts of the document indicate that it is the intention to allow cross-licensing of programs. In other words, the government may provide a license to one company to use the software created by another company.

This proposal created a storm when it was introduced. Trade representatives expressed serious reservations about the bill. The Ministry of Education has submitted an alternative bill and, at last report, both bills have been temporarily withdrawn while the agencies maneuver. The basic conflict as to whether software should be covered by a copyright, or a trademark, or a patent under Japanese law is still open.

# 4. Some Commercial Observations

## 4.1 Overall Assessment

It is clear that the U.S. will dominate basic computer science research over the next decade. The Japanese are aware of their deficiencies, and they are reassessing their position. The Japanese are highly competitive and strive for excellence. They enjoy the reputation of being "the best," and this attitude could lead to more emphasis on research.

Japan has been successful in focusing efforts on a national scale in order to penetrate and ultimately dominate a particular market. Should a decision be made that more basic research was required in computer science, one could easily imagine an effort under the auspices of MITI to redress any deficiencies.

## 4.2 Observations

The body of the report contains observations, conclusions, and forecasts. We do not expect any immediate commercial upsets because U.S. research investments are still sufficiently greater than Japan's except in the area of supercomputers.

Resources for supercomputer development in the United States are considerably less than those in Japan. The National Superspeed Project, with a ten year budget of approximately one hundred million dollars, has two to three times this amount available from commercial sources. To these amounts, the internal funding efforts of Fujitsu, Hitachi, and Nippon Electric must be added. In the United States, the only organizations devoted exclusively to supercomputer development are CRAY and ETA both of which are small organizations with relatively small budgets.

## 4.3 Recommendations

There are several broad areas in which the United States can strengthen its position in the computer science field.

- Encourage venture capital to invest in new high-technology in the interest of speeding up the movement of products from the laboratory to the market.

- Increase research efforts in both public and private universities with added government and industry support.

- Create incentives that encourage basic researchers to follow their ideas through advanced development and product engineering.

- Plan now for the training of competent computer science personnel who will be required at all levels in the near future.

- Better and expanded training is required in secondary schools, technical training institutions, community colleges, universities, and the extensive training facilities maintained by the armed forces.

The U.S. government should continue to press the Japanese government to open computer and communication markets to American vendors. The U.S. government should also remain alert to issues that may act to restrict trade such as the pending Japanese software legislation.

U.S. industry and government needs better methods of monitoring foreign technical information. Because of language, timely information from Japan is particularly difficult. In addition to encouraging and supporting technical studies in our schools, we need to reinstitute and reemphasize the study of foreign languages especially Japanese. State and industry scholarships in Japanese language is one means of achieving this goal.

In the meantime, expert technical translation services from Japanese to English are required now. If American industry requires current Japanese technical data, it must establish the means to obtain the data and pay the costs.

It is finally being recognized that the cost of research and advanced development are too heavy for any one corporation. Therefore, more R&D coalitions such as MCC are needed.

The U.S. government should consider adapting the Japanese practice of requiring a prototype for any government funded research project. This practice would help speedup the movement of advanced development.

# 5. References

**Section 3.1**

1. Y. Matsumoto and others, "SWB System: Software Factory" Software Engineering Environments, H. Huenke, ed., North-Holland Publishing Company, 1981, pp. 305-318.
Matsumoto describes the software factory.

"A Software Design Methodology: Bridge from Requirements Specification to Software Design" Japan Annual Reviews in Electronics, Computers and Telecommunications: Computer Science and Technologies, T. Kitagawa, ed., North-Holland Publishing Company, 1982, pp. 175-192.
Describes the design methodology.

2. Database Engineering, IEEE Computer Society, 6, No. 1, (March 1983).
This is a special issue devoted to the latest reports about the Japanese commercial databases.

**Section 3.2.1**

1. Hisao Yamada, "Japanese Text Input Methods in Flux," Japan Annual Review in Electronics, Computers & Telecommunications: Computer Science & Technologies, T. Kitagawa, ed., OHM, Tokyo, 1983, pp. 101-117.

2. Ken-ichi Mori, Tsutomu Kawada, Shin-ya Amano, "Japanese Word Processor," Japan Annual Review in Electronics, Computers & Telecommunications: Computer Science & Technologies, T. Kitagawa, ed., OHM, Tokyo, 1983, pp. 119-127.

**Section 3.2.2**

1. Hideki Kasuya, "Japan's Drive for Speech Recognition," Speech Technology, September/October 1982, pp. 10-20.

2. R. Dixon, T. Martin, eds., Automatic Speech and Speaker Recognition, IEEE Press, 1979.

3. Osamu Ishii, "Outline of the PIPS Project and Prototype of the Total System," Japan Annual Review in Electronics, Computers & Telecommunications: Computer Science & Technologies, T. Kitagawa, ed., OHM, Tokyo, 1982, pp. 6-22.

4. Toru Ifukube, Katsuhiko Shirai, eds., "Special Issue on 'Speech'," Journal of the Robotics Society of Japan 2, (1), February 1984.

5. J. Flanagan, and others, Application of Automatic Speech Recognition in Severe Environments, Washington, D.C.: National Academy Press, 1984.

Section 3.2.3

1. Jonathan Slocum, "Machine Translation: its History, Current Status, and Future Prospects," Association for Computational Linguistics, Proceedings of the Conference, 1984, Stanford University, California, July 1984, pp. 546-561.

2. Jun-ichi Nakamura, Jun-ichi Tsujii, Makoto Nagao, "Grammar Writing System (GRADE) of Mu-Machine Translation Project and its Characteristics," Association for Computational Linguistics, Proceedings of the Conference, 1984, Stanford University, California, July 1984, pp. 338-343.

3. Jun-ichi Tsujii, Jun-ichi Nakamura, Makoto Nagao, "Analysis Grammar of Japanese in the Mu-Project: A Procedural Approach to Analysis Grammar," Association for Computational Linguistics, Proceedings of the Conference, 1984, Stanford University, California, July 1984, pp. 267-274.

4. Makoto Nagao, Toyoaki Nishida, Jun-ichi Tsujii, "Dealing with Incompleteness of Linguistic Knowledge in Language Translation: Transfer and Generation Stage of Mu Machine Translation Project," Association for Computational Linguistics, Proceedings of the Conference, 1984, Stanford University, California, July 1984, pp. 420-427.

Section 3.2.4

Kaufman, The Handbook of Artificial Intelligence, Volumes 1-3, Los Altos, California, 1981 & 1982.

Paul Kinnucan, "Computers That Think Like Experts," High Technology, January 1984, pp. 30-42.

Hitoshi Aizawa, ICOT Journal, Tokyo, 1983.

ICOT Technical Report Series, Tokyo, n.d.

T. Moto-oka, ed., "Fifth Generation Computer Systems," JIPDEC, North Holland Publishing Company, 1982.

E.A. Feignenbaum and P. McCorduck, The Fifth Generation. Reading, MA: Addison Wesley Publishing Company, 1983.

Section 3.2.5

Koichi Furukawa, Akikazu Takeuchi, Susumu Kunifuji, Mandala: A Concurrent Prolog Based Knowledge Programming Language/System, Technical Report, TR-029, ICOT Research Center, November 1983.

D.H.D. Warren, L.M. Pereira, F. Pereira, "PROLOG - The Language and Its Implementation Compared to LISP," SIGPLAN Notices 12, (8), 1977, pp. 109-115.

P.H. Winston and B.K.P. Horn, LISP. Reading, MA: Addison-Wesley Publishing Company, 1981.

**Section 3.2.6**

Dwight B. Davis, "English: The Newest Computer Language," High Technology, February 1984, pp. 59-64.

Makoto Nagao, "A Survey of Japanese Language Processing -1980 1982-," Japan Annual Review in Electronics, Computers & Telecommunications: Computer Science & Technologies, T. Kitagawa, ed., OHM, Tokyo, 1983, pp. 94-100.

Hideki Yasukawa, LFG in PROLOG: Toward a Formal System for Representing Grammatical Relations, Technical Report, TR-019, ICOT Research Center, August 1983.

**Section 3.4.5**

1. Yuji Yamadori, "Office Automation in Japan," Science & Technology in Japan, (10), April/June 1984, pp. 24-26.

**Section 3.5.1**

1. "Computer Education in Technical Schools," JIPDEC [Japan Information Processing Development Center] Report, No. 53, 1983, pp 54-61.

2. K. Oshima, "Science and Technology Education in Japan - Prospects & Future Tasks," Journal of Japanese Trade & Industry, No. 5, 1983, pp. 18-20.

3. "Data Processing Education at Universities," JIPDEC [Japan Information Processing Development Center] Report, No. 53, 1983, pp 38-53.

# Appendix: Literature Support to Computer Science Panel*

In addition to technical assessments, an important objective of the JTECH Program is the identification of relevant and timely Japanese S&T source material for the technical areas addressed by the panels. Thus, panel members are requested to evaluate, from their technical perspective, the usefulness of the various types of source materials provided. Panel members are also encouraged to give recommendations on Japanese source material in their technical areas which they feel should be acquired and dissiminated to the U.S. technical community on a continuing basis.

In this appendix, we summarize:

- the JTECH approach in providing literature/translation support to the panels, and a description of the general types of material provided,
- open-source Japanese literature available in the computer science area,
- literature provided to the computer science panel by the JTECH staff, and
- summary and recommendations

## A.1 Approach

The following approach has been followed in providing panel members with pertinent source material:

1. Initially, at the panel "kickoff" meeting, the JTECH staff provides general source material for panel members' review. This material is mainly in English and usually of an overview nature. Also presented and discussed is background information on the various types of Japanese publications, technical society

---

* This appendix written by the JTECH Program Staff; the staff member responsible for literature and translation support is Dr. Y. Kim.

and working group meetings, organizations, etc. which are potential sources of information for particular technical topics.

2. Each panel member then ascertains particular source material (including papers in Japanese) needed for their technical topic.

3. The JTECH staff then collects, evaluates, and translates those requests by the panel members.

(Steps 2 and 3 are iterative, with several interactions taking place during the panel assessments together with trips to Japan by Y. Kim to collect information.)

4. Toward the end of the panel life, the panel members, together with the JTECH staff, assess the source materials found to be most useful for the technical area. Also, panel members are encouraged to offer recommendations related to more effective methods of acquisition/translation/dissemination for the U.S. technical community.

Under the JTECH literature support approach, two factors are emphasized:

A. Timely, Pertinent Source Material. In addition to providing panelists with readily available "open" material (such as excerpts from Japanese technical journals and periodicals), the emphasis of the approach is in providing very recent "semi-open" sources of Japanese research accomplishments and plans (such as results of Japanese ad hoc seminars and working groups). (These different categories of source material are discussed in the next section.)

B. Selectivity. To be effective under practical time and budget constraints, it is, of course, important that some selectivity be exercised in the source material provided. In providing the panelists with translated material, the following "selective" approach is used: (1) The titles of papers from a Japanese document are translated and distributed to panelists; (2) for seemingly interesting titles, a panelist requests that certain abstracts be translated (this step is usually done by telephone); and, (3) based on the abstracts, the full translation of the most relevant papers is performed. (In applying this procedure for the Computer Science Panel, only about 10% of the total Japanese source material collected required full translation, as indicated later in Section A.5.)

## A.2 Categories of Source Material

In the JTECH approach for providing literature support to panelists, we consider the Japanese S&T source material in terms of three categories. These categories are defined below, and the material provided to the computer science panel in each category is listed in the next section.

### A. General Material (mostly in English)

These are materials initially provided to assist panel members in identifying specific source material needed for their particular area of expertise. This material is typically shallow in technical depth, but provides a general program background. Included are popular overview articles, previous technology assessment studies, and translations of Japanese Government publications on overall programs, research objectives, participating organizations, funding levels, etc.

### B. Open Material

This category consists "openly available", regularly published technical publications. A list of the technical journals pertinent to computer science is given in the next Section.

C. Semi-Open or Closely-Held Material

The emphasis under the JTECH program is in providing panelists with this category of source material (described below), which is of a more timely benefit, as illustrated in Figure A-1.

Japanese academic societies hold biannual or annual meetings, and participants are provided with the abstracts of the meeting. Many manuscripts of these abstracts are hand-written; they are not reproduced in regular journal publications. Frequently these abstracts contain very current and useful information, but they have to be accessed through privately-held material. In this sense, the meeting abstracts should be regarded as semi-open.

In the Japanese high tech community, frequently ad hoc seminars and conferences are organized to pool and disseminate the state-of-the-art research information. Such meetings are attended usually by the invitees only, and the conference proceedings are printed in Japanese with most manuscripts being handwritten to allow entry of the latest information. Some of the more formal technical information contained in these proceedings eventually work their way through the appropriate professional journals published in English, but with a typical delay of six to eighteen months.

Japanese industries participate in national R&D projects under a "Research Association" type of arrangement. (The ongoing national projects in the computer science related area are: Supercomputers, Optoelectronics, Future Electronics Devices, and Fifth Generation Computers.) Technical progress made in these projects are disseminated first in the technical committees composed of member companies, and refined (edited) technical papers reach the public domain at a much later time. In a highly competitive R&D development, timely access to the raw data is important.

Most Japanese high-technology industries maintain in-house R&D centers that are staffed with leading technical expertise. Many companies publish periodical reports that contain some useful technical information. Of course, commercially sensitive technical information is closely-held as proprietary.

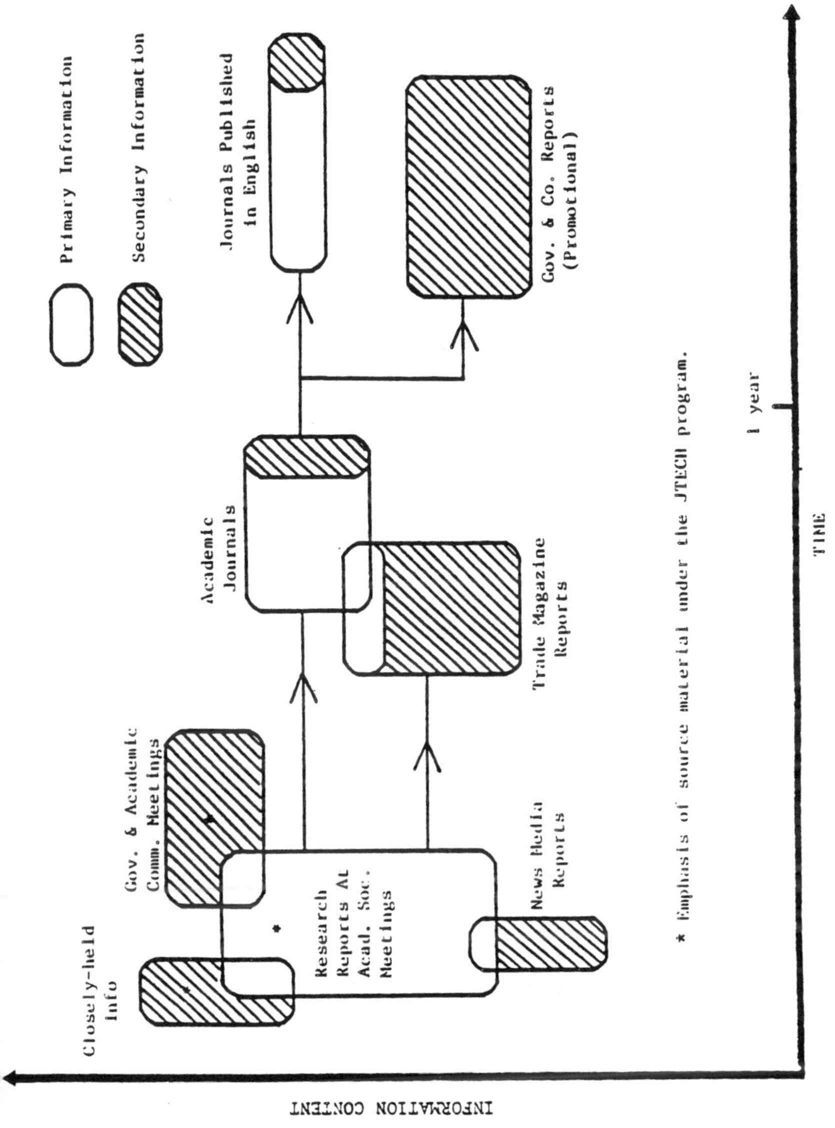


Figure A-1. Flow of Technical Information from Japan.

## A.3 Open Sources of Japanese Literature in Computer Science

### 1. Professional Societies

Japanese literature in computer science is disseminated mostly through the two main professional societies.

#### A. Information Processing Society of Japan (IPSJ)

The IPSJ, which is similar to the U.S. Association of Computing Machinery, publishes a bimonthly Japanese language journal of referred articles called "Transaction of Information Processing Society of Japan."

IPSJ also publishes a monthly journal called "Joho Shori," which is less formal than the Transaction and contains many special issues on current topics, tutorial articles, reviews on overseas development, and other society news.

IPSJ publishes a quarterly English-language journal of refeered articles called "Journal of Information Processing." In addition to technical articles, it contains: 1) English translations of the abstracts of the articles appearing in the Transaction, 2) Abstracts from the various Working Group preprints, and 3) English translation of content pages (article titles and authors) of "Joho Shori."

#### B. Institute of Electronics and Communication Engineers of Japan (IECE)

The IECE, similar to the U.S. IEEE, publishes a monthly Japanese language journal called "Transactions of the Institute of Electronics and Communication Engineers of Japan." The 1984 volume consists of 5 different parts: A. Theory, B. Electronic Circuits, C. Communication, D. Information, and E. Articles in English.

The articles relating to computer science are contained mostly under part D. The English-language publication, Part E., is much like the JIP English Journal, but its coverage spreads to broader subjects.

IECE publishes a monthly Japanese language tutorial journal called "Journal of IECE." It serves the function of "Joho Shori," but covers broader areas. The volume numbers of this Journal run parallel with those of the Transactions.

Each of these societies holds Working Group research meetings (ten meetings per year), and the meeting reports contain the more current information. However, these reports are usually distributed to members only and should be regarded as semi-open source material. The breakdown of these meetings into various subfields and subtopics is given below:

(I) Computer Architecture

(A) IPSJ
- (1) Computer architecture
- (2) Microcomputers
- (3) Control and evaluation of computer systems
- (4) Distributed systems
- (5) Automated design

(B) IECE
- (1) Computers

(II) Software Technology

(A) IPSJ
- (1) Software theories
- (2) Software engineering
- (3) Database system
- (4) Symbols process
- (5) Knowledge engineering and artificial intelligence
- (6) Natural language
- (7) Graphics and CAD
- (8) Numerical analysis

(III) Network

(B) IECE
- (1) Communication methodology
- (2) Exchange (telephones)
- (3) Information network

(IV) User Friendly Interface

(A) IPSJ
- (1) Computer vision
- (2) Japanese language input

(B) IECE
- (1) Graphics engineering

(V) Digital Transmission

(B) IECE
- (1) Antenna and signal propagation
- (2) Satellite communication

(VI) Devices

(B) IECE
- (1) Semiconductors, transistors
- (2) Electron devices
- (3) Superconducting electronics
- (4) Quantum electronics
- (5) Electronic parts and materials
- (6) Magnetic recording

(VII) Theory

(B) IECE
- (1) Automaton and language
- (2) Pattern recognition and learning
- (3) Educational material

2. Society Publications

(A) IPSJ
- (1) Transactions of IPSJ (bimonthly in Japanese and quarterly in English), listed in IA
- (2) Joho Shori (review and tutorial), listed in IA
- (3) Working Group Research Meeting Reports
- (4) Symposium Proceedings
- (5) Biannual IPJS Meeting Proceedings

(B) IECEJ
- (1) Transactions of IECE
  A. Theory, B. Electric Circuits, C. Communication, D. Information, E. English
- (2) Journal of IECE (Review & tutorial articles), listed in IB
- (3) Working Group Research Meeting Proceedings
- (4) Symposium Proceedings
- (5) IECE Annual Conference Proceedings
- (6) IECE Biannual Topical Conference Proceedings

(C) Other Society Journals
- (1) Software Society (newly established)
- (2) Television Society
- (3) Electron Graphics Society
- (4) Robot Society
- (5) Office Automation Society

3. Research Reports

(A) National Projects
- (1) ICOT
- (2) Superspeed Computer for scientific computing
- (3) Research Association for new functional devices
- (4) Research Association for optical devices and control systems

(B) Laboratory Research Reports
- (1) Electrotechnical Laboratory Reports (ETL Reports)
- (2) ETL Research Reports
- (3) Electrocommunication Application Laboratory Reports
- (4) NTT Electrocommunication Laboratory Reports

(C) In-house Industry Laboratory Reports
Most Japanese industrial laboratories publish periodical research reports. Some examples are: Hitachi Review, NEC Technical Reports, Toshiba Review, Sony Central Research Laboratory Review.

4. Popular Magazines & Promotional Publications

- (1) Nikkei Electronics
- (2) Nikkei Computer
- (3) Electronics (Ohm Sha)
- (4) Business Communications (Planning Center)
- (5) Japan Information Processing Development Center (JIPDEC) Reports

(6) Computer White Paper, annual publication by JIPDEC
(7) INS (Information Network System)
(8) Frontier Technologies

5. Japanese Literature in Computer Science Published in English or Translated into English

(1) IPSJ English-language Quarterly & Section E (English) of Transactions of IECE
(2) JIPDEC Reports
(3) ICOT Journals and Technical Reports
(4) Systems*Computers*Controls (published by Scripta Publishing Co.) Translations of articles selected from the Journal of IECE.

6. Japanese Science & Technology Literature Available in the United States

The Library of Congress has an impressive holding of Japanese S&T literature, including almost all of the open-source material mentioned above. Most periodicals appear to arrive with a four to six month delay. The cataloging of Japanese S&T literature is done mostly in terms of romanization of Japanese titles. Therefore, it is rather difficult to make use of these materials unless the researcher is familiar with the Japanese language.

A.4 Literature Provided to JTECH Computer Science Panel

A. General

1. James C. Abegglen and Akio Etori, Japanese Technology Today, 1983. n.p.

2. James F. Decker and others, Recommended Government Actions to Retain U.S. Leadership in Supercomputers. Report to the Federal Coordinating Council on Science, Engineering and Technology Supercomputers Panel, August 1983.

3. Richard Dolan, "Japan's Fifth Generation Computer Project," Scientific Bulletin, pp. 63-97.

4. Kazuhiro Fuchi, Fifth Generation Computer: Some Theoretical Issues. Tokyo, Japan: Institute for New Generation Computer Technology [ICOT], n.d.

5. Christian Galinski, "VLSI in Japan," IEEE Computer, March 1983, pp. 15-21.

6. A. Carl Haussman, Very High Performance Computing: National Needs and Options in Energy R&D. Lawrence Livermore National Laboratory, Prepared Statement and Testimony before the Committee on Science and Technology, U.S. House of Representatives, 15 June 1983.

7. Barry Hilton, "Government Subsidized Computer, Software and Integrated Circuit Research and Development by Japanese Private Companies," Scientific Bulletin, Department of the Navy Office of Naval Research Far East: October-December 1982, pp. 1-22.

8. George E. Lindamood, "The Rise of the Japanese Computer Industry," Scientific Bulletin, pp.55-72.

9. S. . Okamatsu, "Japanese Computer Industry & Government Policy," Journal of Japanese Trade & Industry, Issue 2, (1982), pp.22-35.

10. K. Oshima, "Science & Technology Education in Japan," Journal of Japanese Trade & Industry, Issue 2, 1982. pp. 18-22.
This is a quarterly publication sponsored by MITI. It promotes international understanding of Japanese positions in world trade.

11. Harold S. Stone, Computer Research in Japan: A Comparison and Appraisal, Amherst, Mass.: University of Massachusetts, ECE Department, June 1983.

12. John Walsh, "Supercompeting over Supercomputers," Science Vol. 220, 6 May 1983, pp. 581-584.

13. Jack Worlton, Japanese Supercomputer Initiatives: Challenge and Response. Los Alamos, New Mexico: Los Alamos National Laboratory, n.d.

14. Preliminary Report on Study and Research on Fifth Generation Computers 1979-1980. Japan Information Processing Development Center (JIPDEC), Fall 1981.

15. "Publications List," ICOT Journal. Tokyo, Japan: Institute for New Generation Computer Technology (ICOT), pp. 32-38.
Contains list of technical reports and memorandums; abstracts of technical reports 1 through 23; technical memrandums 8, 9, 23, & 24.

B. Open

1. K. Kawanobe, "Present Status of Fifth Generation Computer Systems Project," pp. 15-23. "ICOT News," pp. 24-25. ICOT Journal Number 3, (1984).

2. Teijiro Kubo, "Trends in Personal Computer Software," JIPDEC Report, Number 57, 1984, pp. 10-24.

3. Mitsui Takahashi, "A User's View of Personal Computers," JIPDEC Report, Number 57, 1984, pp. 25-35.

4. Yuji Yamadori, "Office Automation in Japan," Science and Technology in Japan, Volume 3, Number 10, April/June 1984, pp. 24-26.

5. "Current News," JIPDEC Report, Number 57, 1984, pp. 41-45.

6. Digest of Japanese Industry and Technology (DJIT), Number 198, 1984, pp. 38-39, 42-64.

7. "Explanation on Patents Related to IC," Electronics Industries Monthly, pp. 41-42.

8. Journal of Information Processing. Vol. 6, No. 1-4, 1983.

9. "New Developments for an Information-Oriented Society,--Outline of White Paper on Science and Technology 1983," Science and Technology in Japan, pp. 27-30.

10. "Outlook of Policies Related to Information," Electronics Industries Monthly, Volume 26, Number 3, (March 1984), pp. 3-9.

11. Prospects of Demand for Electronic Industries. Japan Electronic Engineering Association, July 1983.

12. "Speech Research in the University and Laboratory," Journal of the Robotics Society of Japan: Special Issue on Speech, Volume 2, Number 1, (February 1984), pp. 40-46.

13. "Speech Synthesis and Recognition," Journal of the Robotics Society pp. 47-53

14. "Trends in Electronics Technology Reflected in U.S. Patents," Electronics Industries Monthly," pp. 31-40.

15. White Paper on Computers: Towards the Realization of New Media. JIPDEC, [Japan Information Processing Development Center] 1983.

C. Semi-Open and Closely-Held

1. H. Kashiwagi and K. Miura, Japanese Super-Speed Computer Project. Electrotechnical Laboratory and Fujitsu Limited, Japan, 1982.

2. Computational Scheme/Algorithm Workshop. Superspeed Computer Research Association, Tokyo University, March 1983.

3. Japanese Computer Related R&D Supported by MITI: 1983 Research Results Report and 1984 Research Plans. n.p., n.d.

4. Japanese 5G Computer System: Summary of R&D Results During the First Period and the Planning for 1984 FY. n.p., n.d.

5. Present State of Information Processing in Japan. Investigation on Actual State Information Processing. MITI, December 1983.

6. "Understanding 5G," Report on Ultramodern Computers (Fields of Application-Influences). ICOT, March 1983, pp. 239-250.

7. "America's Viewpoint Towards 5G," Report on Ultramodern Computers, pp. 251-255.

8. Report on Ultramodern Computers (Technological Trends) Research Reports of Advanced Computers. Japan Institute of Machine Industry & New Generation Computer Technology Development Organization, 1983.

9. R&D on Superspeed Computer Systems. Summary of Progress in 1982. Electrotechnical Laboratory, March 1983, pp. 38-44.

10. R&D on Superspeed Computer Systems. Summary of Progress in 1983 FY and Plans for 1984 FY. n.p., n.d.

11. Research and Development of Element Technologies for Next Generation Computers. Research association for Fundamental Computer Technologies, June 1984.

12. Second Symposium on New Functional Devices. New Functional Device Research Association & MITI, July 6-7, 1983. This publication contains abstracts covering R&D results in three areas: superlattice, radiation hardened ICs, and three dimensional circuits.

13. "Trends in Computer Industry in U.S. & Europe," Survey of Computer Industry in U.S. and Europe. 4th Report of Research Committee on the Information Processing Industry, n.d., pp. 5-18.

14. "Report on IBM," Survey of Computer Industry in U.S. and Europe, pp. 126-137.

15. "Report on SRI," Survey of Computer Industry in U.S. and Europe, pp. 184-191.

## A.5 Statistical Summary of Literature Provided

Table A-1 summaries in various categories provided to the computer science panel.

Table A-1. Statistical Summary of Literature Support to JTECH Computer Science Panel

| | Number of Source Material Items | Amount of Material (pages) Source Material in English | Source Material in Japanese | Translated Material |
|---|---|---|---|---|
| General | 15 | 417 | | |
| Open-Source | 15 | | 144 | 65 |
| Semi-Open | 10 | | 1035 | 66 |
| Closely-Held | 5 | | 99 | 39 |
| TOTAL | 45 | 417 | 1278 | 150 |

A.6 Summary and Recommendations

To remain competitive with Japan, the JTECH Computer Science Panel strongly recommends the establishment of a national information center, or several regional centers, perhaps, through which the U.S. industry, academia, and national laboratories can keep abreast of the current Japanese R&D in computer science. If established, such a center must be structured to avoid the many difficulties we are presently facing in collecting and disseminating JSTI in a meaningful way.

In the area of Japanese literature acquisition and collection of current technical information, the Panel's experience can be summarized as follows:

A. Open-source Literature

In Section A.3, a complete list of available Japanese literature in computer science was given. As noted, some Japanese computer science literature is published in English. The following journals can be taken as important examples:

(1) IPSJ English-language Quarterly Section E (English) of IECE Transaction

(2) JIPDEC Reports

(3) ICOT Journals and Technical Reports

This material is published in English (through Japanese initiative) and presents a snapshot picture of a large body of Japanese computer science literature. It reaches the American audience, however, with the standard delay involved in translating Japanese research results into English. Even with this limitation, few U.S. research libraries presently have access to this material on a continuing basis.

Systems*Computers*Controls is published bimonthly by Scripta Publishing Co., in Silver Spring, MD. It contains translations of Japanese articles selected from Transactions of IECE (Parts A, B, C, and D), approximately 24 weeks after the publication of its Japanese-language equivalents. Apparently, the selection of Japanese articles for translation is done by the Japanese editor and authors of IECE Transactions, but they do not assume the responsibility for the translated versions. This publication is offered to the public on a subscription basis, but the technical information is published in English 12 to 18 months from when the same result research is made known in Japan.

While the Japanese computer science literature in English listed above serves some useful purpose in keeping the U.S. computer science community in touch with developments in Japan, time and coverage are scarcely adequate for providing useful input in critical assessment studies, such as carried out in the present program.

In fact, with the exception of ICOT publications, none of the open source literature used by the present panel was contained in the above list.

B. Acquisition of Current Technical Information

As has been indicated in Figure A-1, most of the primary information resulting from Japanese high tech R&D is reported at various academic society meetings, where presentations are given, and meeting abstracts are published, in Japanese. Government or Academy-sponsored committee meetings are frequently held to disseminate important primary research information for synergetic use. The computer science discipline is no exception to this general picture; in fact, most members of the present panel were surprised to learn of the existence of semi-public Japanese Working Groups (listed in Section A.3) similar to the U.S. DARPA Principal Investigator meetings.

The technical information derived from such meetings is of critical importance in assessing the state of current Japanese R&D, and the present panel make heavy use of such information. The information available from this source is of a semi-open or closely-held nature, and its acquisition and translation must involve appropriate U.S. technical experts who are familiar in both Japanese language and the Japanese R&D environment on the subject matter. In order to perform this task effectively on a continuing basis, we must observe the rules of tact and protocols already mentioned by the panel at the end of Section 3.

In summary, the Panel recommends the establishment of some appropriate mechanism of monitoring Japanese R&D in computer science on a continuing basis, taking into consideration the following specifics:

(1) Encourage maintenance and expansion of the presently available pool of Japanese computer science literature in English and its dissemination to a wider audience.

(2) Establish a mechanism of monitoring the current Japanese R&D information which are of semi-open or closely-held nature. This activity must involve appropriate U.S. technical experts who are familiar with the Japanese scene both in language and R&D environment.

(3) In the longer range view, it is necessary to encourage the study of the Japanese language for some of the U.S. students in computer science and other related disciplines.

# Part II

# Opto- and Microelectronics

The information in Part II is from *JTECH Panel Report on Opto- and Microelectronics,* prepared by H. Wieder, W. Spicer, R. Bauer, F. Capasso, D. Collins, K. Hess, H. Kroger, R. Scace, W.-T. Tsang, and J. Woodall. The study was prepared for Science Applications International Corporation under contract to the U.S. Department of Commerce, May 1985.

# Acknowledgements

The Japanese Technology Evaluation Program (JTECH) is indebted to the panel members for their efforts in completing the assessment in the short time allotted while continuing to meet other commitments and pursuing ongoing research interests.

We appreciate the contributions made by Mr. George Mu from the Department of Commerce, International Trade Administration, Office of Japan. His help and guidance as the DoC Contracting Officer was invaluable. We express our gratitude to the National Science Foundation for their supplemental financial support of this study.

# About the Authors

Harry H. Wieder

Harry Wieder is Adjunct Professor, Electrical Engineering and Computer Sciences Department, University of California, San Diego. From 1949 to 1953 he was a physicist at National Bureau of Standards. He was Head Physicist, Dielectrics and Semiconductors Branch, Naval Weapons Center, Corona, California from 1953 to 1970. In 1970 he joined the Naval Electronics Laboratory Center, San Diego, as Head, Semiconductor Physics Branch. From 1973 to 1981 he was Head, Electronic Material Science Division, Naval Ocean Systems Center, San Diego. Harry Wieder is a member of several professional organizations and has published extensively in his field.

William E. Spicer

William E. Spicer graduated from William and Mary, M.I.T., and University of Missouri (Ph.D. 1955). RCA Laboratories 1955-62 (two achievement awards). Stanford University: Associate Professor 1962-65; Professor of Engineering 1965 to present; Stanford W. Ascherman Chair of Engineering since 1978; Guggenheim Fellow; Oliver Buckley Prize - American Physical Society, 1980; Scientist of the Year - R&D Magazine, 1981; Medard W. Welch Award - American Vacuum Society, 1984. A founder of photoelectron spectroscopy. Principal present research interest: physics and chemistry and electrical properties of metal-semiconductor interfaces.

Robert S. Bauer

Dr. Robert S. Bauer is Manager of the Integrated Opto-Electronics Area at the Palo Alto Research Center of Xerox Corporation. His group conducts R&D for specialty opto-electronic-based components. Dr. Bauer's own research has been directed at understanding the formation and electronic properties of semiconductor interfaces and surfaces. Recent focus has been on bonding and growth of heterostructures using molecular beam epitaxy (MBE) combined with soft x-ray photoemission. Dr. Bauer has over 85 publications on semiconductor interface science, surface physics, and electronic and optical properties of solids. In 1978, he served on the Editorial Board of the JVST and from 1979 was Associate Editor for Electronics Materials. Bauer was elected to the Executive Committee of the EMP Division of AVS in 1979. He has served on the Organizing and Program Committees for numerous conferences, including the 27th national AVS Symposium and the Annual Conferences on the Physics and Chemistry of Semiconductor Interfaces (PCSI) since 1977. He was chairman of the 1982 IUPAP/UNESCO Semiconductor Symposium on "Surfaces and Interfaces: Physics and Electronics." In addition, Dr. Bauer was chairman of the Stanford Synchrotron Radiation Laboratory User's Organization in 1981, and during July 1982 served on the DARPA Materials Research Council. He is currently on the Electronic Materials Advisory Board for the Berkeley Center of Advanced Materials and a Director of the AVS.

Federico Capasso

Dr. Capasso received the doctorate degree summa cum laude in physics from the University of Rome, Italy, in 1973. In 1974 he joined Fondazione U. Bordoni, Istituto Superiore Poste e Telecomunicazioni, where he worked on the theory of nonlinear effects in optical fibers. At the end of 1976 he joined Bell Laboratories, where he has since been engaged in research on impact ionization 1.3-1.6-μm detectors, liquid-phase epitaxy of III-V materials, surface passivation studies, deep level spectroscopy, high-field transport in semiconductors, novel avalanche photodiode structures, bipolar transistors, and superlattices. He introduced the technique of "band gap engineering" and has used it in the design of a new class of superlattice and variable gap heterostructures, including new photodetectors and bipolar transistors. In 1984 he received the AT&T Bell Laboratories Distinguished Member of Technical Staff Award and the Award of Excellence of the Society for Technical Communication. He has given thirty invited talks at international conferences and coauthored over seventy papers. Dr. Capasso is a Member of the American Physical Society, the New York Academy of Sciences and the IEEE.

**Douglas M. Collins**

Douglas M. Collins received his B.S. degree in Electrical Engineering from Montana State University in 1971. He received his M.S. (1972) and Ph.D. (1977) degrees in Electrical Engineering from Stanford University. He currently manages the Materials Department in Hewlett-Packard's High-Speed Devices Laboratory. This department is responsible for optical and electrical characterization of III-V Compound Semiconductor Materials in addition to the Molecular Beam Epitaxy and Organo-Metallic Vapor Phase Epitaxy growth of Gallium Arsenide and Aluminum Gallium Arsenide. Dr. Collins is a member of Tau Beta Pi, Sigma XI, Phi Kappa Phi, the American Physical Society, the American Vacuum Society, the Institute of Electrical and Electronics Engineers, and the Electrochemical Society.

**Karl Hess**

Karl Hess was born in Trumau, Austria. He studied physics and mathematics at the University of Vienna in Austria where he received his Ph.D. degree in 1970. In 1969 he became associated with the University of Vienna and the Boltzmann Institute as a Research Assistant. In 1971 he was promoted to Assistant Professor and in 1977 Universitätsdozent. He was a Visiting Assistant Professor at the University of Illinois at Urbana-Champaign during the academic year 1973/74 and returned to the University of Illinois in 1977 where he is currently Professor of Electrical Engineering and Research Professor in the Coordinated Science Laboratory. In 1982 he was named a Beckman Associate in the Center for Advanced Study. Professor Hess has done extensive research on the high frequency conductivity of semiconductors, the acoustoelectrical properties of semiconductors and is currently involved in research on the electronic properties of quantum wells heterostructures and superlattices. Professor Hess is an internationally recognized authority in this field.

**Harry Kroger**

Harry Kroger received a B.S. degree from the University of Rochester in 1957 and a Ph.D. degree from Cornell University in 1962. Both degrees were in Physics. From 1962 to 1983 he was employed at the Sperry Research Center in Sudbury, Massachusetts where he worked on ultrasonics, microwave semiconductor amplifiers and oscillators, microwave bandwidth avalanche photodiodes, silicon and gallium arsenide digital circuits, and analog and digital Josephson devices. Since 1983 he has been employed at MCC (Microelectronics and Computer Technology Corporation) in Austin, Texas as the technical director of the Packaging/Interconnect Program.

**Robert Scace**

Robert I. Scace is Deputy Director, Center for Electronics and Electrical Engineering, at the National Bureau of Standards. He holds BS and MS degrees in Physics from Purdue University and did further graduate work at Rensselaer Polytechnic Institute. Prior to joining NBS in 1975, he was with the General Electric Co., where he did research on the SI-C phase diagram, developed a number of power semiconductor devices, and managed advanced development for the Semiconductor Products Department. At NBS, he has done technology assessment studies in semiconductors and was chief of the Semiconductor Materials and Processes division from 1981 until taking up his present duties in late 1984. Mr. Scace has been developing industry standards for the semiconductor industry for many years. He is Chairman of ASTM Comittee F-1 on Electronics and active in SEMI and DIN standards work. He has received numerous awards from ASTM, DIN and the Department of Commerce for this work.

**Won-Tien Tsang**

Dr. Won-Tien Tsang is a member of the technical staff of AT&T Bell Laboratories. He obtained his doctorate in Electrical Engineering from the University of California at Berkeley. He has been awarded the Adolph Lomb medal of the Optical Society of America. He is an internationally recognized authority on heterojunction lasers and epitaxial growth of semiconductors and the inventor of a number of important laser structures.

**Jerry M. Woodall**

Dr. Woodall was born and grew up in the metropolitan area of Washington D.C. He attended MIT where received his B.S. degree in Metallurgy in 1960. In 1982 re received a Ph.D. in Electrical Engineering from Cornell University. In 1960 he joined the Clevite Transistor Products in Waltham, Massachussetts where he worked on germanium crystal growth and device processing. In 1962 he joined the IBM Thomas J. Watson Research Center in Yorktown Heights, New York. During the early phase of his career he studied the crystal growth of high puriety bulk gallium arsenide. Subsequently he developed the use of liquid phase epitaxy method for fabricating silicon doped p-n junction gallium arsenide light emitting diodes and gallium aluminum arsenide/gallium arsenide LED's and lasers. In the late 60's he invented the gallium aluminum arsenide/gallium arsenide high efficiency solar cell, and with co-workers, IBM developed this structure into a state-of-the-art high efficiency solar cell. Recently he has turned his attention to the studies of gallium arsenide surfaces and interfaces with particular emphasis on the fundamental aspects of the formation of both Schottky barrier and ohmic contacts and the use of photochemical techniques for device fabrication and surface pacification. Also he has been studying the role of fusion cell chemistry on epitaxial layers grown by the molecular beam epitaxy method.

# Executive Summary

An evaluation and comparative assessment of research and development in Japan and the U.S. concerning solid state electronics and opto-electronics reveal trends, which when extrapolated, suggest that quantum mechanically-based dimensionally constrained devices and integrated circuits may represent the most significant future electronic and electro-optic data and signal processing options. Specific emphasis is placed on the synthesis, growth, fabrication, and applications of III-V compound semiconductor quantum wells and superlattice heterojunction structures used for high speed transistors, low threshold lasers, and high speed photosensors in electronic and electro-optic integrated circuits. The Japanese are aggressive in acquiring, improving, and implementing these technologies, whose conceptual aspects were developed in the U.S. In opto-electronics, in particular, the Japanese have made major, original contributions and, while their adaptive ingenuity can be expected to continue to produce market oriented products, their original creative contributions to this field are expected to increase steadily in the future. Of particular significance are the collaborative interactions between industrial organizations, and between these organizations and the Japanese government, with the Opto-electronics Joint Research Laboratory as a clear example.

The Japanese government, which plays an important role in promoting its industries through such means as protectionists trade policies, does not dictate selection and development of specific technological options. Industrial organizations in Japan have made long term commitments to these technologies, and they are not deterred from pursuing them even if the payoff is not immediate or is not generic in character. A clear example of this is the Japanese R&D of the superconductive Josephson junction integrated circuit technology abandoned by IBM after it appeared that its potential use for general purpose computers has become questionable. It is pursued in Japan with the goal of developing a more limited purpose superspeed computer. In summary, while the U.S. scientific and technological efforts concerned with future generation solid state electronics represent a slim lead, the Japanese are making a determined effort to maintain their adaptive product-oriented R&D

while at the same time encouraging original creative fundamental and applied research on all aspects of solid state electronics and electro-optics.

We recommend continuous monitoring of trends in microelectronics and periodic assessment of the status of scientific as well as technological issues. Although we agree, in general, with the statements made by Roland W. Schmitt in his paper on National R&D Policy ["An Industrial Perspective", Science, Vol. 224 (1984) p. 1206]: ".... The Japanese are agressive in aquiring, improving and implementing technology that they did not develop ...", there is a substantial base of original work underway in Japan which is likely to increase with the passage of time. It is important, therefore, to foster and to continue to increase present day information exchanges for mutual benefit. This, indeed, is the recommendation of each one of our panel members in their respective sections contained in this report.

While the U.S. may never attain equity in numbers compared to Japanese who are fluent or at least conversant with technical and scientific aspects of the English language, we must try to provide every opportunity for our scientists and engineers to master not only the rudiments of the Japanese language but to reach a level compatible with that of attendance and participation in technical and scientific meetings where the papers are presented in Japanese. (An example is the Symposium on Future Electron Devices, July (1985) sponsored by MITI and the Japan Industrial Technology Association).

# 1. Introduction

The future of solid state electronics, opto-electronics and of their integrated circuit technologies involves a variety of scientific and technical issues which represent a radical departure from the present-day dominance of silicon-based devices and of the silicon integrated circuit technology.

Alternative approaches are being investigated in the U.S., Japan and a number of European countries. These investigations suggest that a very important role is to be had by different non-silicon based options and that we are now seeing the outline of what may well become the solid state electronics and electro-optics of the next century. These extrapolated technologies are concerned with development of electronic and photo-electronic digital and analog signal and data processing systems based on electronic and photo-electronic devices and integrated circuits (IC's) operating at much higher rates and with lower dissipation than those of the extant silicon-based technology. In view of the ever increasing demand for high volume, high throughput data and signal processing systems such alternative options are being intensively investigated worldwide. Such research is of importance in view of the physical limits which we are all reaching, probably within the next decade, on shrinking the size (scaling) the dimensions of field-effect transistors. Higher signal processing efficiencies, higher reliability, and lower costs have been achieved in past decades by scaling down the size of transistors while increasing their density.

The new and emerging technological options are based principally on the inherent electronic properties of compound semiconductors synthesized from various combinations of binary, ternary, and quaternary alloys of elements selected from columns III and V of the periodic table of the elements. The synthesis, characterization, and device applications of these materials are considerably more complex than those employing silicon. They provide the means for making not only high speed and lower power transistors than those based on silicon but also visible and infrared lasers and corresponding detectors which can be integrated on the same semi-insulating substrate. Such compound semiconductor IC's can be used in conjunction with extant systems based on silicon but they also provide some hitherto unavailable options such as those involved in the generation, switching and detection of optical radiation pulses. Such optical signals can be generated and processed within a chip or can be transmitted between chips which handle data by conventional procedures. Optical interconnections represent a potentially important alternative to conventional conductive interconnection between devices which may circumvent many of the present impediments of large scale very high speed device integration.

Research and development of compound semiconductors and of their applications require a multi-disciplinary approach. Conventional demarcation lines between solid state physics, electrical and electronic engineering, chemistry and chemical engineering become somewhat blurred.

Such investigations require a well developed, sophisticated industrial base for the supply of high purity raw materials. Solids many orders of magnitude purer than conventional reagent grades are required for the synthesis of these compound semiconductors grown in the form of single crystal layers on appropriate substrates. Organo-metallic gases, many of them still in the research phase, are required for advanced research and industrial development.

Sustained long term research, development and industrial application of advanced IC technology based on compound semiconductors require a well-educated and properly motivated scientific and technical workforce capable of assimilating and adapting discoveries made worldwide. Rapid and accurate communication of technical and scientific results is mandatory. It is vital all the more because the time delay between conceptualization of a device, component or circuit configuration and its practical implementation continues to become shortened by the synergistic interaction between scientists, engineers, and technicians with complementary backgrounds constituting goal-oriented action teams.

A cornerstone of compound semiconductor devices is the heterojunction, i.e., the junction between dissimilar semiconductors whose cyrstalline lattice constants and thermal expansion coefficients match. Multiple thin heterojunctions made of alternating sequential layers of two different fundamental bandgaps constitute a superlattice. If the thickness of these layers is comparable to the wavelength of an electron, then the charge carriers are confined within such layers and constitute quantum wells. The specific electronic properties of such quantum wells and superlattices provide hitherto unavailable options for making high speed low power dissipation transistors, IC's, lasers, and photo detectors.

The properties and applications of quantum wells and superlattices depend on the methods available for their synthesis and controlled growth. The principal methods are molecular beam epitaxy (MBE) and metal-organic vapor phase epitacy (MOVPE). MBE was developed in the 1960's at the Bell Laboratories. It is based on the epitaxial growth of thin layers using molecular beams directed towards a substrate where they combine and condense in the form of a single crystal a few hundred atoms in thickness at a rate of approximately one atomic layer per second.

MOVPE is a technique in which the constituents of a semiconducting compound are introduced in the gas phase into a reaction chamber where they combine, by pyrolisis, onto an appropriate substrate. MOVPE is particularly well suited for industrial applications and has a somewhat greater versatility than MBE in dealing with a variety of semiconducting compounds not suitable for MBE. The ability to grade abruptly as well as gradually the composition of the interfaces between compound semiconductor structures used for superlattices and quantum wells by computer-controlled shuttering of molecular beams has profound implications for a new type of device and integrated circuit engineering defined as bandgap engineering. Band gap engineering is the latest of the great advances in solid state electronics which can be traced to a combination of advances in conceptual clarification, improved methods of material synthesis, characterization, and the development of new device principles.

The organization of science and technology and the subsequent input of the results therefrom into production are noticeably different in Japan from that in the United States. To give the reader perspective on the

Japanese organization, a section is provided by W. E. Spicer of Stanford University, Stanford, California.

In this report an evaluation and comparative assessment of Japanese and U.S. research and development on the synthesis of compound semiconductors is provided by D. M. Collins of Hewlett-Packard Laboratories, Palo Alto, California. His conclusions are that the U.S. maintains leadership or equity with respect to the development of MBE, OMVPE, and bulk cyrstalline growth of compound semiconductors. However, the collaborative and cooperative interaction between industrial organizations and their government provides the Japanese with important advantages in commitment to long range goals of product development keyed to specific markets.

K. Hess, of the University of Illinois, describes in the following pages research on quantum wells and superlattice structures in Japan and the U.S. He emphasizes that the Japanese recognize the long term significance of such research for the development of "functional" devices to be used for artificial intelligence system hardware components and have committed substantial amounts of money for this purpose. He indicates, furthermore, that although the Japanese excel at technological adaptivity and are engaged in intensive device and IC development of quantum wells and superlattice structure concepts developed in the U.S., the original, creative contributions of Japanese scientists and engineers must not be underestimated. Technical and scientific information exchange between the U.S. and Japan is, and will continue to be, of considerable mutual benefit.

A bird's eye view and assessment of Japanese research and development of semiconductor near-infrared heterojunction lasers, light emitting diodes, and integrated optics is provided in the following pages by W. T. Tsang of AT & T, Bell Laboratories. A capsule summary of his conclusions is:

1. In the component development area, the Japanese have surpassed the U.S.
2. In the system development area, the U.S. is still leading.

3. Competition for dominance in opto-electronics is primarly between Japanese industrial organizations rather than competition from abroad.
4. The driving force for Japanese opto-electronics R&D is based on demand of their domestic markets with emphasis on a large share of the world market to be obtained, eventually.
5. The quality of their research is high, their R&D is efficient and their technology transfer from laboratory to pilot plant production is effective and efficient, particularly in process-intensive areas.

Japanese research on heterojunction avalanche photodetectors and opto-electronic integrated circuits with a spectral range compatible with the radiation emitters discussed by W. T. Tsang is treated in this report by F. Capasso of AT & T Bell Laboratories, with particular emphasis on the impact on this research on solid state devices and IC technology of the future. His conclusions, in agreement with those of Collins, Hess and Tsang are:

1. Opto-electronic research in Japan has a broad gauge and it is long term in character.
2. Cooperation between industrial laboratories and between industry and government is the hallmark of Japanese R&D.
3. While Japanese scientists and engineers have made ingenious adaptions of basic research and applied research performed primarily in the U.S., they have demonstrated and continue to exhibit creative solutions in the development of photo-electronic building blocks and associated technological issues.

In this report J. Woodall of the T. J. Watson IBM Research Laboratory, N.Y. deals with a comparative assessment of Japanese and U.S. research concerning ohmic contacts and Schottky barriers. These are crucial issues related to the fabrication of compound semiconductor devices and ICs. He concludes that the perception of the nature of the metallurgical problems (this involves the approach to potential solutions) is approximately similar in the U.S. and Japan. He accentuates the painstaking care to detail and solid engineering accomplishments of the Japanese which have led them, step by step, to develop economically competitive products proceeding systematically from concepts, developed elsewhere, through feasibility study and then to prototype development. He also emphasizes the long range aspects of investment in future technologies such that embodied in the principal issues described in this report.

A brief evaluation and comparison of research on surfaces and interfaces in Japan and the U.S. pertinent to technological issues discussed in this report is presented in a separate section by W.E. Spicer, Stanford University, Palo Alto, California.

In his assessment of Japanese plans and programs on solid state lasers in the near infrared and visible portions of the spectrum, R. S. Bauer of the Xerox Research Laboratories, Palo Alto points out that integrated optoelectronics is a more than $100 million research cooperative and is specifically targeted by the Japanese Ministry of Trade and Industry (MITI). Including optical devices, office equipment and fiber optics systems, the Japanese production is expected to rise from a base of $0.34 billion in 1980 to a projected $50 billion by the turn of the century.

An evaluation of equipment required for the synthesis of heterojunction, superlattices and quantum wells specifically manufactured in Japan is provided in this report by R.I. Scace of the National Bureau of Standards. He points out that, as a rule, several years pass from the initial announcement of equipment embodying a significant improvement, to its appearance in the domestic market (in Japan) and thereafter its appearance in the international market place. Furthermore, there is some indication of collaboration or at least a close interaction between the equipment designer, manufacturer and its user (in Japan). While much of the original equipment for material synthesis by MBE and MOVPE was purchased from abroad, there is evidence of continuing research and development on such equipment intended initially for their own (Japanese) use and thereafter for export.

A different approach for the implementation of high speed large scale signal and data processing technologies is that based on superconductive Josephson junctions described in this report by H. Kroger of the MCC, Austin, Texas. The main advantage of the Josephson junction technology for large scale integrated circuits is the superconductive interconnection which makes use of lossless and dispersionless transmission lines connecting fast switching binary digital devices with low cross-talk between them. In Japan, such research (which started, initially, by imitating the IBM-developed Josephson device technology) has now developed a distinctive flair of its own. Each facet of this program is aimed at product development, not basic research. This work is now exploring more advanced technologies than those which were being investigated by IBM.

Although IBM has now abandoned development of Josephson junction integrated circuits for general purpose computer applications, the Japanese are not deterred. Their intent is to aim such work towards a superspeed computer and they are making steady progress in that direction.

Future electronic and electro-optic systems would benefit from the availability of a technology analogous to the complementary silicon integrated circuit technology employing insulated gate field effect transistors with performance characteristics superior by far to devices in present use. Presently available data obtained primarily in the U.S. and Japan suggests that this is feasible using III-V compound insulated gate field effect transistors and integrated circuits. A comparative assessment of such work performed in the U.S. and Japan is provided in this report by H. H. Wieder of the University of California, San Diego in La Jolla, CA. Although the conceptual device aspects of such transistor were developed primarily in the U.S., the work underway in Japan is of a comparable level both in research and development in universities as well as in industrial laboratories. Japanese industry, in particular, is poised to exploit rapidly the potential of insulated gate III-V compound field effect transistor integrated circuits provided that the remaining technological proglems related to the dielectric-semiconductor interface are solved and a comparison with alternative technologies justify this in terms of performance, cost, and reliability.

In summary, this report presents many of the available technological options of solid state electronics and electro-optics emphasized in Japan and the United States. It would be desirable to supplement this report with a subsequent evaluation of the potential of the planar large scale integrated circuit technology employing Schottky barrier gates and p-n junction gates to gallium arsenide field effect transistors made by selective area ion implantation into semi-insulating GaAs. Following many years of research on basic device aspects and small and medium scale circuit integration in the U.S., Japan and Europe there is a present thrust, worldwide, to adapt such circuits to large scale system integration in Japan.

# 2. Organization of the Japanese Effort on Non-Silicon Based Opto- and Microelectronics

W.E. Spicer, Stanford University

## 2.1 Introduction

This section is directed towards the support and direction given by the Japanese government to the non-silicon based opto- and micro-electronics industry.

There have been significant changes in government funding for the Japanese electronics industry and related areas in the past decade.

## 2.2 The Ministry of International Trade and Industry (MITI)

MITI is the force spearheading the Japanese into dominance in new technological areas. With guidance from senior members of the Japanese technical and scientific communities, MITI allots large amounts of funds for well-defined projects, It organizes "joint laboratories" with a predefined and relatively short lifetime to develop the generic technology and scientific knowledge necessary for such projects. In addition, research is commissioned in industrial laboratories and is also carried out in government laboratories as part of such projects.

The first such joint laboratory in electronics (The VLSI Cooperative Laboratory) was formed about a decade ago and was disbanded after its predetermined life expired around the turn of the decade. The success of this effort can perhaps best be judged by the growth of the Japanese integrated circuit (IC) industry in the last decade.

While well defined and suitable generic problems are attacked by the joint laboratories, simultaneously work is carried out for the project by government laboratories and commissioned in industry. In the Optoelectronic

Project, the joint laboratory has responsibility for the common basic technology with emphasis on the materials and closely related problems. The additional work in the government laboratories and the commissioned work in industry seems to be more concerned with complete systems and their overall design than is the joint laboratory.

## 2.3 The Optoelectronics Joint Research Laboratory of MITI

To illustrate how such joint laboratories operate and to focus on the scope of this report, let us examine the Optoelectronics Joint Research Laboratory. Opto-electronics means a wide range of technology in which optical as well as electrical functions are carried out. It includes such apparently diverse areas as optical communications and fast computing. The emphasis is on reaching very high frequencies. This plus the fact that Si is not a useable light emitter, automatically brings concentration on 3-5 compound semiconductors, e.g., GaAs.

Figure 2-1 indicates the original organization of the Optoelectronics Laboratory. It is especially important to note the six research groups into which the laboratory was organized since this gives a good overview of the areas of concentration. Note two things: First, the organization is strongly built around materials growth and processing; second, it goes from the most basic aspect, the growth of "ideal" large single crystals, to such advanced topics as Maskless Ion Beam Doping, and Compound Semiconductor Device Fabrication. Again it is important to emphasize that the titles of the various groups indicate emphasis on the generic problems of the technology with detailed development of systems to be manufactured being left up to the individual companies.

By examining the way in which the MITI joint labs are staffed, one can get an insight into the joint government - private industry partnership represented by these laboratories. Figure 2-2 gives a list of the companies which are members of the Opto-electronics Joint Laboratory. The research staff of approximately fifty professionals comes from the permanent staffs of these companies, and they will return to their parent company either during

Figure 2-1. OUTLINE OF OPTOELECTRONICS JOINT RESEARCH LABORATORY*

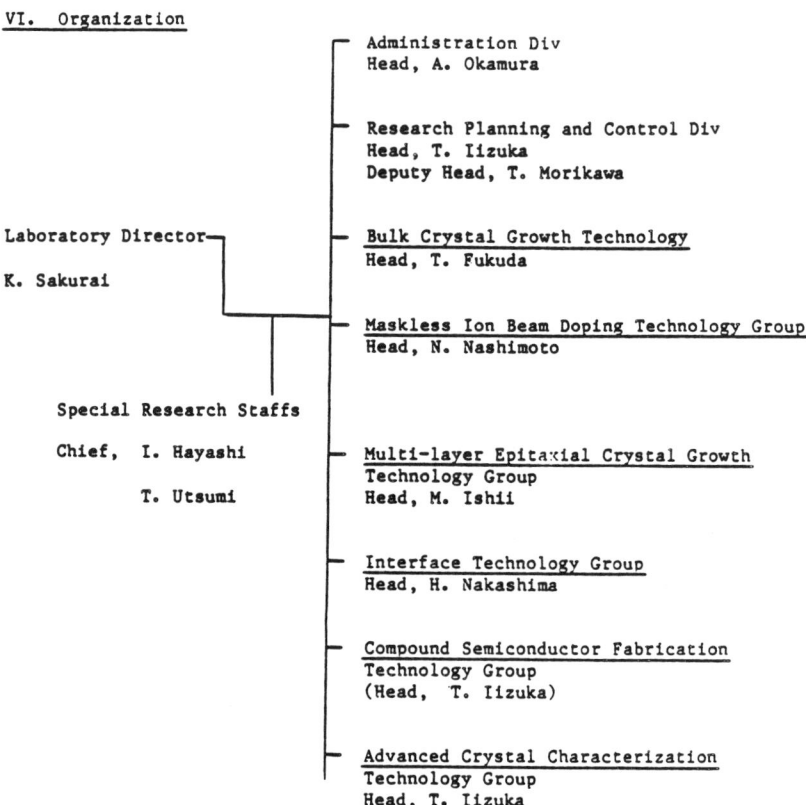

VI. Organization

- Laboratory Director, K. Sakurai
  - Special Research Staffs
    - Chief, I. Hayashi
    - T. Utsumi
  - Administration Div
    - Head, A. Okamura
  - Research Planning and Control Div
    - Head, T. Iizuka
    - Deputy Head, T. Morikawa
  - Bulk Crystal Growth Technology
    - Head, T. Fukuda
  - Maskless Ion Beam Doping Technology Group
    - Head, N. Nashimoto
  - Multi-layer Epitaxial Crystal Growth Technology Group
    - Head, M. Ishii
  - Interface Technology Group
    - Head, H. Nakashima
  - Compound Semiconductor Fabrication Technology Group
    - (Head, T. Iizuka)
  - Advanced Crystal Characterization Technology Group
    - Head, T. Iizuka

* Total number of Research Staffs is about 50.

Figure 2-2. MITI JOINT OPTOELECTRONICS LABORATORY

- MEMBER COMPANIES

  OKI ELECTRIC INDUSTRY CO., LTD.

  OPTOELECTRONICS INDUSTRY AND TECHNOLOGY DEVELOPMENT ASSOCIATION

  SHIMADZUI SEISKUCHO, LTD.

  SUMITOMO ELECTRIC INDUSTRIES LTD.

  TOSHIBA CORPORATION

  NIPPON SHEET GLASS CO., LTD.

  NIPPON ELECTRIC CO., LTD.

  HITACHI, LTD.

  FUJITSU LTD.

  FUJI ELECTRIC COMPONENTS

  RESEARCH AND DEVELOPMENT LTD.

  FUJIKURA CABLE WORKS, LTD.

  FURAKAWA ELECTRIC CO., LTD.

  MATSUSHITA ELECTRIC INDUSTRIAL CO., LTD.

  MITSUBISHI ELECTRIC CORPORATION

  YOKOGAWA ELECTRIC WORKS, LTD.

- Each company provides professionals to make up research staff of ~ 50 for Joint Lab.

- Laboratory to be dissolved in 1986 and staff returned to companies.

the life of the Joint Lab or when it is dissolved in about 1986. The opto-electronics thrust started in 1979; however, the Joint Laboratory was still receiving new equipment when I visited it in September 1983.

Summary:

1) These laboratories are funded by the government.

2) The research staff comes from and returns to the companies which are members of the laboratories.

3) Each research staff member has responsibility to transfer the results obtained by the laboratory back to his company. The companies may then use this knowledge for rapid development of new product lines.

4) The Joint Laboratories are dissolved after a fixed period (perhaps eight years) and all research staff members return to their respective companies.

The Joint Laboratories resemble laboratories such as the Radiation Laboratory at M.I.T. which as formed during World War II and was dissolved at the end of that war. Questions have been raised in recent years about U.S. Government Labs which may have outlived their mission. The Japanese Joint Laboratories may serve as one example of an alternative approach.

We have concentrated on the Joint Opto-Electronics Laboratory as an example of one part of the Japanese strategy to move forward quickly and efficiently in electronics. Figure 2-3 attempts to put the joint laboratories into perspective within the whole Japanese structure in electronics. We have concentrated on the joint laboratory since it is one of the most innovative parts of this structure.

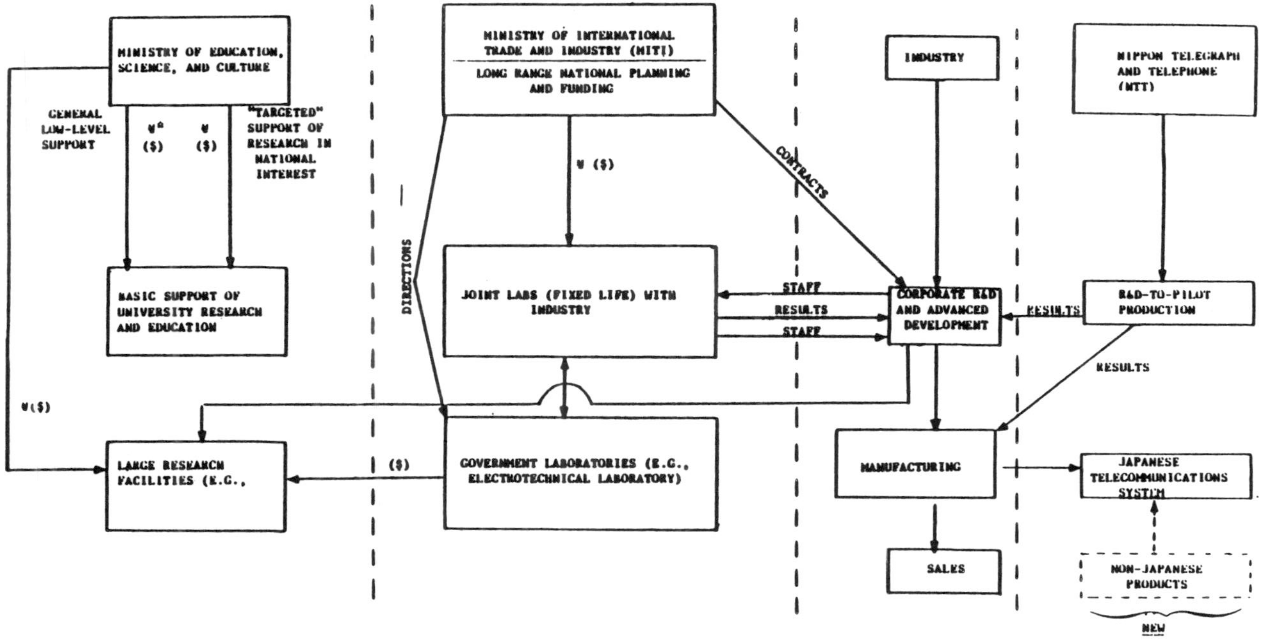


Figure 2-3. A View of the Organization of Japanese "Electronics" and Related Areas.

* Yen indicates flow of funds.

## 2.4 The MITI Optoelectronics and Supercomputer Programs

The Joint Optoelectronics Laboratory is one part of the overall MITI thrust in Optoelectronics which started in 1978 and will be finished in 1986. The MITI funding is $75 million for the complete program. It is not clear how much the industrial funding will add to this in the overall research and development phase which extends into the period where the Joint Laboratory closes and the industrial staff move back to their respective laboratories; however, a fourfold increase in funding over MITI's original investment is probably not unrealistic. It should not be assumed that the work in industrial laboratories starts when the joint laboratory closes. It should be considered a continuous process stimulated by information flow between the joint laboratory and the member companies. Remember that each member of the staff at a joint lab has the responsibility of keeping communications flowing with his own company. Because of lower salaries and longer working hours, the actual productivity per dollar may be as much as twice that in the United States. The inflated value of the dollar with respect to the yen is another factor. This reduction in costs is somewhat offset by capital equipment expenditures which seem to be considerably higher per researcher than in many U.S. laboratories. I found the Joint Optoelectronics Laboratory to be one of the best equipped per staff member that I have seen. Table 2-1 gives the MITI projection of the Japanese share for the Optoelectronics market. Note the estimate is $50 billion by the year 2000.

The supercomputer thrust is another area of MITI involvement (see Table 2-1). In addition to the work in computer science and architecture, strong efforts are being made in Si, 3-5 and Josephson junction hardware under this MITI program. I do not have any actual numbers on the 3-5 expenditures, but I suspect they are comparable to that in the Optoelectronics area. It is to be expected that there is a strong element of mutual support between these areas, particularly after information is transmitted back to the companies involved.

Table 2-1. ORGANIZATION OF JAPANESE

ADVANCED OPTO- AND MICROELECTRONICS

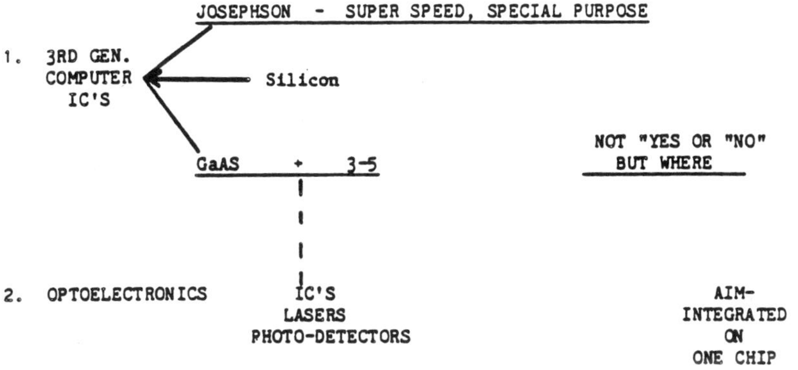
JOSEPHSON - SUPER SPEED, SPECIAL PURPOSE
1. 3RD GEN. COMPUTER IC'S
Silicon
GaAS + 3-5
NOT "YES OR "NO" BUT WHERE
2. OPTOELECTRONICS
IC'S
LASERS
PHOTO-DETECTORS
AIM-
INTEGRATED
ON
ONE CHIP

OBJECTIVES AND PROJECTIONS

OPTOELECTRONICS - TO DEVELOP INTO OEIC-(OPTOELECTRONIC INTEGRATED CIRCUITS)

JUNE 1983 NO EXISTING OEIC, BUT JAPANESE BELIEVE IT WILL BE DEVELOPED AND GROW TO BECOME KEY ELEMENT IN FUTURE ADVANCEMENT OF OPTICAL COMMUNICATIONS, INFORMATION PROCESSING AND SENSING SYSTEMS

MITI JOINT OPTOELECTRONICS LABORATORY $75M 1979-86

SIZE JAPANESE OPTOELECTRONICS INDUSTRY

| | | | | | |
|---|---|---|---|---|---|
| 1978 | $0.1B | 1983 | $1.8B (E) | 1990 | $ 8.4B (E) |
| 1982 | 1.1B | 1985 | 4.2B (E) | 2000 | 50.0B (E) |

(E) = ESTIMATE

## 2.5 The Overall Organization of Japanese Electronics

Figure 2-3 is divided into four parts. The Ministry of Education and Culture (MEC) and MITI are directly funded by the government. In recent years it has appeared that the MITI sector is the fastest growing. However, the work and facilities supported by MEC cannot be ignored. Of particular importance for this report is the funding for research for national strength. Through this means, strong funding for areas such as 3-5's is made available for the academic community. When an area is considered of national importance, funding is not only coupled into the industrial sector by MITI as outlined above but also into university research by special funds from MEC. This should be reflected by university research results cited in this report.

Another function of the MEC is the support of large national facilities. For example, the $60-100M synchrotron facility called the Photon Factory is heavily used in electronics work. This facility is just coming into operation and is being used by the Optoelectronics Joint Laboratory, university research, and NTT (see Figure 2-3). I suspect it is also being used by industry.

The Japanese system of technical and scientific education differs from that in the United States. Partially because of the highly competitive exams, Japanese high school students work harder than their U.S. counterparts, and they come to the University better prepared. From their freshman year, the students concentrate more strongly on their area of technical or scientific interest than do American students. By the end of the junior year the Japanese student is far ahead in his technical or scientific area of speciality. The senior year is spent as a member of a research group containing master and doctoral candidates. As a result, the Japanese undergraduate is in an advanced position. Many of the best are recruited directly into industry, and a certain fraction of these obtain their doctoral degree (D.Sc.) on the basis of research and publication in industry without further work at the university. This provides a rather unique "bonding" between the D.Sc. and his company. The other path to the D.Sc. is essentially the same as that in the U.S. Whichever route is taken, a Doctor of Science degree is the same.

Note the last column in Figure 2-3, Nippon Telephone and Telegraph, (NTT). Historically NTT has played much the same role in Japan as ATT has in the U.S. As of April 1, 1985 NTT was divested somewhat as has happened to ATT. Since most of the material presented in this report predates that divestment, it has little effect on this report. However, the future effects may be quite significant. One parallel action, the opening of the Japanese telecommunication industry (the operating branch of NTT) to imported equipment has been a disappointment to non-Japanese companies. One reason for this will be apparent as we describe NTT.

Two historic differences between ATT and NTT should be noted. First, the work of NTT's research laboratories has been much more directed toward applications than that of Bell Labs. Second, it has no manufacturing and is equivalent to the old ATT without Western Electric. The results from the NTT R&D efforts have been fed into electronic manufacturing companies such as NEC, Toshiba, and Hitachi and manufacturing contracts are let to these companies to supply the equipment needed by NTT. The specifications for the equipment were closely dependent on NTT work in a detailed way. This is probably one of the reasons why non-Japanese companies have found so much difficulty with the specification required for for equipment to be bought by the operations arm of NTT.

NTT has the largest electronics R&D effort in Japan, and it appears to be making a concentrated effort to exhibit its work at U.S. meetings, open up its laboratories for U.S. visitors, and, in general, build better relations with the United States.

NTT has just completed the new Atsugi R&D laboratory. It consists of several buildings which I estimate are over 100,000 square meters of floor space. When I visited it in September 1983, I was told that it had a research staff of 400 members plus support. It was not clear whether or not it was yet fully staffed. The research emphasis at Atsugi are: (1) R&D on Si VLSI, (2) GaAs and Josephson VLSI, and (3) optoelectronic semiconductor devices based on 3-5 materials. The parallel with the MITI thrust mentioned above is striking. A specific difference from U. S. work was close contact between the Si and GaAs work with frequent transfer of personnel from GaAs to Si and vice versa.

Particular pride was taken in 50,000 square feet of continuously connected Class 10 clean room space. I was shown a pilot production line for IC's such as the Si 256K RAM. Each processing step was being automated. The wafers were transported via an enclosed cassette system by an overhead trolley between the approximately twenty-three processing stations. The wafers were never exposed to the Class 10 atmosphere. By using computers and remote control, the ultimate objective was to remove all personnel from the processing room thus achieving, with the cassette transport, better than Class 1 processing environment.

I also noticed in the same clean space as the pilot line, two MBE machines for GaAs-based research - the latest model Riber plus a Japanese made unit. I estimate that I saw only one-third to one-fourth of the clean room space.

The Atsugi Laboratories are making a concentrated effort to mount a significant program using synchrotron radiation at the large National Light Source ($60 million plus) facility just coming into operation north of Tokyo.

I was impressed with the magnitude and focus of the electronics effort and the quality of the scientists and engineers at NTT. The only United States analogy I can think of is one that one would have all the talent at Bell Laboratories sharply focused for the last twenty years into problems directly dealing with IC, opto-communication, and closely related areas.

The "industrial" column in Figure 2-3 will not be discussed in any detail as it is quite similar to its U.S. counterparts. However, the reader's attention is directed to the various arrows between the industrial sector and the government and NTT sectors. In general these are quite different from anything in the U. S. although some parallels exist.

# 3. Synthesis of Compound Semiconductor Materials

Douglas M. Collins, Hewlett-Packard Laboratories

## 3.1 Introduction

The importance of compound semiconductors in micro- and optoelectronics is primarily based on: (1) the particularly suitable electrical or optical properties of the material (e.g., band-gap, electron mobility, etc.), (2) the ability to tailor these properties to those desired by forming ternary or quaternary compound semiconductor alloys, and/or (3) the ability to form heterojunction structures of two or more different compounds. In the least complex applications (an example of which is the GaAs MESFET), nothing more than high-quality bulk substrate material may be required. For more complex device structures (e.g., heterojunction high-speed devices such as the MODFET or light emitting devices such as the multiquantum-well laser), it is necessary to use sophisticated epitaxial growth techniques such as molecular beam epitaxy (MBE) or organometallic vapor phase epitaxy (OMVPE) to produce complex, well-controlled, highly uniform heterojunction material structures.

The Japanese and U.S. sources of technical information which have been surveyed cover aspects of compound semiconductor material synthesis which are related to these device and integrated circuit applications. Thus, the emphasis will be on epitaxial materials grown by MBE and OMVPE as well as on the growth of the bulk, liquid-encapsulated Czochralski (LEC) GaAs substrate material which is required for many of these applications. There will be a subsection devoted to each of these technologies guided by the following objectives: (1) to provide an objective assessment of the technical information disseminated by the Japanese, in both English-language and Japanese-language publications, in comparison to the open technical literature published as a result of research efforts in the United States; and (2) to indicate areas in which (a) U.S. efforts lead Japanese efforts, (b) there is equity between U.S. and Japanese efforts, and (c) Japanese efforts lead U.S. efforts.

An additional major objective of this report is to identify the sources of information which are the most valuable for keeping abreast of recent Japanese contributions in these three areas of research and development. This topic is addressed in the final subsection of this report.

## 3.2 Molecular Beam Epitaxy

The technique of molecular beam epitaxy (MBE) was pioneered at Bell Laboratories in the U.S. [1]. It was also at Bell Laboratories that the two most promising materials structures were invented. These are the multiquantum-well heterostructure [2] which has led to important applications in the area of optoelectronic devices (e.g., the quantum-well laser [3-5]) and the modulation-doped superlattice [6] which has led to important high-speed device and integrated circuit applications (e.g., the MODFET [7-9]). Currently, MBE is being widely developed in both the U.S. and Japan for application to optoelectronic and high-speed devices and integrated circuits.

In optoelectronics, two major areas of research and development will be discussed here: (1) MBE materials for short-wavelength (650 - 870 nm) emitters and detectors, and (2) long-wavelength (1,300 - 1,600 nm) emitters and detectors. The short-wavelength work was pioneered in the U.S. in the GaAs/AlGaAs material system [10,11]. For a time, the U.S. held a significant lead in this technology. However, most of the U.S. MBE activity has subsided in favor of OMVPE (see the next subsection). In contrast, the Japanese have been pursuing this technology enthusiastically. In the GaAs/AlGaAs materials system, Fujitsu [12-14], NTT [15,16], Mitsubishi [17], and the Optoelectronics Joint Research Laboratory (OJRL) [18] have pursued quantum-well laser R&D while the Electrotechnical Laboratory has reported a surface emitting laser [19]. In addition, at least two Japanese laboratories are developing AlGaInP/GaInP materials for use in visible light emitters. NEC has reported a double-heterojunction laser with a GaInP active layer operating at 683 nm [20] and NTT has reported lasers with AlGaInP active layers operating between 660 and 680 nm [21,22]. There is also R&D, primarily in Japan, on the MBE growth of II-VI materials for short-wavelength emitters.

One major application of these materials is thin-film electroluminescent displays.

There are a number of laboratories which are pursuing the MBE growth of InP, InGaAs, InGaAsP or related materials for long-wavelength optoelectronic applications. These include Bell Laboratories in the U.S. [23,24], British Telecom in the U.K. [25], and NTT [26], Tokyo Institute of Technology [27], and Sumitomo Electric in Japan [28-30]. The work at Bell Laboratories is the most advanced in the sense that workers there have reported MBE growth of InGaAsP [24]. This was accomplished through the use of gas sources (arsine and phosphine) which are cracked by passing them through a high temperature zone as they are admitted into the MBE system. Because this technique results in As and P atoms in the incident group V flux it has thus far been the only method to offer the control over the simultaneous incorporation of As and P required for MBE growth of this alloy. Conventional MBE As and P sources result in sublimation of the As and P tetramer molecules and, when followed by a "cracking" furnace, As and P dimer molecules result. Because in both of these latter cases As molecules displace the P molecules at the growth interface, workers have been unsuccessful in controlling the relative amounts of As and P in the MBE film with enough precision to routinely achieve suitable lattice-match to the InP substrate. Another way of getting around this problem is through the use of quaternary alloys which contain three group III elements and one group V element. Workers at Tokyo Institute of Technology have reported the growth of InGaAlAs with band-gaps corresponding to the 1,000 - 1,650 nm wavelength range [27]. There is also R&D, primarily in the U.S. and in Europe, on the MBE growth of II-VI materials for long-wavelength devices. One major application of these materials is very long-wavelength detectors.

There are two primary heterojunction materials systems which are being developed by MBE for high-speed device and integrated circuit applications: (1) AlGaAs/GaAs (on GaAs substrates) and (2) InAlAs/InGaAs (on InP substrates). In both cases, the principal applications are MODFET's and MODFET integrated circuits. In the AlGaAs/GaAs material system, Japanese and U.S. laboratories are on a roughly equal footing. Advanced MBE technology which is being largely dedicated to this application exists at Bell

Laboratories [31-33], Rockwell [34,35], TRW [36], Hewlett-Packard [37], Honeywell [38], IBM [39], the University of Illinois [40,41], and Cornell University [42,43] in the U.S. and at Fujitsu [44-48], NEC [49], and Hitachi [50] in Japan. Overall, the U.S. laboratories hold a slight lead in speed (ring oscillators with 12.2 ps gate delay at 300 K (Rockwell [35]), frequency dividers operating at 10.1 GHz at 77 K (Bell Laboratories [32]), and discrete MODFET devices with f(t) = 80 GHz at 300 K for a 0.2 μm gate length (Cornell [43]) and f(t) = 33 GHz at 300 K for a 0.8 μm gate length (Hewlett-Packard [37])). The Japanese hold the lead in the complexity of MODFET integration (4 K SRAM with a 2 ns access time at 77 K (Fujitsu [46])) as well as in MESFET integration (16 K SRAM with a 4.1 ns access time at 300 K (NTT [51])). A potentially important advance reported recently by NEC [49] is the substitution of an AlAs/GaAs superlattice structure with the Si Doping in the GaAs layers (to overcome the deep-donor or deep-level effects in Si-doped AlAs and AlGaAs) for the normal Si-doped AlGaAs layer in the MODFET structure. NEC's promising results have already been repeated at Bell Laboratories [33]. The key problems which continue to face laboratories in both countries are the same: the elimination of oval defects (the lowest densities are ~200 cm^-2 in both the U.S. and Japan [52]) and improvement in GaAs substrate materials.

There is much less activity in the MBE growth of InAlAs/InGaAs. The majority of the work on these materials has been carried out in the U.S., primarily at Bell Laboratories [53,54] and Cornell University [55]. Related work in Japan includes the MBE growth of AlGaInAs lattice-matched to InP at Tokyo Institute of Technology [27] and the chloride vapor phase epitaxial growth of high mobility (μ = 9,400 cm^2/V-s at 300 K and μ = 71,200 cm^2/V-s at 77 K) modulation-doped structures at Fujitsu [56]. In addition, CNET in France has reported the growth of InAlAs/InGaAs by MBE [57].

There are a number of novel techniques related to MBE which are being pursued in both the U.S. and Japan. However, most of the more ambitious, longer range work such as reactive MBE (Electrotechnical Laboratory [58]), cluster-ion beam epitaxy (Kyoto University [59]), radical beam epitaxy (Hiroshima University [60]), and maskless ion-beam doping (OJRL [61,62]) were all pioneered and are currently most advanced in Japan. A major area which was pioneered in the U.S. and which has been described above is the use of gas sources for the growth of InGaAsP (i.e., cracked arsine and phosphine [24,63]). In addition, very recently a number of laboratories have begun using organometallics such as trimethylgallium, trimethylindium, trimethylarsenic, and trimethylphosphorous as gas sources of Ga, In, As, and P in MBE systems. This technique, which is being called MO-MBE, is being studied at Tokyo Institute of Technology [64,65], Bell Laboratories [66], and AAchen Technical University [67].

Other items which bear mentioning are (1) the development in the U.S. of an indium-free mount for substrates during MBE growth [68,69], (2) work in the U.S. [70-72] and Japan [73,74] to develop passivating overlayers to facilitate MBE regrowth following processing, and (3) efforts in Japan [75,76] to understand the causes and effects of oval defects in MBE films.

Finally, it is important to note that most of the advanced MBE R&D being performed in both the U.S. and Japan uses commercially manufactured MBE systems. The most common systems are manufactured in the U.S. (Varian Associates (~7 systems in Japan) and Perkin-Elmer (~1 system in Japan)) and Europe (Riber (~30 systems in Japan) and Vacuum Generators (~8 systems in Japan - mostly for Si MBE)) [77]. The MBE systems manufactured in Japan by Anelva have not yet been widely accepted by Japanese researchers. However, Anelva's most recently introduced system (January, 1984) has all of the important features of MBE systems manufactured in the U.S. and Europe [78]. It will be important to monitor the acceptance of this system in the market place.

### 3.3 Organometallic Vapor Phase Epitaxy

Organometallic vapor phase epitaxy (OMVPE), alternatively referred to as metal-organic chemical vapor deposition (MOCVD), is being developed primarily for the growth of the III-V compound materials GaAs, AlGaAs, InP, InGaAs, and InGaAsP for optoelectronic device applications. There is also some R&D activity on AlGaAs/GaAs and InP/InGaAs heterojunction structures for high-speed device and integrated circuit applications as well as on other compounds such as the II-VI's.

The leading OMVPE technology is probably in the western countries. The most advanced results for visible quantum-well lasers have been achieved at the Xerox Palo Alto Research Center [79,80]; whereas for long wavelength emitters using the InP based III-V alloys, the work at Thomson-CSF in France [81] is currently the most advanced. The use of OMVPE for the growth of modulation-doped GaAs/AlGaAs heterojunctions has been most actively pursued at Thomson-CSF [82], Hewlett-Packard [83], and Cornell University [84]. All of these laboratories have succeeded in producing two-dimensional electron gas mobilities close to 100,000 cm^2/V-2 at 77 K, thus being suitable for MODFET device applications.

While western countries currently hold the lead in OMVPE technology, the gap is narrowing. Sony has reported the OMVPE growth of superlattice structures with very abrupt interfaces [85] as well as multiquantum-well lasers operating at 709 nm [86]. Hitachi has reported a rather unique, undoped, self-aligned enhancement-mode MODFET device [50]. The OJRL has reported the growth of high purity OMVPE GaAs (μ(77 K) = 100,000 cm^2/V-s) [87]. NEC has reported the low pressure (~70 Torr) growth of very high uniformity GaAs (+-2.6% in thickness and +-3.5% in carrier concentration over 3" diameter GaAs substrates) [88]. And, lastly, OKI has reported extraordinary results in the growth of GaAs on Si substrates [89,90].

This latter accomplishment warrants further discussion. OKI has reported several "firsts" in their GaAs/Si work. They have reported the growth of GaAs on Si substrates by using (1) an interfacial layer of Ge deposited by cluster ion beam epitaxy, and (2) an interfacial layer consisting of an AlGaAs/GaAs superlattice. In addition, they have reported vanadium doping of OMVPE GaAs to yield semi-insulating (> 10^8 ohm-cm) buffer layers. Finally, they have fabricated 17-stage E/D MESFET ring oscillators on the GaAs/Si films with gate delays of 66.5 ps for 0.5 μm gate lengths.

GaAs/Si R&D is also being pursued in the U.S. The most successful work to date has been reported by MIT Lincoln Laboratory in which double-heterojunction lasers have been successfully fabricated on MBE grown AlGaAs/GaAs layers grown on Ge coated (~0.15 μm of Ge) Si substrates [91].

While the Japanese are still playing catch-up in most aspects of OMVPE technology, the gap is closing quickly. In addition to the novel GaAs/Si work at OKI, the Japanese (i.e., Sumitomo Chemicals) supply the highest purity alkyl source materials for use in OMVPE. Thus OMVPE activities in Japan surely warrant close observation.

Essentially all OMVPE work reported to date has been carried out in reactors which were custom designed and built in the laboratories in which they are used. Although commercially manufactured reactors have been available for several years (vendors include Metals Research in England and Crystal Specialties and Spire in the U.S.) these have not been widely accepted by established OMVPE researchers. Recently SPC Electronic Corporation in Japan has announced their entry into the OMVPE reactor market. The initial specifications for this reactor look impressive and is a product to watch [92].

## 3.4. Liquid Encapsulated Czochralski GaAs

Ultimately, the promise of using III-V materials for high-speed LSI or VLSI circuits as well as for complex integrated optoelectronic circuits will depend on the ability to produce high-quality bulk substrate material. This will be required in order to produce high-quality epitaxial materials from which high yield integrated circuits can be made. This is well recognized by the Japanese as is evident by the prominence of the bulk substrate growth program within the MITI sponsored Optoelectronic Joint Research Laboratory [93]. The importance of substrates has not been overlooked by researchers in the U.S. either. In fact, numerous activities are underway throughout the U.S. and Japan [93-105]. Much of this work is being carried out in commercial LEC crystal pullers. The most common commercial LEC pullers are manufactured by Cambridge Instruments in England. These are high-pressure pullers and are in use at Rockwell [94,95] and Westinghouse [96,97] in the U.S., Cominco [98] in Canada, and the OJRL [99,102] and Sumitomo Electric [103,104] in Japan. In addition, a number of laboratories are using commercial or internally developed low pressure LEC pullers. These include Texas Instruments [105] and Hewlett-Packard [106] in the U.S.

There are four main approaches being reported for the improvement of bulk LEC GaAs. These are (1) indium doping (~1-3% indium in the melt), (2) growth in a magnetic field, (3) improved thermal design to reduce thermal gradients, and (4) computer controlled growth (especially automatic diameter control). Most of the laboratories in both the U.S. and Japan are investigating two or more of these approaches in the same puller.

The indium doping of LEC GaAs, which increases the critical resolved shear stress for plastic deformation in the crystal, was pioneered by Westinghouse [97]. Other laboratories are following this lead, in particular Sumitomo Electric [104] which is one of the world's leading suppliers of GaAs substrate material. Sumitomo is expected to be the first company to market (probably in 1985) semi-insulating In-doped GaAs substrates with very low dislocation densities. Sumitomo has used an improved thermal geometry in addition to the indium doping.

The application of a magnetic field to the LEC (i.e., MLEC) growth of GaAs was pioneered by the OJRL in Japan [99-102]. They have reported significant improvements in the homogeneity of the crystals grown under fields greater than 1300 G. NTT has also reported the growth of MLEC GaAs [107] and recently reported the use of indium-doped, MLEC, GaAs to achieve 99.8% bit yields in 16K MESFET SRAM's [51]. To date, no other laboratories have reported significant magnetic LEC activities.

In addition to the MLEC work, the OJRL has made other advances in LEC GaAs technology. This includes in situ melt purification using a distillation which occurs when the Ar pressure in the puller is abruptly decreased from ~20 atm to ~1 atm. The conductivity of the melt is monitored during this process and significant reduction in impruities, particularly Si, has been reported [99].

The most recent progress report of the OJRL [102] suggests that their overall LEC GaAs program is quite successful. Two-inch diameter undoped semi-insulating GaAs with dislocation densities less than 1,000 $cm^{-2}$ is reported in addition to indium-doped semi-insulating material which is dislocation-free.

Overall, the sophistication of the R&D activity in the U.S. and in Japan is comparable. However, the Japanese (especially Sumitomo Electric) are in a good position to be the first to commercialize this technology and remain the only country with a long range objective of dominance in this market and a plan to meet this objective.

### 3.5 Resource Assessment

In the course of this survey it has become clear that there is a vast resource of literature published by the Japanese. This is neither surprising nor are the specific journals which are most useful surprising. The journals which are most popular for the U.S. MBE, OMVPE, and bulk GaAs researchers (e.g., the Journal of Applied Physics and Applied Physics Letters) are also popular among Japanese researchers. Also, Japan's English-language journals, primarily the Japanese Journal of Applied Physics and the Japanese Journal of Applied Physics Letters are very valuable sources in these areas of research.

English-language conferences in the U.S. and abroad provide additional, and more timely, information about Japanese technological developments. Recent examples of such conferences include the 2nd International Conference on Molecular Beam Epitaxy (Tokyo, Japan, August 27-30, 1982), the 5th U.S. Molecular Beam Epitaxy Workshop (Atlanta, Georgia, October 6-7, 1983), the 3rd International Conference on Molecular Beam Epitaxy (San Francisco, California, July 31-August 3, 1984), the 2nd International Conference on Metal-Organic Vapor Phase Epitaxy (Sheffield, England, April 10-12, 1984), the 3rd International Conference on Semi-Insulating III-V Materials (Warm Springs, Oregon, April 24-26, 1984), the annual Device Research and Electronic Materials Conferences, the International Electron Device Conference (IEDM), the International GaAs Integrated Circuit Symposium, and the International Conference on GaAs and Related Compounds.

In addition, other valuable and timely sources of information about Japanese technological advances have been made available to this panel. These are translations of abstracts, figure captions, summary reports, etc., of conferences held under the auspices of the Japanese Applied Physics Society and of program reviews or meetings held by Japanese government agencies (e.g., the symposium held by the Research and Development Association for Future Electron Devices in July, 1984 [47]). A key difference between Japanese conferences and U.S. conferences is that many Japanese researchers have a good command of the English language and can profit by attendance at English-language conferences; whereas most U.S. researchers have little knowledge of the Japanese language and Japanese-language conferences of little use. However, this may not be as large a problem as it first appears. First, translations similar to those provided to this panel could be more widely distributed. Second, U.S. companies and government laboratories could identify individuals within their organizations who understand Japanese and sponsor their attendance at these conferences in order to bring back information. Information-gathering is simplified at Japanese conferences because it is common practice for attendees to photograph visual material presented by the speakers. Very complete reports can result.

Personal visits by active Japanese and U.S. researchers to each other's laboratories can also provide opportunities to learn of Japanese technological achievements. Naturally, the amount of valuable technical information gleaned from these visits varies widely from visit to visit. The use of good judgment by the visitor and the host can insure that technical exchanges during these visits remain on a level which does not compromise proprietary or classified information while facilitating a mutually beneficial technical exchange.

## 3.6 Summary and Interpretation

There is extensive active research and development in the U.S., Japan, and Europe in the areas of MBE, OMVPE, and LEC growth of GaAs and related III-V compound semiconductors. In the case of the epitaxial growth techniques MBE and OMVPE, the pioneering work was done in the U.S. As with any emerging technology, the open technical literature and open technical conferences played an important role in disseminating advances in these areas which stimulated additional R&D in the U.S. and new efforts in Japan and Europe.

In the case of LEC growth of bulk GaAs, the demand for higher quality substrates for use in MBE and OMVPE, as well as other applications, became clear to researchers throughout the world and stimulated original work concurrently in the U.S. and Japan. Again, reports of technical advances in the open technical literature and at open technical conferences stimulated additional work and cross-fertilized ongoing work, resulting in rapid progress in both the U.S. and Japan.

There is no question that the Japanese benefited from the availability of MBE and OMVPE research results published by pioneering researchers in the U.S. In addition, other U.S. companies have benefited from the availability of these early publications, as well as publication of later work in Japan and Europe. Thus, the mechanisms that exist for the dissemination of technical information which are outlined in the previous subsection seem to provide an equitable technical exchange between the U.S. and Japan as well as other countries active in R&D in these technologies. This type of exchange has played a major role in the rapid growth of all high-technology industries.

In contrast to the seemingly equitable exchange of technical information between the U.S. and Japan, key differences between the U.S. and Japanese governments' approach to R&D funding of these technologies became apparent during the course of this survey. These differences are the extent of government and industrial committment (1) to the commercialization of these tehcnologies and (2) to the establishment of long-range goals and plans needed to meet this committment.

One need only take a cursory look at the differences between the U.S. and Japanese R&D activities surveyed in this report to see one of the most important impediments to their commercialization in the U.S. as compared to Japan. Most Japanese companies which accept Japanese government R&D funding are in an excellent position to apply the developed technologies to their expanding commercial markets. For example, Fujitsu receives Japanese government funds for semiconductor laser R&D (potential commercial application: high-speed optical data links for their computer systems), NTT receives Japanese government funds for short and long-wavelength laser R&D (potential commercial application: optical communication systems), and Sony receives Japanese government funds for visible laser R&D (potential commercial application: optical video and audio disc players). In contrast, most U.S. government R&D funds support work at companies that have no objectives for the commercial application of these technologies. It would appear that this is not the most efficient way of converting U.S. government R&D investments into products which result in commercial market dominance and thus corporate profits for U.S. industry.

However, it is not just the technology that the Japanese have gleaned from the U.S. literature nor the product oriented use of government R&D expenditures which have led to the Japanese domination of markets based on that technology. Additionally, the Japanese establishment of and committment to specific long-range goals is an important factor. The establishment of the OJRL for the development of optoelectronic integrated circuits typifies this long-range planning. U.S. industrial managers and U.S. government policy-makers might well look to the Japanese example in the establishment of long-term goals and the long-range plans as an effective means to commercialization of technologies pioneered in the U.S.

## 3.7 References


1. A.Y. Cho and J.R. Arthur, Prog. Solid State Chem. 10, 157 (1975).
2. R. Dingle, W. Wiegmann, and C.H. Henry, Phys. Rev. Lett. 33, 827 (1974).
3. W.T. Tsang, C. Weisbuch, R.C. Miller, and R. Dingle, Appl. Phys. Lett. 35, 673 (1979).
4. R.D. Dupuis, P.D. Dapkus, R. Chin, N. Holonyak, and S.W. Kirchoefer, Appl. Phys. Lett. 34, 265 (1979).
5. N. Holonyak, R.M. Kolbas, R.D. Dupuis, and P.D. Dapkus, IEEE J. Quantum Electronics QE-16, 170 (1980).
6. R. Dingle, H.L. Störmer, A.C. Gossard, and W. Wiegmann, Appl. Phys. Lett. 33, 665 (1978).
7. T. Mimura, S. Hiyamizu, T. Fujii, and K. Nanbu, Jap. J. Appl. Phys. 19, L225 (1980).
8. D. Delagebeaudeuf, P. Delecluse, P. Etienne, M. Laviron, J. Chaplart, and N.T. Linh, Electron. Lett. 16, 667 (1980).
9. S. Judaprawira, W.I. Wang, P.C. Chao, C.E.C. Wood, D.W. Woodard, and L.F. Eastman, IEEE Electron Device Lett. EDL-2, 14 (1981).
10. A.Y. Cho and H.C. Casey, Appl. Phys. Lett. 25, 288 (1974).
11. W. Tsang, Appl. Phys. Lett 34, (1979).
12. T. Fujii, S. Hiyamizu, O. Wada, T. Sugahara, S. Yamakoshi, T. Sakurai, and H. Hashimoto, J. Crystal Growth 61, 393 (1983).
13. T. Fujii, S. Hiyamizu, S. Yamakoshi, and T. Ishikawa, Abstracts of the 3rd International Conference on MBE, p. 136 (San Francisco, (1984).
14. Yoon Soo Park, ONR/Far East, Semiconductor Lasers and Crystal Growth Technology Report from a Topical Meeting of the Japan Society of Applied Physics (JSAP), Scientific Bulletin, Volume 9, Number 2, pp. 79-85 (1984)
15. H. Iwamura, T. Saku, H. Kobayashi, and Y. Horikoshi, J. Appl. Phys. 54, 2692 (1983).
16. H. Iwamura, T. Saku, T. Ishibashi, M. Naganuma, and H. Okamoto, Collected Papers of MBE-CST-2, p. 47 (Tokyo, 1982).
17. K. Mitsunaga, K. Kanamoto, M. Nunoshita, and T. Nakayama, Abstracts of the 3rd International Conference on MBE, p. 70 (San Francisco, 1984).

18. Yoon Soo Park, ONR Far East, "III-V Compound Semiconductor Research at the Optoelectronics Joint Research Laboratory", Scientific Bulletin, Volume 9, Number 1, pp. 152-157, (1984).

19. M. Ogura and T. Yao, Abstracts of the 3rd International Conference on MBE, p. 139 (San Francisco, 1984).

20. Yoon Soo Park, ONR/Far East, Semiconductor Lasers and Crystal Growth Technology Report from a Topical Meeting of the Japan Society of Applied Physics (JSAP), Scientific Bulletin, Volume 9, Number 2, pp. 79-85 (1984)

21. Yoon Soo Park, ONR/Far East, Semiconductor Lasers and Crystal Growth Technology Report from a Topical Meeting of the Japan Society of Applied Physics (JSAP), Scientific Bulletin, Volume 9, Number 2, pp. 79-85 (1984)

22. H. Asahi, Y. Kawamura, K. Wakita, and H. Nagai, Abstracts of the 3rd International Conference on MBE, p. 25 (San Francisco, 1984).

23. W. T. Tsang, Appl. Phys. Lett. 44, 288 (1984).

24. M.B. Panish and H.T. Temkin, Abstracts of the 3rd International Conference on MBE, p. 45 (San Francisco, 1984).

25. E.G. Scott, D. Wake, A.W. Livingstone, D.A. Andrews, G.J. Davies, Abstracts of the 3rd International Conference on MBE, p. 49 (San Francisco, 1984).

26. Y. Kawamura, H. Asahi, and H. Nagai, J. Appl. Phys. 54, 841 (1983).

27. K. Masu, T. Mishima, S. Hiroi, M. Konagai, and K. Takahashi, J. Appl. Phys. 53, 7558 (1983).

28. Abstracts of "Symposium on New Functional Devices (Future Electronics Devices)", held in July 1983. Consists of three sessions on Superlattice, Hardened IC's, and Three-Dimensional IC's.

29. Mitsuo Kawashima, "Present Status of Future Electron Devices", Abstracts of the Third Symposium on Future Electron Devices, July 4-5, 1984.

30. Y. Matsui, H. Hayashi, K. Kikuchi, K. Yoshida, Abstracts of the 3rd International Conference on MBE, p. 15 (San Francisco, 1984).

31. R.H. Hendel, S.S. Pei, R.A. Kiehl, C.W. Tu, M.D. Feuer, and R. Dingle, IEEE Electron Device Lett. EDL-5, 406 (1984).

32. S.S. Pei, N.J. Shah, R.H. Hendel, C.W. Tu, and R. Dingle, IEEE GaAs Integrated Circuit Symposium Technical Digest, p. 129 (Boston, 1984).

33. C.W. Tu, J. Chevallier, R.H. Hendel, and R. Dingle, Abstracts of the 3rd International Conference on MBE, p. 156 (San Francisco, 1984).

34. S.J. Lee, C.P. Lee, D.L. Hou, R.J. Anderson, and D.L. Miller, IEEE Electron Device Lett. EDL-5, 115 (1984).

35. C.P. Lee, D. Hou, S.J. Lee, D.L. Miller, and R.J. Anderson, GaAs IC Symp. Tech. Dig. 1983, p. 162 (1983).

36. J. Berenz, K. Nakano, K. Weller, 1984 IEEE Microwave and Millimeter Wave Circuits Symposium, p. 83 (San Francisco, 1984).

37. M. Hueschen, N. Moll, E. Gowen, and J. Miller, IEDM (San Francisco, 1984).

38. N.C. Cirillo, J.K. Abrokwah, and M.S. Shur, IEEE Electron Device Lett. EDL-5, 129 (1984).

39. P.M. Solomon, C.M. Knoedler, and S.L. Wright, IEEE Electron Device Lett. EDL-5, 379 (1984).

40. T.J. Drummond, H. Morkoc, and A.Y. Cho, J. Crystal Growth 56, 449 (1982).

41. T.J. Drummond, S.L. Su, W. Kopp, R. Fischer, R.E. Thorne, H. Morkoc, K. Lee, and M.S. Shur, IEDM Tech. Dig., p. 586 (1982).

42. W.I. Wang, C.E.C. Wood, and L.F. Eastman, Electron Lett. 17, 36 (1981).

43. L. Camnitz and L.F. Eastman, 11th International Symposium on GaAs and Related Compounds (Biaritz, France, 1984).

44. S. Hiyamizu and T. Mimura, J. Crystal Growth 56, 455 (1982).

45. S. Hiyamizu, J. Saito, K. Kondo, T. Yamamoto, T. Ishikawa, and S. Sasa, Abstracts of the 3rd International Conference on MBE, p. 52 (San Francisco, 1984).

46. S. Kuroda, T. Mimura, M. Suzuki, N. Kobayashi, K. Nishiuchi, A. Shibatomi, and M. Abe, IEEE GaAs Integrated Circuit Symposium Technical Digest, p. 125 (Boston, 1984).

47. Abstracts of the Third Symposium on Future Electron Devices, July 4-5, 1984. Research & Development Association for Future Electron Devices, sponsored by MITI. (Compilation of titles and short English abstracts.

48. Mitsuo Kawashima, "Present Status of Future Electron Devices", Abstracts of the Third Symposium on Future Electron Devices, July 4-5, 1984.

49. T. Baba, T. Mizutani, and M. Ogawa, Jap. J. Appl. Phys. 22, L627 (1983).

50. Y. Katayama, M. Morioka, Y. Sawada, K. Ueyanagi, T. Mishima, Y. Ono, T. Usagawa, and Y. Shiradi, Jap. J. Appl. Phys. 23, L150 (1983).

51. Y. Ishii, M. Ino, M. Idda, M. Hirayama, and M. Ohmori, IEEE GaAs Integrated Circuit Symposium Technical Digest, p. 121 (Boston, 1984).

52. Private communications.

53. T.P. Pearsall, R. Hendel, P. O'Conner, K. Alavi, and A.Y. Cho, IEEE Electron Device Lett. EDL-4, 5 (1983).

54. K. Alavi, A.Y. Cho, and W.R. Wagner, Abstracts of the 3rd International Conference on MBE, p. 34 (San Francisco, 1984).

55. T. Griem, M. Nathan, G.W. Wicks, J. Huang, P.M. Capani, and L.F. Eastman, Abstracts of the 3rd International Conference on MBE, p. 86 (San Francisco, 1984).

56. M. Takikawa, J. Komeno, and M. Ozeki, Appl. Phys. Lett. 43, 280 (1983).

57. L. Goldstein, M.N. Charasse, A.M. Jean-Louis, G. Leroux, M. Quillec, M. Allovon, and J.Y. Marzin, Abstracts of the 3rd International Conference on MBE, p. 107 (San Francisco, 1984).

58. S. Yoshida, CRC Critical Reviews in Solid State and Materials Sciences, 11, 287 (1984).

59. T. Takagi, I. Yamada, K. Matsubara, and H. Takaoka, J. Crystal Growth 45, 318 (1978).

60. S. Miyazaki, H. Hirata, and M. Hirose, Extended Abstracts of the 16th (1984 International) Conference on Solid State Devices and Materials, p. 447 (Kobe, 1984).

61. "R&D Progress Report of Optoelectronics Joint Research Association", published in June 1984.

62. Y. Bamba, E. Miyauchi, K. Kuramoto, A. Takamori, and T. Furuya, Jap. J. Appl. Phys. 22, L331 (1983).

63. A.R. Calawa, Appl. Phys. Lett. 38, 701 (1981).

64. E. Tokumitsu, Y. Kudou, M. Konagai, and K. Takahashi, J. Appl. Phys. 55, 3163 (1984).

65. E. Tokumitsu, Y. Kudou, M. Konagai, and K. Takahashi, Abstracts of the 3rd International Conference on MBE, p. 89 (San Francisco, 1984).

66. W.T. Tsang, Abstracts of the 3rd International Conference on MBE, p. 90 (San Francisco, 1984).

67. N. Putz, E. Veuhoff, H. Heinecke, M. Heyen, H. Luth, and P. Balk, Abstracts of the 3rd International Conference on MBE, p. 91 (San Francisco, 1984).

68. L.P. Erickson, G.L. Carpenter, D.D. Seibel, P.W. Palmberg, P. Pearah, W. Kopp, and H. Morkoc, Abstracts of the 3rd International Conference on MBE, p. 20 (San Francisco, 1984).

69. T. Hierl, G. Ross, G. Muraoka, Varian Associates, Inc., private communications (June - September, 1984).

70. S.P. Kowalczyk, D.L. Miller, J.R. Waldrop, P.G. Newman, and R.W. Grant, J. Vac. Sci. Technol. 19, 255 (1981).

71. D.L. Miller, R.T. Chen, K. Elliott, and S.P. Kowalczyk, Abstracts of the 3rd International Conference on MBE, p. 30 (San Francisco, 1984).

72. Y.-J. Chang and H. Kroemer, Abstracts of the 3rd International Conference on MBE, p. 8 (San Francisco, 1984).

73. Abstracts of "Symposium on New Functional Devices (Future Electronics Devices)", held in July 1983. Consists of three sessions on Superlattice, Hardened IC's, and Three-Dimensional IC's.

74. N.J. Kawai, T. Nakagawa, T. Kojima, K. Ohta, and M. Kawashima, Electron. Lett. 20, 47 (1984).

75. Y. Suzuki, M. Seki, Y. Horikoshi, and H. Okamota, Jap. J. Appl. Phys. 23, 164 (1984).

76. M. Shinohara, T. Ito, K. Wada, and Y. Imamura, Jap. J. Appl. Phys. 23, L371 (1984).

77. D.S. Morsetts, Perkin-Elmer, Physical Electronics Division, private communication (June, 1984).

78. T. Ishida and K. Katano, Anelva Corporation, private communication (1984).

79. R.D. Burnham, C. Lindstrom, T.L. Paoli, D.R. Scifres, W. Streifer, and N. Holonyak, Appl Phys. Lett. 42, 937 (1983).

80. C. Lindstrom, T.L. Paoli, R.D. Burnham, D.R. Scifres, and W. Streifer, Appl. Phys. Lett. 43, 278 (1983).

81. M. Razeghi, R. Blondeau, K. Dazmierski, M. Krakowski, B. De Cremoux, and J.P. Duchemin, Appl. Phys. Lett. 45, 784 (1984).

82. S.D. Hersee, J.P. Hirtz, M. Baldy, J.P. Duchemin, Electron. Lett. 18, 1076 (1982).

83. Y.-M. Houng and A.F. Sowers, Abstracts of the 26th Annual Electronic Materials Conference, p. 112 (Santa Barbara, 1984).

84. L.F. Eastman, Cornell University, private communication (1984).

85. Mitsuo Kawashima, "Present Status of Future Electron Devices", Abstracts of the Third Symposium on Future Electron Devices, July 4-5, 1984.

86. Yoon Soo Park, ONR/Far East, Semiconductor Lasers and Crystal Growth Technology Report from a Topical Meeting of the Japan Society of Applied Physics (JSAP), Scientific Bulletin, Volume 9, Number 2, pp. 79-85 (1984)

87. S. Takagisha and H. Mori, Jap. J. Appl. Phys. 22, L795 (1983).

88. "Technical and Research Reports", Japan Science & Technology News, Volume 3, Number 3, June 1984. pp. 20-36.

89. Mitsuo Kawashima, "Present Status of Future Electron Devices", Abstracts of the Third Symposium on Future Electron Devices, July 4-5, 1984.

90. Mitsuo Kawashima, "Present Status of Future Electron Devices", Abstracts of the Third Symposium on Future Electron Devices, July 4-5, 1984.

91. T.H. Windhorn, G.M. Metze, B.-Y. Tsaur, J.C.C. Fan, Appl. Phys. Lett. 45, 309 (1984).

92. JST News 3, 54 (1984).

93. "Japanese National R&D Projects in Microelectronics".

94. R.T. Chen and D.E. Holmes, J. Crystal Growth 61, 111 (1983).

95. D.E. Holmes and R.T. Chen, J. Appl. Phys. 55, 3588 (1984).

96. R.N. Thomas, H.M. Hobgood, G.W. Eldridge, D.L. Barrett, and T.T. Braggins, Solid-State Electron. 24, 387 (1981).

97. H.M. Hobgood, D.L. Barrett, L.B. Ta, G.W. Eldridge, and R.W. Thomas, Abstracts of the 25th Annual Electronic Materials Conference, p. 8 (Burlington, Vermont, 1983).

98. Private communications.

99. "Collection of Research Papers Published by Optoelectronics Joint Research Laboratory", Jan-Dec 1983.

100. K. Terashima and T. Fukuda, J. Crystal Growth 63, 423 (1983).

101. T. Shimada, K. Terashima, H. Nakajima, and T. Fukuda, Jap. J. Appl. Phys. 23, L23 (1984).

102. R&D Progress Report of Optoelectronics Joint Research Association", June 1984.

103. M. Sekinobu and K. Matsumoto, J. Electronic Engineering, p. 32 (September, 1983).

104. K. Tada, S. Murai, S. Akai, and T. Suzuki, IEEE GaAs Integrated Circuit Symposium Technical Digest, p. 49 (Boston, 1984).

105. R.L. Lane, Semiconductor International, p. 68, (October, 1984).

106. G. Elliot, C.L. Wei, R. Farraro, G. Woolhouse, M. Scott, and R. Hiskes, J. Crystal Growth, to be published.

107. J. Osaka and K. Hoshikawa, Abstracts of the 3rd Conference on Semi-Insulating III-V Materials, p. 21 (Warm Springs, Oregon, 1984).

# 4. Properties and Applications of Multiquantum Well and Superlattice Structures

K. Hess, University of Illinois

## 4.1 Introduction

New horizons in semiconductor research have opened the possibility of varying the band gaps as well as the doping levels. In addition, light as well as electric currents are used to transport signals in new forms of ultra thin III-V compound layers. This presentation focuses on Japanese semiconductor heterolayer technology which makes use of these new opportunities.

Fig. 4-1 shows an AlAs-GaAs semiconductor heterolayer structure. Combinations of such structures are currently termed superlattices or multiple quantum well structures and will generally be referred to as "superstructures" in the following. Superstructures consist of layers of different semiconductors. The AlAs-GaAs system (or AlGaAs-GaAs system) has been investigated in great detail. Although the interatomic distance in the different semiconductors may differ slightly, a perfect match of lattice constant is desirable because perfect interface conditions are of prime importance. The lattice match is usually also accompanied by a close match of thermal expansion coefficients. In other words the materials should be perfectly compatible and matched to each other. The AlGaAs-GaAs system comes close to this ideal. Although the materials are mechanically very similar, there is still enough difference between them electronically; the electron potential energy differs in AlGaAs-GaAs system up to ~ 0.8 eV depending on the Al content. This potential energy difference allows us to impose boundary conditions on the electron wave function. The advent of MBE and MOCVD technologies has made possible a layer thickness control down to dimension of $10^{-7}$ cm. This is smaller than the electron deBroglie wavelength in GaAs. The size control is therefore on a quantum level.[A-1] III-V compound materials such as AlGaAs and GaAs have long been investigated because these materials are basic for the fabrication of semiconductor lasers and light emitting diodes. Although the optical properties will be discussed by other panel

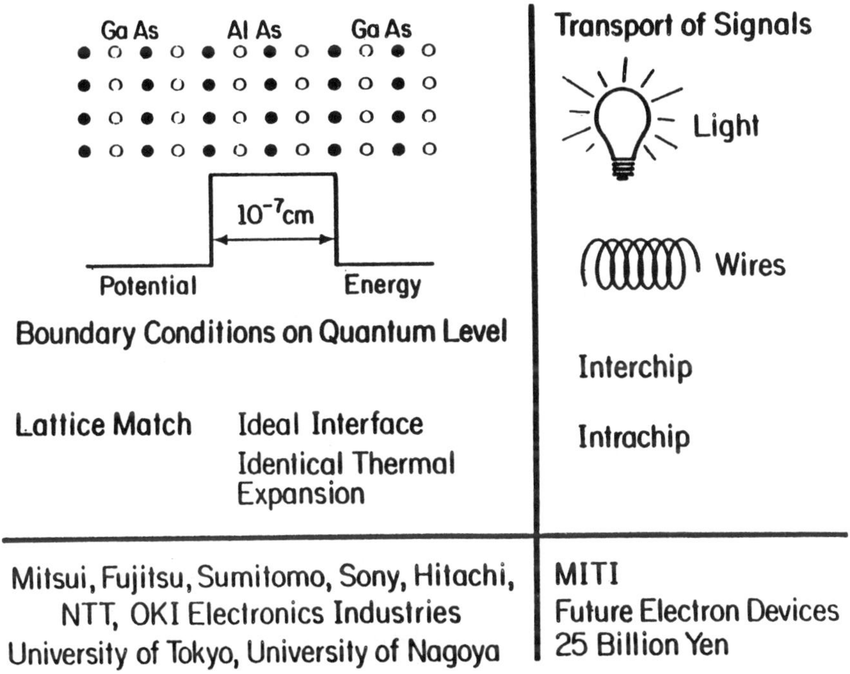


Figure 4-1. AlAs-GaAs Semiconductor Heterolayer Structure.

members, I will include some discussions connected to applications in digital electronics since serious thought is currently given in Japan to the transmission of signals in between and within chips by light.[A.2]

The new possibilities of quantum control of the electronic boundary conditions in semiconductor layers and the possibility of transmitting signals electronically as well as by light have elicited an enthusiastic response in the Japanese semiconductor community.[A.3]

H. Sakaki, University of Tokyo, is promoting the idea of "wave function engineering" [A-3] and Hayashi is one of the major advocates of the new developments in transmitting signals by light in addition to the conventional ways. [A-2] A large number of scientists in Japan are concerned with brainstorming investigations of the functional device concept, [A.4] (to be discussed below) mainly for Artificial Intelligence (AI) applications (see Fig. 4-2).

While ongoing discussions in the U.S. do not seem to have come to a clear conclusion about the technological future of superstructures, the Japanese opinion seems formed and set in favor of superstructure research and development.[A2,3] Japanese objectives (Fig. 4-3) are the production of ultrafast ICs using the high electron mobility transistor (HEMT) technology.[A5] Research on three dimensional circuits is intensively pursued with the declared goal of integrating AI functions in the third dimension.[A-6] Most of the currently funded research in this respect is silicon connected. In general the Japanese scientists attempt to create functional devices by using the basic physics (wave function engineering) of available electronic materials. The optical intrachip connection is discussed also in this context. In the first stage (next three years) research will center on interchip connection.[A-2] This stage is seen as having the two problem areas of packaging and of integrating semiconductor light sources and detectors on silicon chips. While some scientists see a direct factor of two advantage in speed over conventional wire connections (Hayashi, Futai), others see in optical interconnects rather a solution for bottlenecks (Kawago). The second phase of the optoelectronic integrated circuit thrust will address intrachip connections. Japanese scientists are searching for a new device which shows

## Concepts - Ideas - Superstructures

- Wavefunction Engineering (Sakaki, University of Tokyo)
  Boundary Conditions on Quantum Level
  Bandgap Engineering
  } U.S. Origin
- Transmission of Signals by Electrons (light-wires; needs new device)
- Functional Device Concept (AI Functions) (U.S. Origin, Jack Morton, 1965)

Figure 4-2. Concepts - Ideas - Superstructures

## Major Objectives

- *Ultrafast IC's* using High Electron Mobility Transistors *(HEMT's)*
  
  −1984 > 1 billion yen
- *3-dimensional IC's* (tyranny of speed and large systems)
  
  −1984 ≈ 1.8 billion yen
- *Functional Devices* (AI functions etc.)
  
  Several conferences (at least 3 on the specific topic)
- *Optical Inter-Intra Chip Connection*
  
  Intensive preliminary investigation

Figure 4-3. Major Objectives

fast transistor characteristic, emitts light, and can be made very small and integrated in large numbers. It does not seem impossible to construct and fabricate such a device in the near future. The Optoelectronics Joint Research Laboratory has performed intensive preliminary research on these subjects.

Driving forces (Fig. 4-4) in the Japanese semiconductor research community can be deduced from [A.7] and [A.3]:

(1) The collective opinion is that superstructures are useful and will be of great importance in future electronics.

(2) A cooperative and friendly university-industry and industry-industry relationship with well differentiated missions. While the universities attack future concepts, eleven companies (The Research and Development Association for Future Electron Devices) have a coordinated plan of solving difficult technological problems.

(3) The ministry of international trade and industry (MITI) earmarked 25 billion yen for superstructure research.[A.8] This should not be converted to U.S. dollars at the official rate since income per capita is lower in Japan than in the U.S. even for engineers (a division by 100 may be a more appropriate conversion).

(4) The top managers in Japanese electronics are convinced that leading is preferable to catching up. Two quotes from Makoto Kikuchi's book[A-7] on the evolution of Japanese electronics may illustrate this. In the chapter on "the price of maturity" (U.S. maturity) Kikuchi quotes a vice president of Zenith Radio Corp. who asked, "Kikuchi, is it really necessary for a company making and selling television sets, like we do, to have a research department?" Kikuchi continues, "I stared at this man, years my senior, and wondered to myself what on earth had happened to America." (Zenith research was shortly afterwards almost entirely dissolved).

## "Managerial" Driving Forces

- *Collective opinion* about usefulness and *importance of superstructures* has been formed in the Japanese semiconductor community
- *Cooperative*, friendly *university-industry relationship* with well divided missions
- *MITI earmarked 25 billion yen*
- J-S Community seems convinced that leading is preferable to catching up

Figure 4-4. "Managerial" Driving Forces.

In his chapter on the "catching up days" (of Japan) Kikuchi says "The difference between being first and second is the difference between heaven and hell.

4.2 Japanese Superstructure Technology

Figs. 4-5 through 4-9 describe the technological basis and current developments in superstructure research in Japan and in the U.S. Highly developed research exists for the high electron mobility transistor (HEMT) which is illustrated in Fig. 4-5.[B1,2] Especially Fujitsu impresses by outstanding HEMT IC research. Small scale high performance IC's (1k static RAM's, 4k static RAMS) have been fabricated and IC's with $10^4$ gates and superior speed (~ 70 ps gate delay) are expected to be produced in the next 6 years or so.[B.3] The level of sophistication and integration of HEMT's increases about twice as fast as for silicon (based on a few data points only). HEMT integrated circuits are perceived by Japanese researchers as being prime candidates for future supercomputer memories.[B.4] The high performance of the HEMT at 77K makes operation at this low temperatures for special applications probable and possible. In addition, it is anticipated that the HEMT has favorable scaling properties and when scaled to or below 1μm will also have advantages at room temperature operation and therefore be useful for computer graphics and the like.

Device modeling develops in Japan to a sophistication never seen before. Recently they completed a Monte Carlo simulation of "ballistic aspects" (overshoot) of transport in HEMT structures as shown schematically in Fig. 4-6.[B.5] A literature search [B.6] on HEMT related publications reveals that 31% of the world's publications are from Japanese authors. The U.S. holds 38% (with many duplications) and France 23% (also with duplications). The rest of the world (8%) is negligible. There are many acronyms used for the HEMT as shown in Fig. 4-5. The fact that each country has its own, but the U.S. publications use at least three different ones indicates, in my opinion, a not entirely healthy competitive spirit. This is also reflected in a high redundancy of U.S. HEMT research.

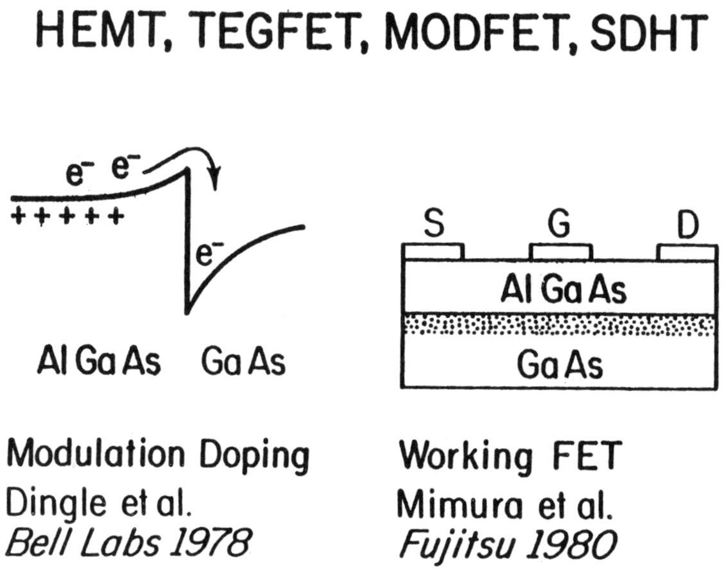


Figure 4-5. HEMT, TEGFET, MODFET, SDHT

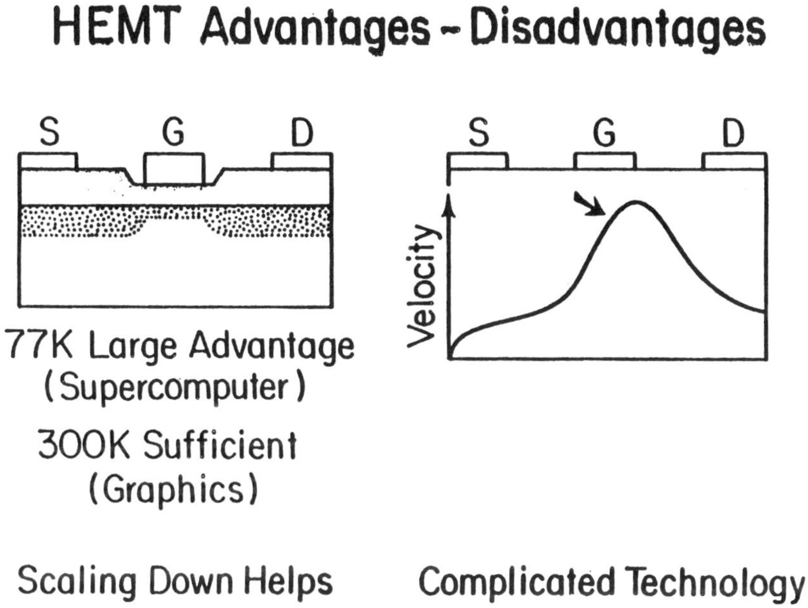


Figure 4-6. HEMT Advantages - Disadvantages.

## Japanese Achievements, Projects and Targets

- HEMT static RAM's
  Dec. 1983 1k
  June 1984 4k (Mimura)
  *Sophistication increases* by a factor of ~2 times *faster than for silicon.*
- Target: $10^4$ gates, 70 ps gate delay (Abe)
- Substrates: Sumitomo, Fujitsu
- Heterolayers: Sony (MOCVD), Sumitomo, Hitachi, Fujitsu (MBE)
- Heterolayer doping: Fujitsu
- Contacts: Fujitsu
- Device fabrication optimization: Hitachi, Fujitsu
  estimated MITI support/year 2.5 billion yen

Figure 4-7. Japanese Achievements, Projects and Targets

## Functional Devices

- Creation of a device function by proper design of the quantum physics *(from physics to function)*
- Example:
  AI function on chip
- Promise:
  In connection with three-dimensional IC concept very complex functions may be generated on very small area
- Lattice matched superstructures are perceived to have extremely high potential

Figure 4-8. Functional Devices.

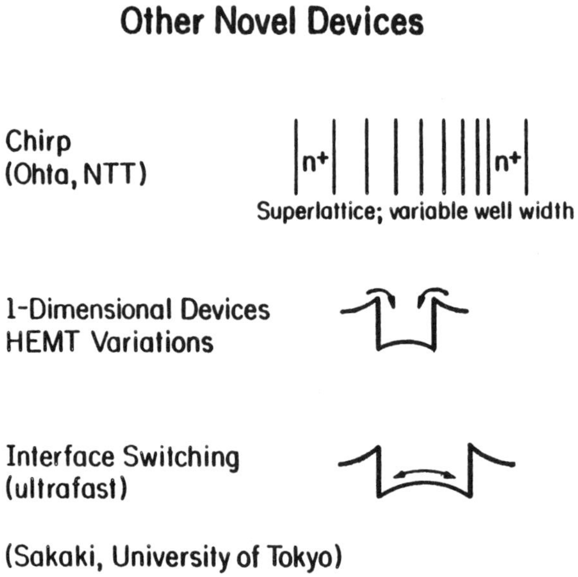


Figure 4-9. Other Novel Devices.

The Japanese work on the HEMT can be characterized as "adaptive" creative. Basic mechanism and structure have been invented in the U.S. and Japanese scientists "only" performed a first reduction to "transistor" - practice. Japan has now a firm lead in HEMT integration, again a sign of their adaptive creativity (Fig. 4-7). The effort on functional devices is also adaptive. Functional devices (Fig. 4-8) were proposed in 1965 by Jack Morton, then Vice President of Bell Laboratories. In his celebrated paper, "From Physics to Function,"[A-4] Morton pointed out that in order to satisfy the tyranny of large systems and high speed, complex device functions should be designed from physical principles. Morton was ahead of his time. With few exceptions such devices could not be made. However, with the possibility of wave function engineering, or (in our terminology) the use of Angstrom boundary conditions such devices seem possible and may be capable of performing the complex functions that are necessary in AI.

In recent years Japanese research goes clearly beyond the adaptive creativity. Leo Esaki (Japanese Nobel Laureate working in the U.S.) is a co-founder of the whole area of superstructure research. New devices and concepts involving superstructures are currently developed in Japan as can be seen from Fig. 4-9. The concept of CHIRP superlattices (Coherent Hetero-Interfaces for Reflection and Penetration), [B.7] and the concepts of Sakaki, (Fig. 4-9) [B.8] demonstrate clearly that creativity flourishes also at the other side of the Pacific.

## 4.3 Summary

The Japanese scientists envision a renewal of electronics by the use of superstructures. Their viewpoint goes far beyond seeing this area as an interesting one for research. Correspondingly, they have plans to invest enormous amounts of money, with the government support alone being around 3-4 billion yen per year for this and closely related areas. A comparison between Japanese and U.S. efforts is given in Fig. 4-10 based on the literature described above. The Japanese viewpoint can be summarized by a managerial principle formulated 20 years ago by Jack Morton (Bell Laboratories): an industry must not be analogous to a single individual in its development (like the steel industry, which is based only on very few basic inventions). It

## Summary of Efforts

| | Japan | U.S. |
|---|---|---|
| • Integration | > | |
| New Device Concepts (NDC) | < | |
| Applications of NDC | > | |
| Basic Research | < | |

• Japanese Government Support

| | |
|---|---|
| Superstructures: | |
| 2.5 billion yen/year | 10 years |
| 3-Dimensional Circuits: | |
| 0.5 billion yen/year | Increasing |
| Optical Interconnections: | Steeply increasing |

Figure 4-10. Summary of Efforts.

should be a dynasty seeking repeated renewal through trips back to the "fountain of youth" of basic science. The basis for the trends in electronics applications of superstructures were listed in Fig. 4-11.

Superstructure research in the United States is performed at Bell Laboratories and at an increasing scale at IBM as well as at several centers which are mainly supported by DOD programs. NSF support is increasing while SRC support is modest. The theoretical advantage of superlattices, i.e., the possibility of designing a final electronic function from basic physical (quantum) principles is not collectively recognized or agreed to by the leaders of U.S. semiconductor research and development.

Given this situation (Fig. 4-12), one may anticipate that Japanese superstructure electronic research will develop vigorously in the coming years. The U.S. will be able to keep pace and may achieve superiority if, and only if, cooperation, faith and monetary support in this area increases.

## Basis for Trends in Applications

- HEMT's proven high performance
- Promise of three-dimensional circuits and new functional devices
- The possibility of optical inter- and intra-chip communication

Figure 4-11. Basis for Trends in Applications.

## Summary

Japan

- *Great enthusiasm* for superstructure electronics
- *Creativity* beyond adaptive creativity
- *Ambitious goals* with respect to circuit speed and integration
- Are determined to be *first in application aspects*

U.S.

- Contributed *major inventions*
- *Leads in* aspects of *fundamental research*
- Attempts to regain supremacy in HEMT IC development
- Is increasingly *developing enthusiasm* for *superstructure electronics application*

The technology of lattice matched heterostructures promises enormous impact within the next five to ten years. Japan is determined to lead. The U.S. will need to increase its effort to regain (and maintain) supremacy.

Figure 4-12. Summary.

4.4 References


A.1 See e.g. K. Hess and N. Holonyak, Jr., Physics Today, October 1980.

A.2 "Survey Reports on Materials for Opto-Information Processing", JEIDA, March 1984.

A.3 1st International Workshop on Future Electron Devices (Heterostructures and Superlattices) Tokyo, February 6-7, 1984.

A.4 J. A. Morton, From Physics to Function, IEEE Spectrum, September 1965.

A.5 A. Abe, M. Mimura, T. Yokoyama and N. Ishikawa, IEEE GaAs integrated circuits symposium, Vol. MTT-30, No. 7, San Diego, 1981.

A.6 Second Symposium on New Functional Devices (sponsored by MITI), Tokyo, July 6-7, 1983.

A.7 Makoto Kikuchi, Japanese Electronics, Simul Press, Inc., Tokyo 107, Japan, 1983.

A.8 "Future Electron Devices, Ongoing Activities and Results in Japan," Mitsui & Co., LTD., Technical Development Division Report, July 1983.

B.1 The basic invention underlying the HEMT was made at Bell Laboratories, R. Dingle, H. L. Stormer, A. C. Gossard and W. Wiegmann, Appl. Phys. Letters 33, 665, 1978. It should be pointed out, however, that the theoretical concept of modulation doping was described by Esaki and Tsu, IBM Research Report No. RC-2418, 1969.

B.2 T. Mimura, S. Hiyamizu, T. Fujii and K. Nanbu, Jap. T. Appl. Phys., Vol. 19, pp. L225, 1980.

B.3 M. Abe, "Why HEMT now," see Ref. A.3.

B.4 Y. Kawatani, G. G. Scarrott, Fifth Generation Computer Project, State of the Art Report, Maidenhead, Berks, England, p. 63, Pergamon Info. Tech., 1983.

B.5 M. Tomizawa, K. Yokoyama and A. Yoshii, IEEE Electron Device Letters, EDL-5, 11, 464, 1984.

B.6 BRS Search, January 1977-May 1984.

B.7 T. Nakagawa, N. J. Kawai, K. Ohta, M. Kawashima, Electronics Lett., 19 (20), 822, 1983.

B.8 H. Sakaki, Superlattices and Microstructures, 1, 5 (1985).

# 5. A Bird's Eye View and Assessment of Japanese Optoelectronics: Semiconductor Lasers, Light-Emitting Diodes, and Integrated Optics

W.T. Tsang, AT&T Bell Laboratories

## 5.1 Background

Within the last decade and a half, the field of optoelectronics has undergone and is still experiencing a tremendous growth in both research and product development. This explosive growth and vitality is fueled by the advent of various major technologies such as lasers, integrated circuits and optics, and lightwave communications and optical storage. It is impossible to survey the entire field completely and in detail. Thus, this study aims at providing overall observations and general trends. This study also provides a comparison between Japanese and U.S. research and the development of semiconductor lasers, light-emitting diodes, and integrated optics.

## 5.2 The Driving-Force of Japanese Optoelectronics

The driving-force for the Japanese optoelectronic R&D is domestic market sales with high hopes for a world market with strong, co-ordinated, and systematic support from the government. The major resources of government support are the well-publicized MITI programs and the NTT's Information Network System (INS). At present, the MITI's major program in optoelectronics in the Japan Optoelectronics Joint Research Laboratory. A summary about this program is given below:

JAPAN OPTOELECTRONICS JOINT RESEARCH LABORATORY

- In operation October 1981
- Monolithic OEIC technology
- Communications, optical measurement and control

- High speed, high quality picture data subsystems, high speed and composite process data subsystems, data control subsystems
- Approximately 50 researchers
- Budget $77M ($5.6M 1981, $25,7M, 1987)
- Floor space 1600 sq. meters, Kawasaki City

The NTT's INS is expected to influence the Japanese optoelectronics industries with $8-13B for research and product development over the next 10-15 years.

The other driving force comes from domestic and world markets. The Japanese optoelectronic markets include the following major areas:

1. COMPONENTS:
   optical fibers
   lasers & LEDs
   photodetectors

2. EQUIPMENTS:
   optical disk units
   laser printers
   electronic fields

3. LARGE & SMALL SYSTEMS:
   telecommunications
   local area networks (LAN)

Sales performance is summarized as follows:

- Past five years sales increased by tenfold, mostly in component area.

- Components represent half of the market

- By 1992, laser applications will be widely used in office and home equipments.

A detailed breakdown of the Japanese optoelectronic market in 1982 and a forecast in 1992 in various area is given below:

THE JAPANESE OPTOELECTRONIC MARKET:

| | | | |
|---|---|---|---|
| Optical Communication: | $85M | (1982) --> | $400 - 800M (1992) |
| LAN : | $70M | (1982) --> | $420M (1992) |
| Laser : | $ 5M | (1982) --> | $18M (1992) |
| LED (comm.) : | $10M | (1982) --> | $29M (1992) |
| Detector : | $0.5M | (1982) --> | $1.25M (1992) |
| Optical disk : | | --> | $1.25B (1992) |
| TOTAL OPTOELECTRONIC MARKET: | | --> | $4-8B (1992) |

One other general observation is that the Japanese government seems to work closely with industry. In the U.S., this appears to occur only for companies which have defense contracts. This communication allows the private Japanese industrial companies easier access and more influence in shaping government policy in terms of trade. Such communication between government and private industries is essential to the economic welfare of the country as a whole and the Japanese appear to be doing particularly well in this respect.

In contrast, the U.S. optoelectronic industry appears to be driven mostly by market forces. Both the private industry and government do not appear to be making any particular effort to communicate and to coordinate general policies. This might affect the future welfare of the U.S. optoelectronics industry.

## 5.3 The Growth of Japanese Optoelectronics, 1969 - Present

The Japanese industry regards optoelectronics as a fast growing technology with a large potential product base. This is reflected by the R&D efforts in a number of industrial companies and their progress through the recent years. Major Japanese optoelectronic industries include NEC, Hitachi, Toshiba, Mitsubushi Electric, Matsushita, Fujitsu, Oki, Sumitomo Electric,

Furukawa, NTT, Sharp, Sanyo, Sony, and KDD. Industries involved in long-wavelength (1.3 μm, 1.55 μm) semiconductor laser-R&D are Hitachi, NEC, Fujitsu, NTT, Mitsubushi Electric, Matsushita, Toshiba, and KDD. Industries involved in developing visible (< ~ 0.78 μm) semiconductor lasers are Sharp, Sony, Sanyo, and Hitachi.

The R&D effort, achievements, and growth in certain fields can be monitored by plotting the total number of scientific publications published per year. Figure 5-1(a) shows the number of published papers (including those in Japanese journals) on semiconductor lasers by different countries from 1969 to 1982. Data for 1983 and 1984 are not available. Figure 5-1(b) shows a smooth version of the same curves in Fig.5-1(a) in order to illustrate the general trend of progress. Some observations are:

1. The field of semiconductor lasers is fast-growing in every industrial country.

2. Before 1976, there was relatively little effort in Japan. After 1976, the Japanese growth appears to be faster than that of the U.S.

3. The various ups-and-downs on the curves can be related to the various important events which occurred, as indicated in the figure.

A similar plot for light-emitting diode (LED) is shown in Figs. 5-2(a) and (b). Some conclusions can be drawn as follows:

1. LED growth appears to peak at 1978 and then suddenly level off. This is due to the takeover by the semiconductor laser as an alternative and more promising light source. The important lesson this teaches us is that: in high technology a fast growing technology can suddenly be replaced by another up-coming more suitable technology without any long term advance warning.

2. By 1976, the Japanese R&D tracks that of U.S. very closely.

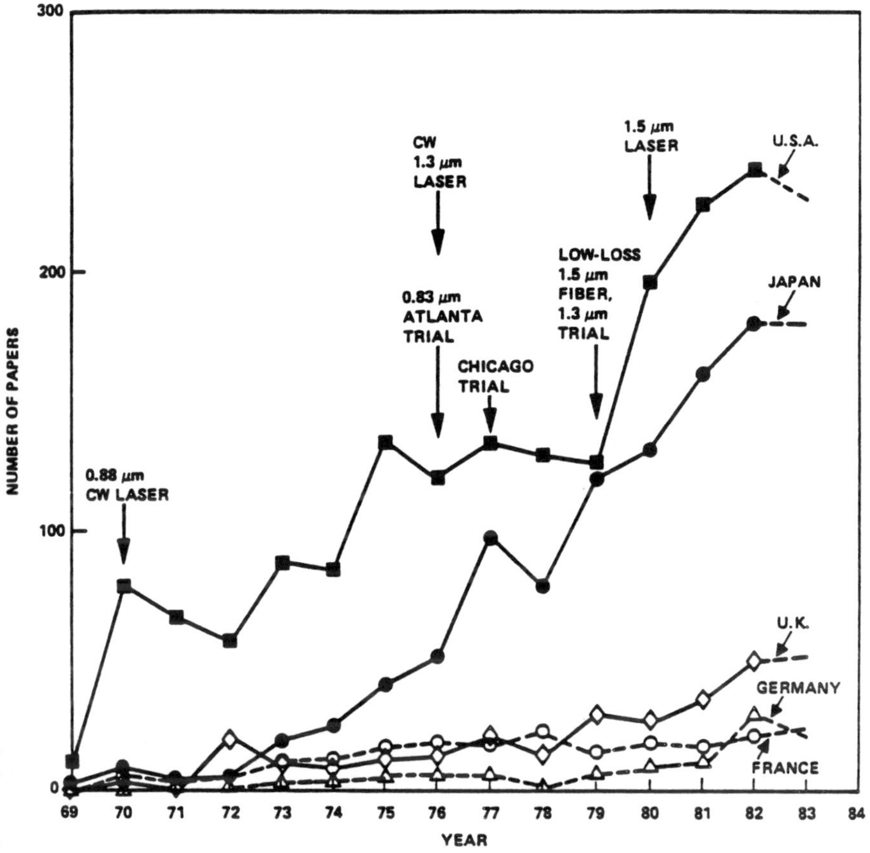


Figure 5-1(a). Published Papers on Semiconductor Lasers by Different Countries.

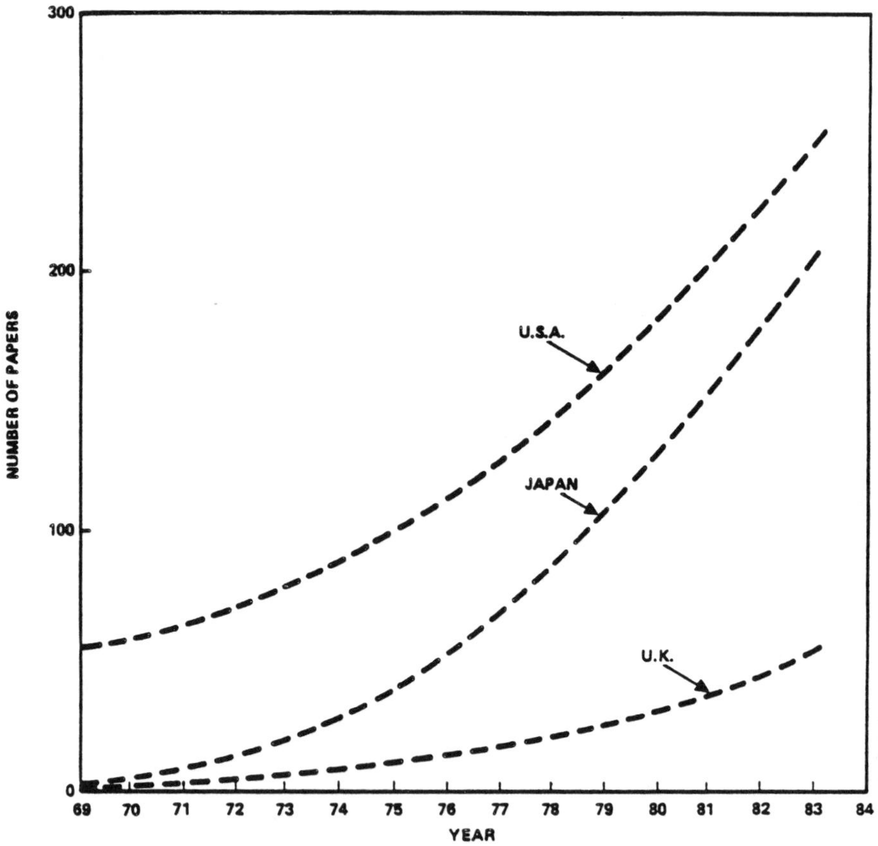


Figure 5-1(b). Published Papers on Semiconductor Lasers by Different Countries.

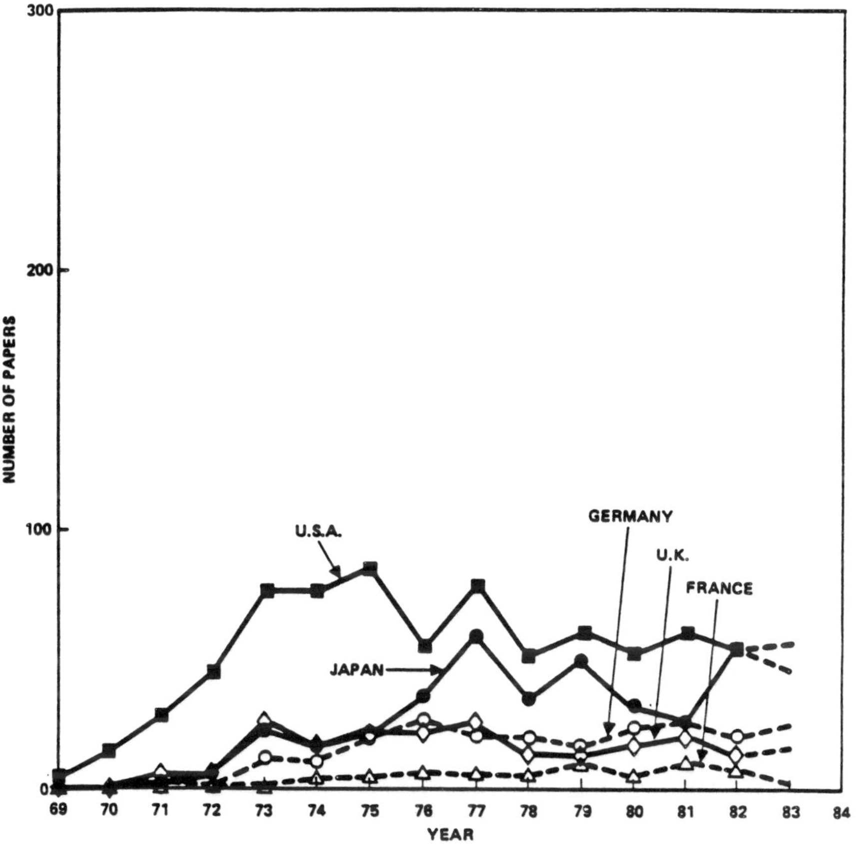


Figure 5-2(a). Published Papers on Light-Emitting Diodes by Different Countries.

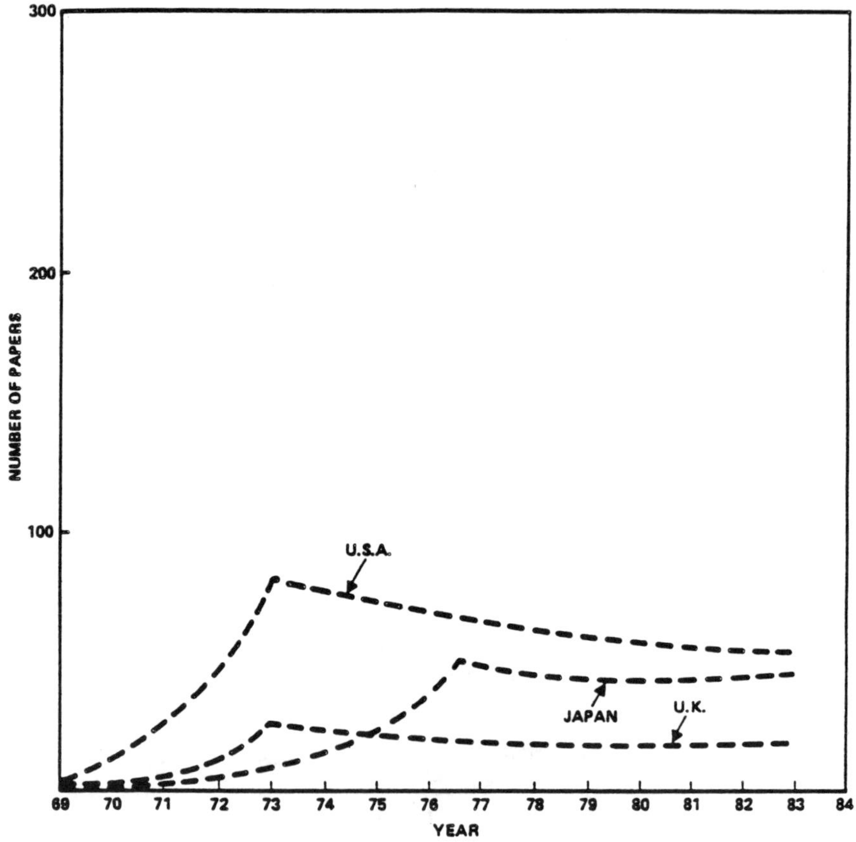


Figure 5-2(b). Published Papers on Light-Emitting Diodes by Different Countries.

3. The effort in LED research is slowly decreasing. Figures 5-3(a) and (b) show similar plots.

On Integrated Optics. Some general observations are:

(1) The field of integrated optics is uneven. The cause of the sudden drop in 1981 is not understood. This drop occurred worldwide.

(2) The Japanese effort in integrated optics did not start until 1976. After 1976, their effort tracked with that of the U.S.

(3) The Japanese put much less effort in integrated optics than in semiconductor lasers, while the U.S. appears to place approximately equal emphasis on both. This suggests that the Japanese do not see as much practical application of integrated optics.

Plotted in Figures 5-4(a) and (b) are the total number of published papers per year for AT&T Bell Laboratories, NTT, Hitachi, Fujitsu, NEC, Mitsubishi, Toshiba. Some observations can be made based on:

(1) The fluctuations of AT&T Bell Laboratories can be related to certain events. The sharp spike around 1975 and the subsequent drop around 1978 are related to the initial decision to base all lightwave systems on 0.83 μm multimode systems, the various difficulties encountered in obtaining high-performance 0.83 μm semiconductor lasers, its subsequent abundance and time-lapse in developing 1.3 μm lasers.

(2) Japan came in late in the field of semiconductor lasers. As a result, they were spared the cost and agony of the 0.83 μm technology for use in lightwave transmission.

(3) There appears to be two characteristic types of developments in Japanese R&D. One typified by NTT and Fujitsu and the other by Hitachi and NEC. The sudden decrease in the cases of Hitachi and NEC might be due to the companies' shift from more research effort to a

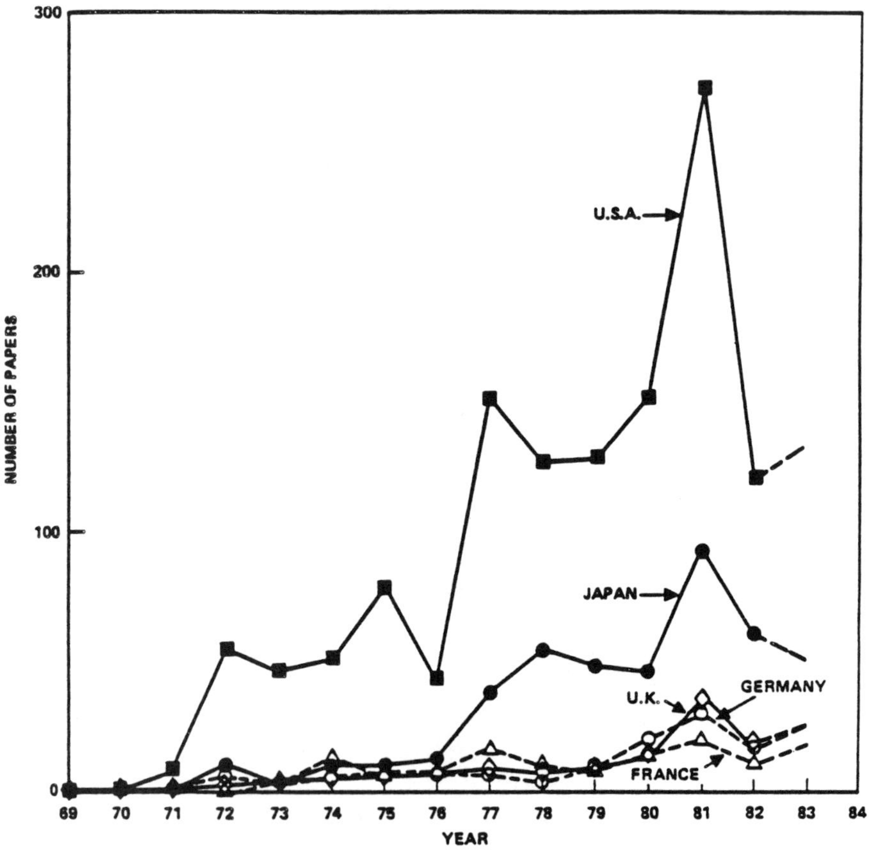


Figure 5-3(a). Published Papers on Integrated Optics by Different Countries.

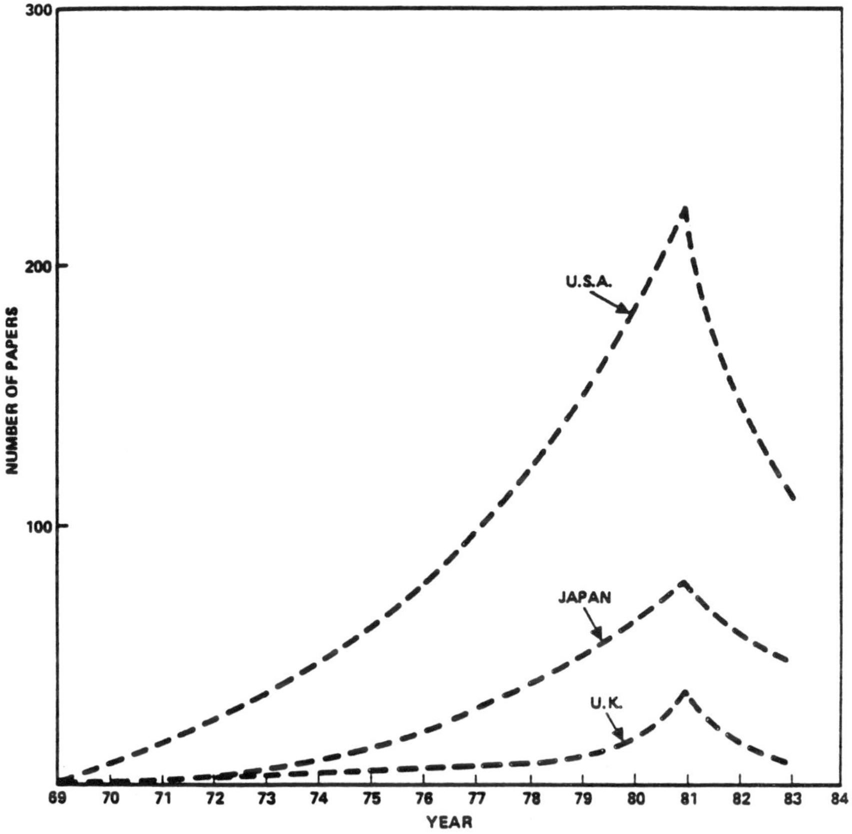


Figure 5-3(b). Published Papers on Integrated Optics by Different Countries.

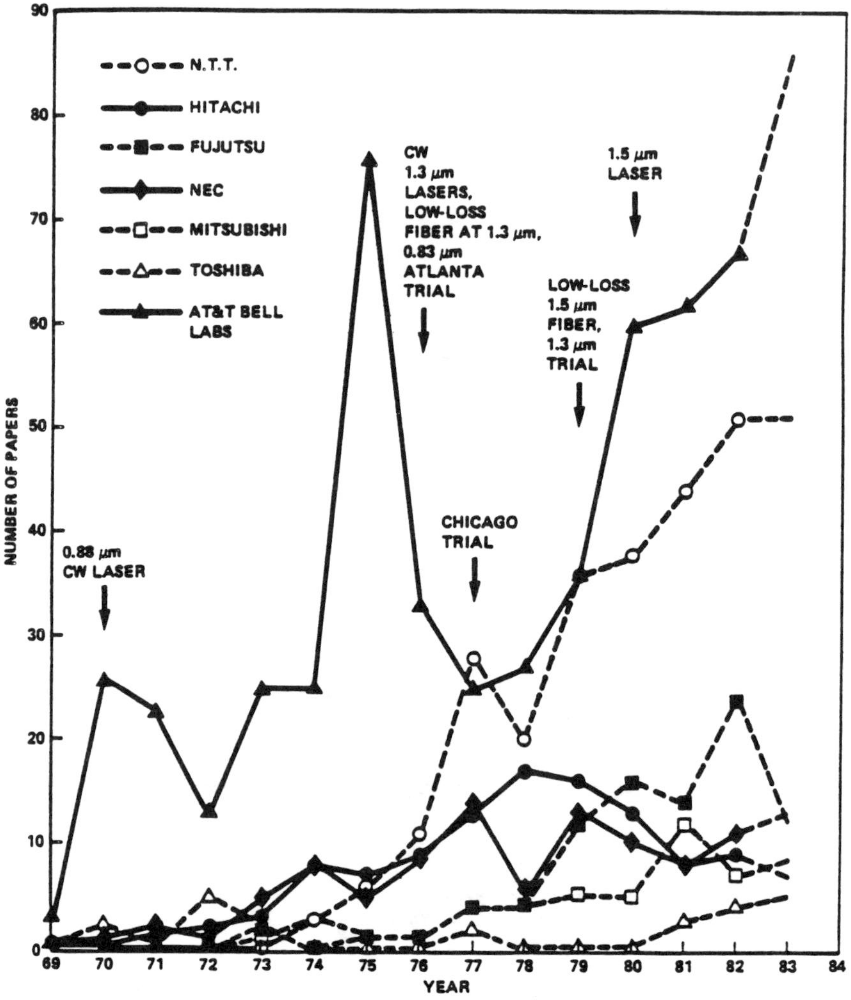


Figure 5-4(a). Published Papers on Semiconductor Lasers by Different Countries.

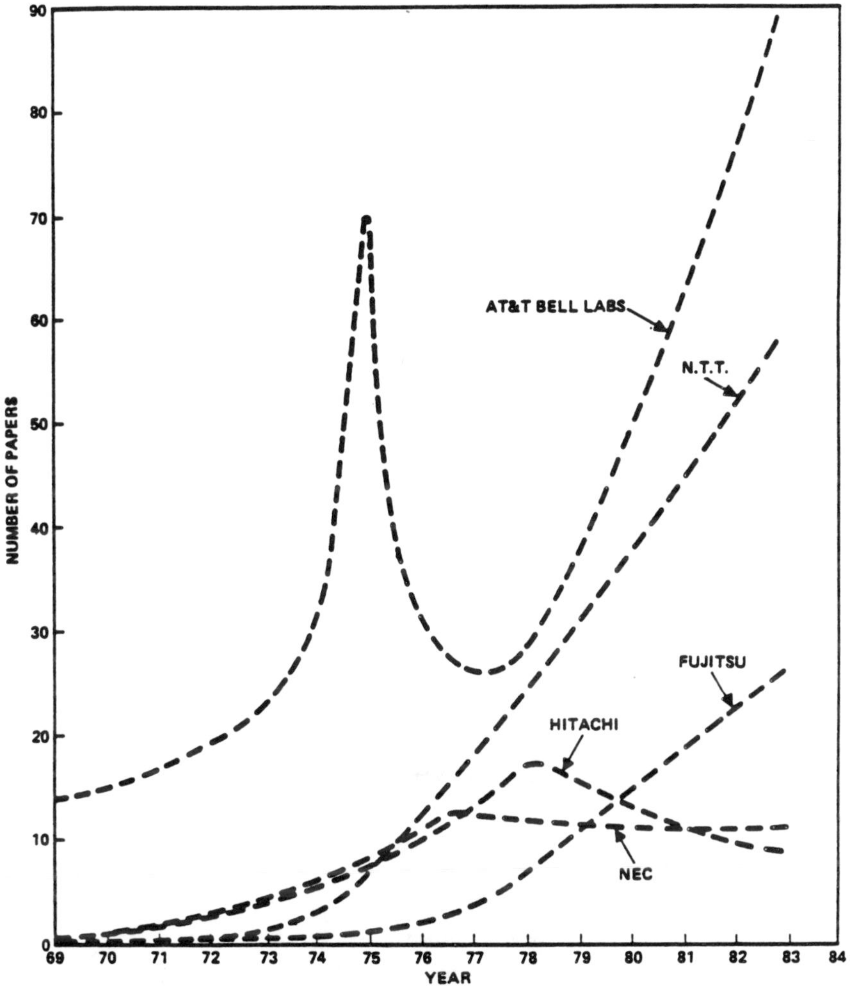


Figure 5-4(b). Published Papers on Semiconductor Lasers by Different Countries.

more product oriented effort, hence, the decrease in the number of publications. This is the likely reason because both Hitachi and NEC are at present heavily involved with very successful opto-electronic products. The price they paid is approximately 5 years of reduced research effort in return for their present dominance in this component market. Less product oriented companies continue to publish heavily.

Finally, Figures 5-5(a) and (b) plot the R&D performance of AT&T Bell Laboratories with respect to U.S. and NTT to Japan on semiconductor lasers. The similarity between AT&T Bell Laboratories and U.S., and NTT and Japan are striking. Both AT&T Bell Laboratories and NTT account for 28 percent of total publications in their respective countries.

## 5.4 Japan and the United States Inventiveness in Opto-Electronics

This is presented by a survey of important inventions in the field of semiconductor laser structures.

The following table shows that all the major laser heterostructures were invented in the West, mostly in the U.S. All these inventions involve fundamental changes of a conceptual nature.

**MAJOR LASER HETEROSTRUCTURES**

| | |
|---|---|
| double-heterostructure (DH) | USSR, USA |
| separate confinement heterostructure (SCH) | USA |
| graded-index SCH (GRIN-SCH) | USA |
| quantum well (QW) | USA |

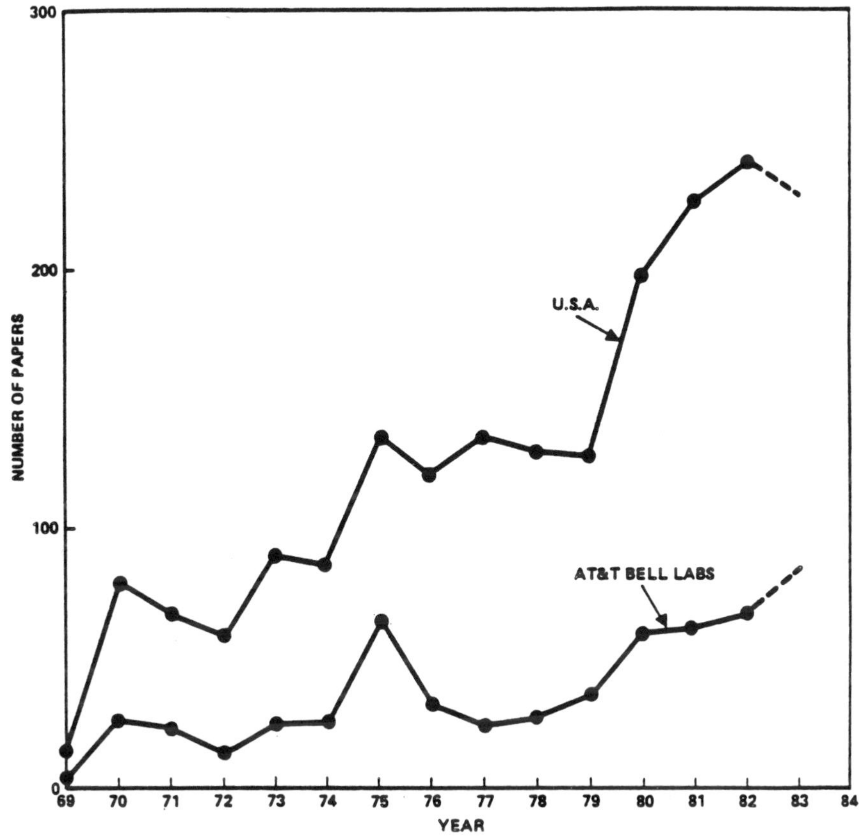


Figure 5-5(a). Published Papers on Semiconductor Lasers.

*AT&T BELL LABORATORIES ACCOUNTS FOR 28% OF TOTAL USA PUBLICATIONS IN 1982.*

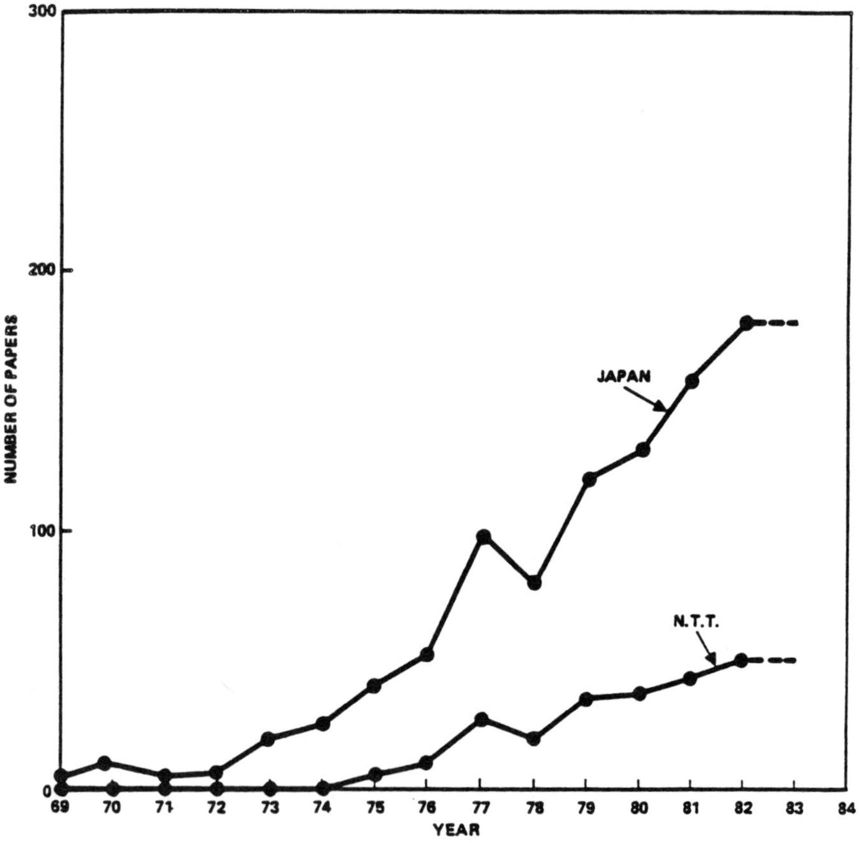


Figure 5-5(b). Published Papers on Semiconductor Lasers.

*N.T.T. ACCOUNTS FOR 28% OF TOTAL JAPANESE PUBLICATIONS IN 1982.*

**MAJOR STRIPE-GEOMETRY LASER STRUCTURES:**

| laser type | invented | commercialized (or active) |
|---|---|---|
| buried heterostructure (BH) | Japan | Japan, USA Europe |
| double-channel planar BH (DC-PBH) | Japan | Japan, USA |
| channel substrate planar (CSP), V-groove | Japan, USA | Japan, USA Europe |
| transverse junctions stripe (TIS) | Japan | Japan |
| strip-buried heterostructure (SBH) | USA | Japan |
| proton-stripe | USA | USA |
| oxide-stripe | USA | USA, Europe |
| ridge-waveguide | Japan | USA, Europe |

**BEST PERFORMANCE LONG-WAVELENGTH FABRY - PEROT LASERS**

in terms of

* low threshold current
* linear light-current characteristics
* high temperature operation
* device-performance stability
* reliability

| laser | invented in | commercialized (or active) |
|---|---|---|
| BH | Japan | Japan |
| DC-PBH | Japan | Japan |
| V-groove | Japan, USA | Japan, USA, Europe |

These inventions are more peripheral and structural than fundamental and conceptual.

All single frequency laser structures were invented in the U.S.

**SINGLE-FREQUENCY LASERS**

| scheme | invented |
|---|---|
| distributed feedback (DFB) | USA |
| cleaved-coupled-cavity (C3) | USA |
| injection-locking | USA |

A more detailed study suggests that the Japanese tend to work on more sophisticated structures in order to obtain better device performance, while the U.S. tends to choose simplicity of fabrication. This may be related to work habits. The Japanese workers traditionally pay more attention to detail than U.S. workers. These observations are listed in the following tables.

**MAJOR DFB LASER STRUCTURES**

| laser | initial work |
|---|---|
| DC-PBH-DFB | Japan |
| BH-DFB | Japan |
| VPT-BH-DFB | USA |
| HRO-DFB | USA |
| Ridge-DFB | UK, USA |

* VPT-BH-DFB -- Vapor Phase Transported Buried Heterostructure DFB

* HRO-DFB -- Hetero-epitaxial Ridge Overgrown DFB

**BEST DFB LASERS**

in terms of

* low threshold current
* spectral stability
* high temperature operation
* high single-frequency output power

| laser | from |
|---|---|
| BH-DFB | Japan |
| DC-PBH-DFB | Japan |
| VPT-BH-DFB | USA |

**BEST DFB LASERS**

in terms of

* simplicity in fabrication
* narrow beam divergence
* spectral stability
* higher yield

| laser | from |
|---|---|
| HRO-DFB | USA |
| Ridge-DFB | UK, USA |

The above study provides the following conclusion:

| | |
|---|---|
| **JAPANESE** | **lead in structural inventiveness,** |
| | **lag in conceptual inventiveness.** |

5.5 The Japanese R&D Efficiency and Technology-Transfer

R&D efficiency and technology-transfer are two of the most important factors in describing the success of a private industrial company. R&D is expensive. Thus, chosing the right issues and the right timing, minimizing unnecessary waste in duplicative efforts, and maximizing productivity are crucial. Efficient and timely transfer of technology to product development is essential. The Japanese industry appears to do both well.

JAPANESE TECHNOLOGY TRANSFER:

- Japanese companies are well organized and managed
- Technology transfer is a one-step process from research to manufacturing especially in process-intensive areas.

- Technology transfer is a transfer of personnel instead of recipes.

JAPANESE R&D EFFICIENCY:

- Their R&D productivity remains high.
- Able to get timely results.
- Only 10-15 percent of their R&D people have Ph.D.s.
- Make efficient use of expensive equipments and facilities.
- R&D results are in general of high quality.
- They work long hours.

## 5.6 Conclusions

From the above studies, some general observations and conclusions can be reached as follows:

- The driving force for the Japanese optoelectronic R&D is at present domestic market sales with high hopes for world market, and strong and coordinated support from the government.
- The strong competition for the Japanese optoelectronic companies comes from other Japanese companies rather from the West.
- In the component area of optoelectronics the Japanese have already surpassed the United States.
- In the system area, the U.S. is still ahead, but Japan is determined to close the gap.
- The Japanese lead in structural inventiveness and lag in conceptual inventiveness.
- The quality of their research is high.

- The Japanese R&D is efficient.

- The Japanese technology transfer is mostly a one-step process directly from research to manufacturing especially in process-intensive areas. This reduces their R&D cost, results in efficient and timely technology transfer, and gives them a lead in product introduction.

- Research in Japanese universities usually complements that in industry. Overlap is avoided. In the U.S., the research topics in universities tend to be the same as those in the industrial laboratories.

# 6. Solid State Lasers in the Near IR and Visible Regions of the Spectrum

Robert S. Bauer, Xerox Palo Alto Research Center

## 6.1 Introduction

This report presents some of the frontiers of technology pursued by the Japanese in short wavelength opto-electronics. Focus is on advances made over the last two or three years in the areas of fundamental GaAlAs materials, devices and processing. Whereas most of the invention is done in the U.S. (we hold the patents for some of the fundamental breakthroughs), there is a great amount of activity in Japan following through on reducing novel phenomena to practical devices. Rather than dwelling on the 1.3-1.55 micron optical fiber window, we discuss how "standard" GaAlAs materials can be pushed into the visible range by using very sophisticated materials growth and device structures. GaAlAs quantum wells have achieved 680 nanometer emission, while at shorter wavelengths, novel III-V materials systems such as InGaAlP, InGaAsP, and AlGaN are applicable.

The visible region of the spectrum is extremely important for a broad range of applications. We seem to focus currently on sources for optical fiber communication because of the very large market potential. But visible lasers allow humans to see the light source in aligning and using real systems. They also have specific applications where spectral overlap with photographic and photosensitive materials is important. These are optimized in the visible region of the spectrum. For example, in printing applications, there are great advantages to working in the visible range. And, finally, for diffraction-limited optical beam applications, the shorter the wavelength, the more the accompanying smaller diffraction limits will result in a narrower spot size. The Japanese are attacking all segments of the market in both short and long wavelength regions of the spectrum.

## 6.2 Japanese Market Targeting

Integrated opto-electronics, which is the integration of light-emitting devices with electronics on a single chip, was a MITI-targeted technology in the late 1970's. Over $100 M of Japanese government funding was commited to a joint industrial research cooperative under the banner of the Joint Optoelectronics Laboratory. This money is crucial not only for its absolute amount but also because of the funding stability that it provides. Being invested over a period of eight years, research can follow technical paths rather than being subjected to shifts in a particular marketplace. This stabilizes otherwise uncertain justifications for industrial R&D investment.

The Japanese themselves project their part of the opto-electronics industry to expand at an enormous rate. From 1980 to the year 2000, the annual growth rate is 28%, which is about double the rate of expansion for the standard electronics industry over the last 10 years. Opto-electronics will grow worldwide to over 100 billion dollars in annual sales by the end of the century. The Japanese openly stated in 1981 that they intend to capture 40 to 50% of that market.

Individual components, such as lasers, and fiber-optic communications systems together represent about 60% of the marketplace. The largest single segment is the office equipment market. This includes such items as optical disks, printers using lasers to write on photosensitive material for computer output, and other applications in the office environment. Office automation environment is exploding and opto-electronics is a key technology in that market.

The important lesson for U.S. commerce is that, when the Japanese decided to pursue opto-electronics in the middle 1970's, the markets for applications of this technology were very small, less than $100 M. Based on technological opportunity, they have created a component market in and of itself. This creates a technology privilege which can be exploited through further innovation to create a larger office systems and communications industry. This steady, long-term investment in market creation through technology push represents a different philosophy than the short-term market pull strategy employed by U.S. industry today.

## 6.3 Advances in AlGaAs Lasers (900 > 1 > 780 nm)

First we discuss "standard" AlGaAs lasers operating in their normal range, which is between 900 and 780 nanometers. The diode laser is composed of a GaAs substrate. Substrates are important in III-V IC's in opto-electronics, and in integration they are crucial. The entire technology of solid state optical sources depends on growing epitaxial layers on top of a GaAs substrate. Any defects that exist in the substrate affect or are grown into the laser material. On top of this substrate, a series of epitaxial layers of AlAs/GaAs is grown. Because they are lattice-matched to the GaAs substrate, almost perfect continuation of the crystal structure is achieved. Through doping during growth, a p-n junction is created near the active layer. This "active" layer is that region where the electrons and holes are brought together to combine and create light emission. This is a very complex structure because one has to fabricate a striped contact to confine electron flow into a very small region. The electrons recombine with holes as they move vertically in this structure, at high enough gain to achieve stimulated photon emission. If there is enough gain compared to the loss in the material itself, then lasing action occurs and a coherent beam of light is produced.

In this section, we discuss the substrates, growth and epitaxy, and then finally some aspects of device structures to show why materials are so important.

### 6.3.1 GaAs Substrates

The Japanese have decided that they will be the world's supplier of GaAs substrates. In 1980 they decided that they were going to build up a massive manufacturing capability. In the period from 1982 to 1985, there was a fourfold increase to eight tons in the production capacity of bulk GaAs by the Japanese. To give you the idea, that is something like 10 billion lasers, if they chose to use all the GaAs for lasers; in fact, larger die will be used for GaAs integrated circuits. It is reported that Sumitomo alone is going to be able to make a hundred metric tons of GaAs by 1988 using Ga supplies from the Peoples Republic of China.

What is abundantly clear is that the only reputable commercial sources of GaAs substrates are in Japan. Therefore, large American companies like Rockwell International grow their own material. But other large companies like IBM and Xerox and start-ups like Gigabit Logic do not have captive material sources; they are dependent on Japanese suppliers.

To improve manufacturability and integration, the Japanese are working very hard on growing GaAs on top of silicon substrates. M. Akiyama of OKI Electric reported MO-CVD growth of GaAs on GaAs/GaAlAs buffer layers to accommodate mismatch to a Si(100) substrate. Silicon is a much more mature technology than GaAs-based microelectronics. If high quality III-V growth on a silicon substrate is achieved, one has an enormous advantage, both in the quality of that starting material and in substrate price and availability.

In Applied Physics Letters, Lincoln Labs reported the first GaAlAs laser grown on a silicon substrate. The properties were poor, but the ability to create stimulated emission for growth of GaAlAs on top of silicon is very exciting. T. Nonaka of OKI has reported fabrication of GaAs MESFET IC's on Si substrates. So even with the great emphasis on GaAs substrates, the Japanese are very active in the field of growing these III-V's on standard Si substrate material.

### 6.3.2 GaAlAs Epitaxy Using MO-CVD

I will emphasize metal-organic chemical vapor deposition (MO-CVD) for epitaxy. Compared to molecular beam epitaxy (MBE), we believe MO-CVD to be more appropriate for large-scale manufacturing. MO-CVD is a field that the Japanese have been late to enter, but in which they presently have a great amount of activity.

MO-CVD is based on a very simple technique. Metal atoms attached to some organic group are brought together with a source of arsine gas at an appropriate temperature. For tri-methyl compounds, decomposed on the GaAs wafer, the reaction simply allows $CH_4$ to be given off with deposition of AlGaAs in the appropriate ratios with which one starts in the gas flow. Control in both doping and composition is achieved by temperature stabilization of the sources, valve and tubing cleanliness, sophisticated flow control, and reactor design in order to heat and rotate the substrates uniformly. Compared to MBE, MO-CVD has demonstrated very similar abilities to control composition, dopant thickness, and abruptness of the interfaces. While it is difficult to control background doping, HEMT devices with high mobility most recently have been grown by MO-CVD.

MO-CVD and MBE are new, novel technologies for growing epitaxial layers which offer real advances over LPE in fabricating artificial microstructures for devices. To achieve its potential, sophisticated and difficult research is required. Looking at the literature, one finds that the Japanese now seem to be leading in the material science of these growth techniques. N. Kobayashi, et al., in Jpn. J. Appl. Phys. 21, L705 (1982) reported basic studies of pressure dependence for MO-CVD GaAs vacancy incorporation. At the Second International Conference on Metal-Organic Vapour Phase Epitaxy in Sheffield, England in April 1985, a number of significant Japanese papers was presented. H. Terao of NEC studied the effect of $O_2$ and $H_2$ introduction during GaAlAs MO-CVD growth; M. Akiyama of OKI reported on Cr doping to achieve semi-insulating GaAs by MO-CVD; H. Ohno and H. Hasegawa of Hokkaido University reported attempts at producing atomic layer doping of MO-CVD GaAs for obtaining novel opto- and high-speed electronic device structures. One can find fundamental work on the effect of relative arsine concentration on GaAs and AlGaAs carrier concentration [Y. Mori, et al., Inst. Phys. Conf. Ser. 63, 95 (1981)] and EL2 deep level concentration [M. Watanabe, et al., Jpn. J. Appl. Phys. 22, 923 (1983)]. These are scientific, not just empirical, studies.

To make this a manufacturable technology, understanding process control variations will be basic. Let me describe one example of the Japanese effort to achieve this. In the work of T. Nakanisi of Toshiba, effects of the arsine source were studied. One can grow MO-CVD material with only 10% arsine in hydrogen. Nakanisi started with a tank of 100% arsine, and divided it into

10 tanks with high-purity hydrogen. A series of growths was conducted where nothing but the small arsine sources were changed. After six runs, when the bottle changed, the luminescent output suddenly dropped by three of four orders of magnitude. Such material would not be suitable for devices. This is a well known phenomenon in limiting yield. What one does after a run or two is to throw the arsine bottle away and start again. To the best of my knowledge, no one really can characterize what it is that is causing the arsine to be "bad." Moisture monitors are used on all MO-CVD reactors and contaminants are checked. In this work from Japan, a very tedious program to characterize the fundamentals of the material sources is being pursued as a necessary condition to harness MO-CVD technology. One does not see much work of this type going on in this country.

## 6.4 Quantum Well Device Structures

Precise control of III-V epi layers is required to achieve uniform, low threshold, efficient opto-electronic devices. By sandwiching a thin GaAlAs region of lower aluminum concentration between higher forbidden band gap alloy regions, a confining region with bound quantum well levels is created. Then electrons which are injected will be captured in the well where they recombine with holes, producing photons. The basis for making efficient devices is to produce very thin wells in which to capture these electrons; the population of carriers can be very high, producing more gain in the active region for better lasing characteristics. There are reports of wells that have barriers within the wells, these barriers being tens of Angstroms in thickness. These can be used for separately optimizing optical confinement by index of refraction changes from gain-producing carrier confinement. The ratio of the various compositions and their spacings is very critical in determining how a device will behave.

One application where one needs to have control, not just of the material in a local region but broad uniformity across a particular wafer, is for high-power lasers. A structure that was invented by D. Scifres, W. Streifer, and R. Burnham [Electron. Lett. 19, 169 (1983)] has achieved 2 1/2 watts of optical power emission out of a single chip. Its basis of operation is to take a large area (250 μm x 400 μm) of material composed of multiple quantum wells and create 40 lasers next to each other through processing. The lasing regions are pumped in parallel and are close enough to couple the optical fields of nearby lasers. These regions all phase-lock into a single lasing mode if the material is uniform and engineered properly.

There is considerable research into the proper device geometry. Material uniformity over 105 square microns is required to avoid filimentary hot spots. Surveying the literature, we find that the Japanese employ both MO-CVD and MBE. At the biannual international laser conference in Brazil in July 1984, Hitachi announced their use of multiple quantum wells for high-power devices. We are seeing that, where the U.S. invented structures just a couple of years ago, the Japanese are very quick to jump on the bandwagon and follow up on these advances.

## 6.5 GaAlAs for Visible Lasers (780 > l > 680 nm)

Most of the lasers that are used in the world are helium-neon gas lasers operating at 632.8 nm; they are used in such applications as supermarket scanners and the original optical videodisks. These will be replaced with solid-state lasers. There already is evidence of this in digital-audio compact disks and videodisk players.

In order to move into the visible range, one needs to work with quantum wells grown by controlled epitaxial materials technologies. The quantum size effect in thin layers provides the ability to achieve visible laser operation [N. Holonyak, Jr., et al., IEEE J. Quantum Electron. QE-16, 170 (1980)]. Using the quantum-size effect, states for electrons inside a well of tens-of-angstroms thickness occur at energies which are directly dependent on the height of the barrier and the width of the well. This effect has been discussed in detail by U.S. researchers [such as Vojak, et al. (1981)]. The narrower the well can be made, the shorter the wavelength of the photon emission because of the higher energy state of the electron level.

One of the nicest demonstrations of this effect was achieved again by a Japanese group [H. Kawai, et al., J. Appl. Phys. 56, 463 (1984)]. A series of GaAs quantum wells was grown 30, 40, 70 and 100 Angstroms thick between 500 A $Al_{0.54}Ga_{0.46}As$ barriers. The corresponding luminescence peaks at 710, 730, 780, and 800 nm, respectively. For GaAs active layers thinner than 50 A, lasing occurs at wavelengths shorter than achievable in standard double heterostructures. Lasing action as short as 680 nm at room temperature has been demonstrated by using this principle. Japanese who are intensely pursuing the quantum size effect for such near IR lasers include researchers at Sony [Y. Mori and N. Watanabe, J. Appl. Phys. 52, 2792 (1981)] and Sharp [S. Yamamoto, Appl. Phys. Lett. 41, 796 (1982)].

## 6.6 Other III-V Visible Lasers (680 nm > l)

The Japanese establish the state-of-the-art in going to III-V materials other than GaAlAs for lasers operating at wavelengths shorter than 680 nanomenters. To achieve shorter wavelengths, one has to use materials that have larger bandgaps than AlAs. Phosphide-based III-V's offer this but the lattice constant for Ga or Al phosphide is very different from GaAs, which might provide a substrate for epitaxial layer growth. GaAlAsP heterojunctions cannot be grown on GaAs without producing a strain-induced effect which keeps the material from working satisfactorily as an opto-elecronic device. Complex quaternary materials of Ga, P, and In with Al or As have larger lattice constants which are capable of achieving lattice matching with GaAs.

The Japanese have a clear lead and are pursuing such quaternaries with substantial resources. References are given at the end of this section. For the phosphorous-based III-V quaternary, lasing in the yellow (579 nm) has been achieved, a safe operating range that should be stable in room temperature diode form. By using reactive MBE, nitrogen allows materials of AlGaN which cover the entire visible spectrum. Using this material, cathode luminescence has been demonstrated into the UV up to six electron volts. The intensity of Japanese publication activity suggests a major thrust by them in this area.

"Organometallic Vapor Phase Epitaxial Growth of $In_{1-x}Ga_xAs_yP_{1-y}$ on GaAs," T. Iwamoto, K. Mori, M. Mizuta, and H. Kukimoto, Tokyo Institute of Technology.

"Fabrication and Visible-Light-Emission Characteristics of Room-Temperature-Operated InGaPAs DH Diode Lasers Grown on GaAs Substrates," S. Mukai, H. Yajima and J. Shimada, Jpn. J. Appl. Phys. 20, L729 (1981).

"$Ga_xIn_{1-x}As_yP_{1-y}Ga_{x'}In_{1-x'}As_{y'}P_{1-y'}$ DH Visible Injection Lasers Grown on (100) GaAs by a Two-Phase-Solution Technique," H. Kawanishi, T. Aota, and T. Iwakami, Kohgakuin University.

"MOCVD-Grown $Al_{0.5}In_{0.5}P$-$Ga_{0.5}In_{0.5}P$ Double Heterostructure Lasers Optically Pumped at 90 K," T. Suzuki, I. Hino, A. Gomyo, and K. Nishida, Jpn. J. Appl. Phys. 21, L731 (1982).

"Yellow Emitting AlGaInP DH Laser Diode at 77K Grown by Atmospheric MOCVD," M. Ikeda, M. Honda, Y. Mori, K. Kaneko, and N. Watanabe, Appl. Phys. Lett., to be published.

"Visible Semiconductor Lasers in AlGaInP System," T. Suzuki, NEC.

"0.66 mm Room-Temperature Operation of InGaAlP DH Laser Diodes Grown by MBE," Y. Kawamua, H. Asahi, N. Nagai, and T. Ikegami

"Properties of $Al_xGa_{1-x}N$ Films Prepared by Reactive Molecular Beam Epitaxy," S. Yoshida, S. Misawa, and S. Gonda, J. Appl. Phys. 53, 6844 (1982).

## 6.7 Summary and Conclusions

The Japanese are striving to perfect semiconductor lasers. They are not driven for immediate economic returns, but they are investing in the future. The Japanese government through MITI provides stability with long-term funding and promotes industry-wide cooperation and pooling of resources. The Japanese expect to dominate this market -- they have openly stated this as their goal. Not only will they continue their dominance of individual opto-electronic components, but they should capture 50% of this growing market by the end of the century.

Japanese GaAs laser activities are very broad-based. Major companies are working in such diverse areas as III-V epitaxial growth techniques, novel processing, new materials alternatives, and device physics. They have sophisticated techniques which allow high-level scientific work.

The Japanese will continue to bring products from the development stage to the marketplace rapidly and they will not hesitate to use U.S. developments as well as their own.

There is an explosion in the information that is coming out of Japan on opto-electronics. They report their latest research results as soon as possible, just as we do. We found no more secrecy among Japanese industrial R&D than is present in the U.S.

The Japanese level of investment in the GaAs field suggests strong activity by them in the future. The gap that is opening up between Japan and the U.S. in III-V opto- and micro-electronics threatens to become a permanent disadvantage for the United States.

Acknowledgment

The collaboration and support of Robert D. Burnham of Xerox PARC is greatly appreciated.

# 7. Avalanche Detector and Optoelectronic Integrated Circuits Research in Japan

Federico Capasso, AT&T Bell Laboratories

## 7.1 Introduction

In the last decade, following the invention of the heterojunction laser, there has been a rapidly growing research effort in the area of optoelectronic devices for fiberoptic communication systems. More recently the development of low loss, low dispersion fibers in the 1.3-1.6 μm region of the spectrum has stimulated intensive studies of III-V alloys and related heterojunction devices (LED's, lasers and detectors) covering the above spectral region.

Other important areas of current research in optoelectronic devices are optoelectronic integrated circuits and the study of complex heterostructures, such as superlattices which we shall call, in general, engineered structures.

In this report I will review the current Japanese research effort in the area of heterostructure avalanche photodetectors and optoelectronic integrated circuits. Particular emphasis will be placed on trends and their potential impact on future technology.

Some general characteristics of the Japanese effort in these areas should be mentioned.

### 7.1.1 Concentrated Mostly In Industrial Laboratories

Industrial laboratories for avalanche photodetectors include, Fujitsu, NEC, NTT and KDD. Work on optoelectronic integrated circuits is, for the material side (growth of high quality substrates), concentrated in the Joint Optoelectronic Lab and, for the device applications, at Fujitsu, NTT, and a few universities. The Optoelectronics Joint Research Laboratory is a typical Japanese creation. Several major Japanese semiconductor electronics industries fund this laboratory and provide the manpower. The benefits of this joint enterprise are then shared by the participating companies.

#### 7.1.2 Wide Spectrum of Device Research, Focused On Both Short-Term and Long-Term Goals

A vigorous research effort on avalanche photodetectors covers a large variety of materials, Si, Ge, GaAs, GaInAsP, $InP/Ga_{0.47}In_{0.53}As$, AlGaAsSb/GaSb. Detector research and development is also pursued in those materials which are far from being optimal for fiber communication such as Ge. In fact, the history of the development of Ge APD's in Japan represents an excellent illustration of the Japanese approach to R&D.

### 7.2 Japanese Research On Avalanche Photodetectors For Long Wavelength (1.3-1.6 μm) Fiber Communication Systems

In the last decade Japanese research on APDs has placed strong emphasis on both Ge and heterostructure detectors.

#### 7.2.1 Ge Detectors

This APD was first realized in the U.S. in the mid 60's;[1] its development was abandoned shortly after due to the relatively high ($\gtrsim$ 10 μA) primary dark currents and to the difficulty of finding a suitable passivation procedure. One additional problem of Ge is the ratio of ionization coefficients ($\alpha/\beta$ ] 1-2), which makes this material unsuitable for low noise long-wavelength APD's.

Despite these difficulties and the fact that it was clear that this detector could not meet the requirements for low noise fibers optic receivers (such as those available at shorter wavelengths with Si APDs), the Japanese persisted. This long range strategy has eventually paid off. Ge APDs are at present the only commercially available long-wavelength avalanche detectors. Many of the technological problems associated with Ge passivation and high dark currents have been solved by the use of appropriate planar-guard-ring structures. For example Fujitsu, in cooperation with researchers at NTT, has developed a planar Ge reach-through APD with primary dark currents in the range of a few tens of nanoamps and demonstrated excellent reliability under high bias and high temperature conditions.[2]

In addition these detectors have performed well in fiber optic transmission experiments. The receiver sensitivity achieved with such detectors obviously does not and cannot match that obtained recently with $InP/Ga_{0.47}In_{0.53}As$ heterojunction APDs at Bell Labs.[3] The reason for that is the higher primary dark currents and the lower ionization rates ratio (compared to InP). Nevertheless, the proven reliability of the Japanese Ge APDs is a valuable characteristic, so that they are being used by the Japanese in prototype long wavelength undersea cable high bit rate fiber optic links. For such applications long term reliability is crucial and has not been demonstrated yet in $InP/Ga_{0.47}In_{0.53}As$ heterojunction avalanche photodiodes.

Fig. 7-1 illustrates a cross section of a typical Fujitsu planar guarding reach-through Ge avalanche photodiode.[2]

The complexity of this structure is evident. Its demonstrated performance results from several breakthroughs in the areas of dielectric deposition, implanted guardings and channel-stop layer.

Very little information is given by the Japanese in their papers on APDs on the deposition of dielectrics. This is one of the crucial steps on the road towards the implementation of passivated reliable and low dark current APDs.

### 7.2.2 Heterojunction Avalanche Photodiodes

In 1976 Japanese researchers of NEC, reported a new low dark current, high gain heterojunction avalanche photodiode.[4] This device, subsequently named SAM APD, consists of an absorption layer (of $Ga_{0.47}In_{0.53}As$ or of GaInAsP of the appropriate composition) and of an adjacent InP pn junction

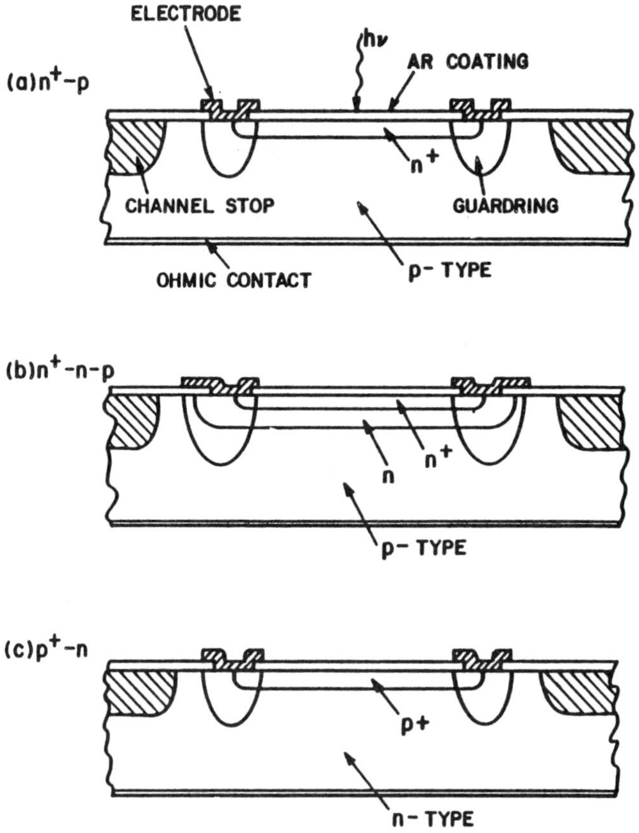


Figure 7-1. Cross-section of typical Fujitsu planar Guard-Ring Reachthrough Ge Avalanche Photodiodes.

where the multiplication process takes place. The band diagram of this structure is sketched in Fig. 7-2. The incident photons are absorbed in the lower gap GaInAs layer. Photoelectrons are collected by the $n^+$ contacts while the photogenerated holes drift toward the wider gap region, surmount the valence band barrier of 0.45 eV and are injected in the n-InP multiplication layer where they undergo avalanche gain because of the high field (4-5) x $10^5$ V/cm.

The crucial advantage of this structure compared to a homojunction APD having the lower gap of the two layers (i.e. $Ga_{0.47}In_{0.53}As$) is that the pn junction (and therefore the maximum electric field) is placed in a wide gap InP region. In a homojunction $Ga_{0.47}In_{0.53}As$ structure, the dark current at the electric field required to achieve sizable gain is very large because of Zener tunneling from the valence to the conduction band.

The InP/$Ga_{0.47}In_{0.53}As$ APD in addition has a larger ionization ratio ($\beta/\alpha = 3$) compared to Ge APDs.

Despite such obvious advantages the design of high performance SAM APDs is critical because of the stringent requirements on doping and layer thicknesses.[5] For example the doping in the gain region must be kept within well-defined limits (typically 1-3 x $10^{16}$) while the background carrier concentration in the GaInAs layers should not exceed $\approx$ 5 x $10^{15}/cm^3$, in order to achieve simultaneously high gain, high quantum efficiency and low dark currents. This was first demonstrated by Forrest et al. at Bell Laboratories.[5]

Another set of problems in this structure is posed by the large valence band barrier ($\approx$ 0.45 eV) present at the InP/$Ga_{0.47}In_{0.53}As$ interface. This leads to the pile-up of photogenerated holes at the bottom of the barrier. Because of the large value of the band discontinuity, these holes are thermionically emitted across the barrier into the InP layer at a relative slow rate. This causes a long tail in the pulse response of such detectors. To solve this basic problem, first recognized by researchers at Bell Labs[6],

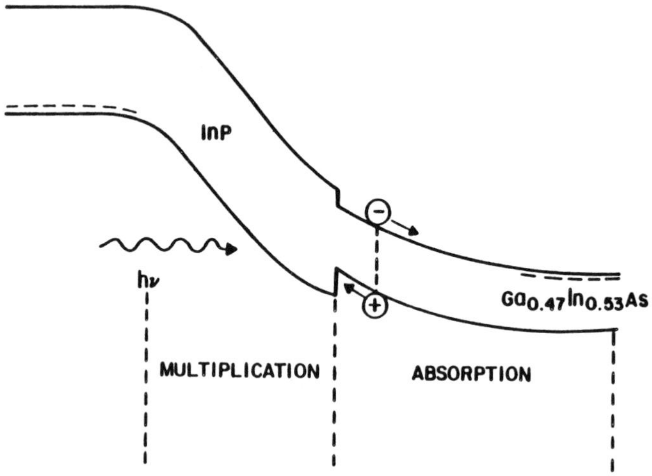


Figure 7-2. Band Diagram of $InP/Ga_{0.47}In_{0.58}As$ SAM APD.

one must grade the interface. This can be done by growing between the multiplication and the absorption layers one or two intermediate grading layers of GaInAsP.

High performance, low noise, and high speed 1.3-1.55 μm InP/GaInAs SAM APDs with a quaternary grading layer have recently been reported by Bell Labs researchers[3] and have led to world record sensitivities in long wavelength fiber optic receivers across the whole data rate range from few tens of megabits to 2 Gb/sec, exceeding by far the sensitivities of receivers employing Ge APD.

An alternative method is to use as intermediate layer a graded gap pseudoquaternary semiconductor consisting of an $InP/Ga_{0.47}In_{0.53}As$ superlattice with varying InP/GaInAs duty factor and constant period.[7] This superlattice simulates a variable gap quaternary layer but is much easier to grow than the random GaInAsP alloy. This technique of growing GaInAsP using only the two end points of the compositional range is very attractive and holds promise to substitute conventional quaternary GaInAsP layers in a variety of applications.

APDs must satisfy the requirements of long term reliability and high device yield per wafer. Such desirable characteristics can be achieved with a suitable planar and guard-ring technology.

In this latter technology, applied to $InP/Ga_{0.47}In_{0.53}As$ APDs, the Japanese companies have extensively worked since the first demonstration of the SAM APD. In fact shortly after the announcement of this detector, four independent efforts were initiated at NTT, NEC, Fujitsu, and KDD (the Japanese phone company).

Particularly broad has been the one at NTT which has included investigations of the physics of these devices, APD planar and guarding fabrication technologies and growth techniques (liquid and vapor phase epitaxy, and hybrid).

The author of this report has had the opportunity of visiting the detector group at NTT in 1981. The group then consisted of about 20 researchers involved in all aspects of heterojunction InP/GaInAs avalanche photodiodes.

The efforts in the other three industrial laboratories (NEC, Fujitsu and KDD) amounts at present to at least sixty researchers and technicians putting the total number of people working on heterojunction APDs to about 80 or more. This estimate is based on visits by the author during two trips to Japan ('81-'83).

This relatively large R&D activity has recently produced many noteworthy results in planar/guarding InP/GaInAs APDs.[8-10] Although the performance of such structures is yet far from that of mesa SAM APDs, especially in terms of advanced receiver sensitivities, the Japanese progress in this area has been impressive. Figs. 7-3 through 7-5 illustrate the APDs structures and some of their characteristics reported recently respectively by NEC, Fujitsu and KDD.[8-10]

The good primary dark currents achieved at the onset of gain (Fig. 7-3) document considerable progress in the area of dielectric deposition and guard-ring technology.

Particularly impressive are the guard ring results. In particular the new buried mesa structure which achieves the equivalent of a guard ring effect, is very ingenious although very difficult to fabricate (Fig. 7-4).

The Japanese clearly lead in planar-guarding APD technology.

We conclude this section by briefly summarizing the U.S. effort in the area of SAM APDs. Shortly after the demonstration of the first SAM APD,[4] several companies and universities undertook investigation of these detectors. In 1980 Lincoln Labs reported high gain, low dark current APDs with InGaAsP absorbing layers and peak response at $\lambda = 1.2$ μm.[11] Bell Labs researchers were the first to achieve high gain and low noise APDs at $\lambda = 1.55$ μm and to elucidate the design rules for such APDs.[5] This was

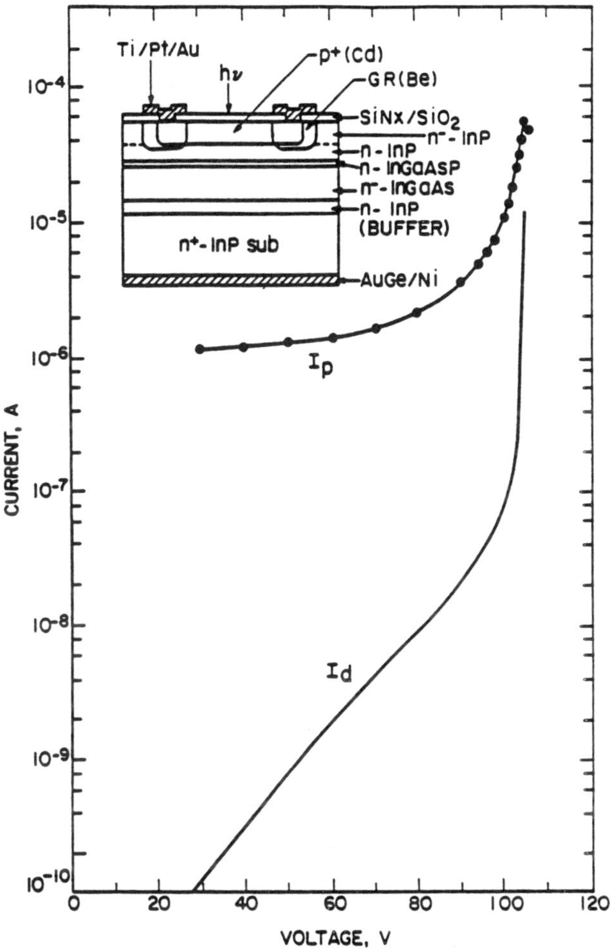


Figure 7-3. Planar SAM APD with Be Implanted Guard-Ring Fabricated at NEC. Shown are also the Dark Current and the Photocurrent vs. Reverse Voltage.

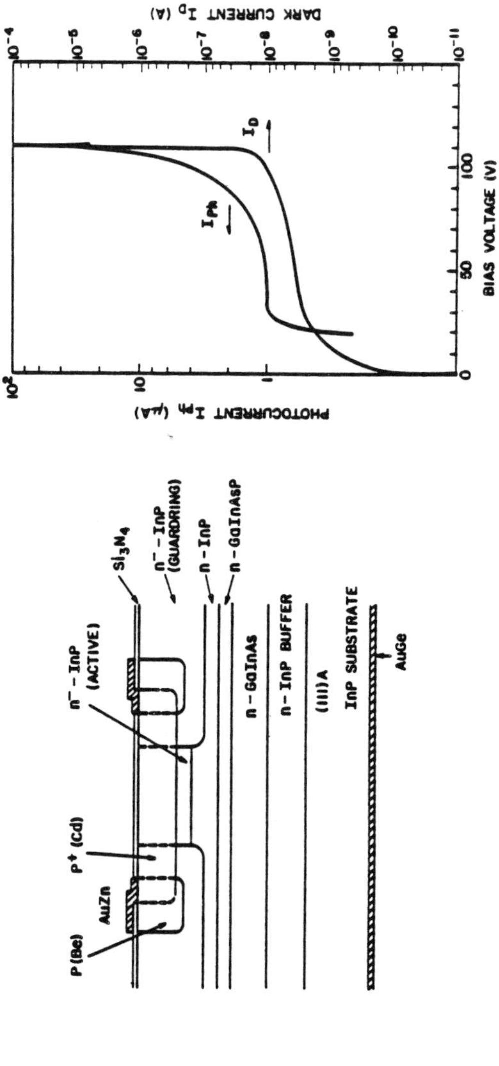


Figure 7-4. Cross-Sectional View of Buried Structure Guard-Ring InP/$Ga_{0.47}In_{0.58}As$ SAM APD Fabricated at Fujitsu.

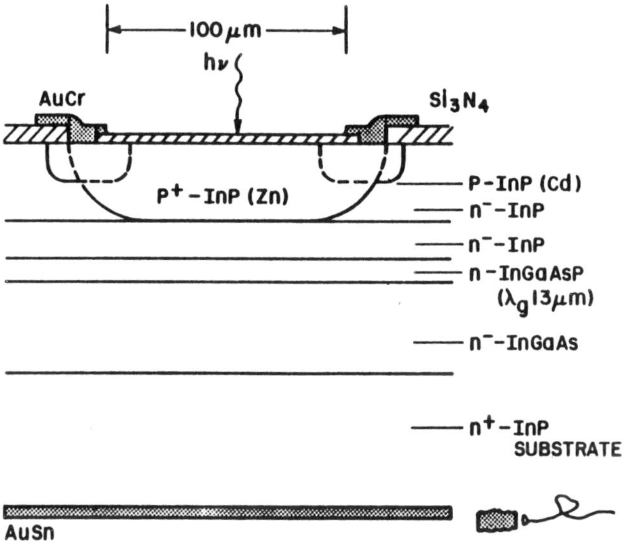


Figure 7-5. Planar InP/$Ga_{0.47}In_{0.58}As$ SAM APD with Cd Diffused Guard-Ring, Fabricated at KDD.

followed shortly after by the discovery, also at Bell Labs, of the pile-up effect of holes and the understanding of its key role in determining the frequency response.[6] The elimination of this effect by the insertion of quaternary grading layers subsequently resulted in the implementation of mesa SAM APDs having world record receiver sensitivities at practically all bit rates from 50 Mbs to 2 Gb/sec.[3] Currently there are two other U.S. companies pursuing R&D in this area: Varian and RCA.

The University of Illinois has carried out basic investigations of the ionization rates for electrons and holes in InP both experimentally and theoretically.[12] Overall in this area U.S. researchers lead in terms of device performance (dark currents, speed, receiver sensitivity, etc.) although the fabrication technology is somewhat less developed. The Japanese lead in planar guard-ring fabrication technology.

In general the U.S. approach has been more innovative, although less systematic.

### 7.2.3 Multilayer avalanche photodiodes and solid state photomultipliers

This area, pioneered by researchers in the U.S., particularly at Bell Laboratories,[13-14] has emerged recently as one of the most exciting of heterojunction research. The aim is to achieve APDs which approach in performance a solid state photomultiplier (Fig. 7-6). Very little in this area has been published so far by the Japanese; this is limited to a proposal by Sakaki at Tokyo University.[15] However, the author is aware of work on superlattice APDs at NTT, NEC and in some universities.

## 7.3 Optoelectronic Integrated Circuits (OEICs)

The goal of this effort is to integrate optoelectronic devices, such as detector and lasers, with electron devices, such as FETs or bipolars on the same chip.

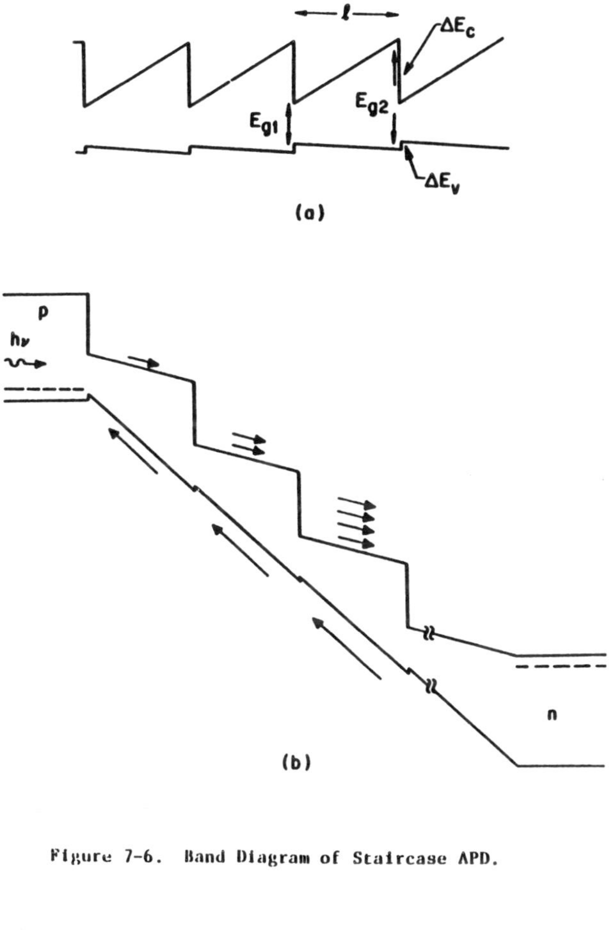


Figure 7-6. Band Diagram of Staircase APD.

The main proponent of this approach in Japan has been Izuo Hayashi at the Optoelectronics Joint Research Lab, coinventor with M. B. Panish of the heterojunction laser.[16] The Japanese envision using such circuits in both optical communication systems and in ultrahigh speed future computers. In the latter, optical interconnections would replace electrical interconnections.

Applications for fiber optics range from simple integrated pin-FET receivers and FET-laser (or bipolar-laser) transmitter to integrated multi-channel repeaters. The latter circuit, illustrated schematically in Fig. 7 integrates on the same chip several receivers (detectors and preamplifiers), amplifiers, laser drivers and lasers (or LEDs).[16]

The OEIC program is a typical example of a cooperative Japanese effort and consists of two-parts. Small scale circuit applications (a few devices at most on the chip) are directly investigated by private companies. They receive for this purpose funding from the MITI. Large scale circuits on the other hand require major material breakthrough in terms of substrates and fabrication technology. This part of the program is carried out by a separate entity, the Optoelectronics Joint Research Laboratory, where roughly fifty researchers from nine industrial companies are organized in a broad materials research program.

For example means of achieving ultra-low dislocation density GaAs substrates ($< 10^2 cm^2$) are investigated. Research on suitable dielectrics and on focussed ion-beam doping is also vigorously pursued.

The Japanese have recognized the need of a strong coordinated materials effort which they view as the key to future device breakthroughs such as OEICs. The same philosophy is also applied to superstructure devices, covered in the contribution of Prof. K. Hess.

The U.S. effort in this area has led to many firsts especially in OEICs with a few devices on the chip. Particularly productive has been the group of Prof. Yariv at Caltech[16,17]. Industrial laboratories have also

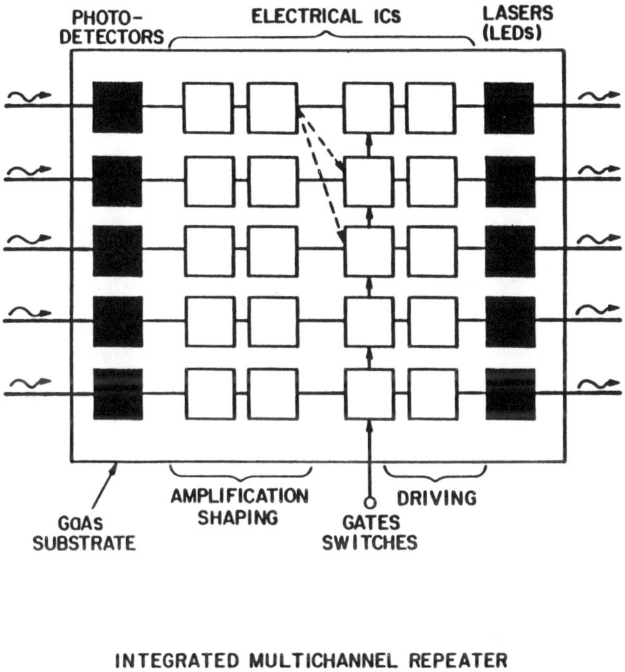


Figure 7-7. Integrated Multichannel Repeater, a Typical Optoelectronic Integrated Circuit; Detectors, Lasers, and Transistors are Integrated on a Single GaAs Substrate.

initiated research activities in this area. Bell Labs work, for example, has been discussed by Forrest in an invited talk at the OFC '85 meeting at San Diego.[17]

There is no concerted materials effort on OIECs in the U.S. comparable to the one in Japan.

## 7.4 Conclusions

One of the most distinctive features of Japanese device research in the optoelectronics area is its broad range, long term character. Technologies that don't promise to produce short or medium term payoffs are patiently pursued. A good example is R&D on Ge APDs at Fujitsu and NITT.

Interestingly the same strategy is pursued in other areas of device technology such as Josephson junctions where a sizeable research activity continues, despite the widely publicized IBM decision to abandon its activity on Josephson computers.

The second distinctive feature is the cooperative nature of research in some of the industrial laboratories such as the Optoelectronics Joint Research Lab.

The last observations concern Japanese creativity and innovation in device research. It is often said that the U.S. invents and Japan copies. Such generalizations are grossly inaccurate and certainly do not favor a genuine understanding of our best competitors.

Dr, Kikuchi, Director of Research at Sony Laboratories, in his book "Japanese Electronics: a Worm's eye view of its evolution"[19] has argued that both the need to catch up technologically with the West after the war, and specific Japanese cultural traits, along with the school system, have favored a kind of "adaptive" creativity; i.e., taking the best ideas generated in the West and adapting them in a creative way to Japanese technological needs.

There are however outstanding exceptions. The tunnel diode was invented by Japanese Nobelist, Leo Esaki, while at Sony. Later on, Esaki became one of the initiators of superlattice research at IBM.

More recently Prof. Sakaki, at the University of Tokyo, has become one of the leaders in innovative heterostructure device research.

In the detector area that we have just examined, Japanese researchers have invented the SAM APD concept and made its first demonstration. Some of the present research on guarding technology shows a high degree of ingenuity.[9]

However, in the photodetector area U.S. researchers have been by far more inventive and more resourceful than their Japanese colleagues. Ninety percent of the novel detector structures have been invented in the U.S. as well as many new creative ways of combining optoelectronic and electron devices on the same chip.

There is a need for a continued and vigorous support of research on dielectrics, passivation, and low etch pit density bulk materials for substrates in the U.S. Such technologies are vital for the implementation of detectors covering the wavelength range from 1 to 10 μm.

This will benefit other areas as well and provide the technological building blocks required for the development of a new generation of heterostructure functional devices and optoelectronic integrated circuits.

## 7.5 References


1. H. Melchior and W. T. Linch, IEEE Trans. Electron. Devices, ED-13, 829 (1966).

2. T. Mikawa, S. Kagawa, T. Kaneda, T. Sakurai, H. Ando, and O. Mikami, IEEE J. Quantum Electronics, QE-17, 210 (1981).

3. J. C. Campbell, A. G. Dentai, W. S. Holden, and B. L. Kasper, Electron. Lett., 19, 820 (1983).

4. K. Nishida, K. Taguchi, and Y. Matsumoto, Appl. Phys. Lett., 35, 251 (1979).

5. O. K. Kim, S. R. Forrest, W. A. Bonner, and R. G. Smith, Appl. Phys. Lett., 39, 402 (1981).

6. S. R. Forrest, O. K. Kim, and R. G. Smith, Appl. Phys. Lett., 41, 95 (1982).

7. F. Capasso, H. M. Cox, A. L. Hutchinson, N. A. Olsson, and S. G. Hummel, Appl. Phys. Lett., 45, 1193 (1984).

8. T. Torikai, Y. Sugimoto, K. Taguchi, K. Makita, H. Ishihaa, K. Minemura, K. Iwakami, and K. Kobayashi. Paper presented at the 1984 OFC Conference.

9. K. Yasuda, Y. Kishi, T. Shirai, T. Mikawa, S. Yamazaki, and T. Kaneda, Electron Lett., 20, 158 (1984).

10. Y. Matsushima, Y. Noda, Y. Kushiro, N. Seki, and S. Akiba, Electron. Lett., 20, 236 (1984).

11. V. Diadiuk, S. H. Groves, C. E. Hurwitz, and G. W. Iseler, IEEE J. Quantum Electronics QE-17, 260 (1981).

12. G. E. Stillman, L. W. Cook, N. Tabatabaie, G. E. Bulman, and V. Robbins, IEEE Trans. Electron. Devices, ED-30, 364 (1983).

13. F. Capasso, W. T. Tsang, and G. F. Williams, IEEE Trans. Electron Devices, 30, 381 (1983).

14. F. Capasso, Laser Focus, July 1984.

15. T. Tanoue and H. Sakaki, Appl. Phys. Lett.

16. Izuo Hayashi, JST News 3, 32 (1984).

17. I. Ury, S. Margalit, M. Yust, and A. Yariv, Appl. Phys. Lett., 34, 430 (1979).

18. S. Forrest. Paper presented at the OFC '85 in San Diego, CA.

19. Kikuchi, "Japanese Electronics: A Worm's Eye View of its Evolution." Simul Press 1982.

# 8. Metal Contacts to III-V Semiconductors

J.M. Woodall, IBM–T.J. Watson Research Laboratory

## 8.1 Science and Technology Development: Historical Perspective

Compared with Japan, the U.S. is leading in GaAs science and device physics and lagging in GaAs technology development. I think the major reason is that the Japanese electronics industry has a stronger belief that GaAs will play an important role in future high speed data processing and communications systems. In addition there are some historical reasons discussed below. The issue of current leadership in product development, manufacturing, and sales is unclear.

The science and technology development of GaAs materials and devices is prototypical of the contrast that has occurred in last two decades between Japan and U.S. industry in the development of electronics. Namely, the U.S. has the "get rich quick and get out" mentality, whereas Japan seems to plan and operate for the long haul. For GaAs technology, the Japanese have taken the best of the U.S. (and European) discoveries, developed economically viable technologies, and successfully made and sold their products.

Some notable examples of U.S./European breakthroughs/early reports include: horizontal Bridgman growth of oxygen and Cr doped semi-insulating GaAs; LPE; VPE; MOCVD; MBE; in the area of materials fabrication, GaAs, GaP, and GaAsP LEDs; GaAs, GaAlAs, GaInAs injection lasers; Gunn devices; FETs; heterojunction bipolar transistors in the area of new devices, and ion implantation, diffusion, lithography and etching, and Au-Ge-Ni ohmic contacts, in the area of device processing. The U.S./European effort in each of these accomplishments (save perhaps the GaAsP LEDs and MBE growth) can be characterized by a period of frenetic research and rapid publication. This was generally followed by a period of almost no activity and then a period of resurgence precipitated by a concern of lagging behind Japanese technology. The consequence of this scenario was that, in addition to the current status of generally lagging the Japanese in nearly all areas of GaAs technology development, the U.S. lost their experts, especially industrial experts, to other areas of R&D. Thus, our current R&D effort is generally dependent on a work force trained within the last decade by a few select universities.

Much of the early and valuable practical knowledge and experience of GaAs technology gained by such leading companies as TI, RCA, GE, IBM, Bell and Howell, Westinghouse has been lost. Fortunately, Stanford, UC Berkeley, Illinois, and Cornell Universities had professors who were versed in early GaAs R&D and trained the people who now manage or influence GaAs R&D. Another factor contributing to a current U.S. lag in GaAs technology development is that ironically most of the companies involved in the early R&D have been profitably marketing products which either do not or only peripherally utilize GaAs devices. Much of the GaAs R&D in the past decade has been spawned by U.S. military needs and hence much of the current U.S. R&D is being done by companies such as Rockwell, Hughes, TRW, and Lockheed which have been manufacturers of military equipment but not the centers of expertise of early GaAs science and technology. The Japanese seem to have committed to long term development plans and have a stable and growing work force devoted to GaAs R&D. Unless the U.S. makes similar committments, it will end up as an importer of GaAs products. This could lead to the position that the U.S. will have to import future generation data processing and communication systems.

## 8.2 Metal Contacts: Science and Technology

It is my opinion that the U.S. and Europe have a current lead in the science of the metal/GaAs interface. The U.S. and Europe are about equal to Japan in the art and technology of metallizing GaAs circuits, except for a possible edge to Japan in high temperature stable Schottky barrier gate metallurgy. The W-Si, and Ti-W-Si gate metallurgy developed in Japan is used with state-of-art performance everywhere. Likewise, the Au-Ge-(Ni) ohmic contact was invented in the U.S., but state-of-the-art performance is successfully practiced everywhere. However, most workers agree that the Au-Ge-Ni technology is dead-ended with respect to LSI applications due to a low eutectic temperature (<400 C) among other things. The U.S. is currently leading in the area of developing possibly better and more thermally stable ohmic contacts.

Furthermore surface science in the U.S. and Europe is starting to impact technological innovation in contacts, especially new concepts in Schottky barrier contacts. The U.S. is in a good position to take advantage of its knowledge in this area. This can be accomplished by encouraging U.S. companies to pursue strong patent protection for contact metallurgy. Also U.S. government contracting agencies should be encouraged to fund R&D in both fundamental surface science and innovation in contacts. Finally, there should be greater incentives to U.S. companies who are not active in GaAs R&D but which have considerable expertise. This is important because, other than GaAs substrate quality and ion implantation technology, the successful commercialization of GaAs high speed device technology will depend on a useful and high yield contact technology. To my knowledge, the Japanese do not yet have a complete contact technology suitable for the high yield manufacture of LSI circuits.

## 8.3 Highlights of the Current State-of-the-Art

### Lessons From Surface Science

The U.S. leads in innovative concepts in contacts to III-V semiconductors. This is largely due to an awareness by the technologists of the comprehensive surface science work on III-V surfaces and interfaces of the past decade which has been reported in Physical Review, Journal of Vacuum Science and Technology and the annual PCSI meetings. The findings can be summarized as follows. The barrier at most but not all metal/III-V semiconductor interfaces is caused by Fermi level pinning. For GaAs this pinning position of about 0.8 eV below the conduction band causes great difficulty in forming both low resistance ohmic contacts and rectifying contacts with precisely controlled barriers. Most device technologists agree that in order to realize the speed promised by GaAs based devices in either a monolithic IC or LSI format, the contact resistance at the source contact must be less than 2 microohm - $cm^2$, and the standard deviation in barrier height of the gate electrode of not more than 10 mV. For an 0.8 eV barrier height, a doping level or a space charge density of about $10^{20}$ $cm^{-3}$ is needed to produce a tunneling ohmic contact that meets the source contact resistance criteria.

This doping level is not easily achieved by most doping methods. A few special reports of high doping and high space charge densities are mentioned below. It is somewhat ironical that the property of the workhorse ohmic contact, i.e. Au-Ge-Ni, which leads to low resistance ohmic behavior is not understood at this time even though it was invented almost twenty years ago! The W-Ti-Si gate contact even with its high temperature stability can produce an unacceptable spread in barrier heights due to both processing and substrate variables.

Even though there is not yet universal agreement on the origin of Fermi level pinning, the research to track its origin has resulted in a recent increase in innovative concepts to form ohmic and Schottky barrier like contacts. Since nearly all of these ideas (discussed below) are the product of U.S. researchers, it appears that Japan has not been aware of the implications of U.S. and European surface science research on contact technology.

Au-Ge-Ni

The Au-Ge-Ni contact is currently the most widely used n-type ohmic contact for GaAs devices and curcuits. Even though its ohmic properties are not understood at this time, it is capable of meeting the source resistance criteria. The contact was invented in the mid 60s at IBM for use in Gunn diode studies. It is characterized as an alloyed contact in that it strongly reacts with GaAs at temperatures above 400°C to form a variety of phases nonuniformly dispersed at the GaAs interface. The resulting interface is nonplanar. This has led to the theory that the observed proportionality of specific contact resistance upon the reciprocal of the doping level is due to the spreading resistance at protrusions in the non planar interface. Both an advantage and disadvantage of this metallurgy is that it is invasive up to several hundred nm from the original interface. This feature is critical to the performance of current HEMT type devices, but is thought to be undesirable for other applications, e.g., lasers and bipolar devices. Also, further device processing at temperatures greater than 400-500°C cause rapid degradation in contact morphology and resistance. However, the biggest drawback is it lateral dimensional instability upon alloying. It is though that as gate lengths and source-to-gate spacing become less than 1 micron this instability will cause both shorting and significant variation in device performance.

There is little evidence that Japan has made significant contributions to the Au-Ge-Ni contact technology. Their main contribution is to successfully employ it in high performance monolithic IC and LSI applications. There have been two minor contributions: a study showing the aging characteristics of Au-Ge-Ni, and a study showing that Ge dopes GaAs n-type by LPE from Au solutions rather than the normally observed p-type behavior of Ge in LPE from Ga melts. My view of Japanese noteworthy accomplishments are listed in Table 8-1.

## 8.4 Recent U.S. Innovations

The following section is a partial list of U.S. innovation in contacts to GaAs, especially ohmic contacts. This is an important part of the assessment of Japanese GaAs technology since it represents an area almost totally dominated by U.S. R&D. It is my view that those who are responsible for setting U.S. R&D policy ought to bear in mind that the work listed below represents a sizable U.S. advantage and that we should take steps to maintain it.

Table 8-1

**NOTEWORTHY JAPANESE ACCOMPLISHMENTS**
**METAL CONTACTS TO GaAs**

- IMPLEMENTATION OF Au-Ge-Ni OHMIC CONTACT TECHNOLOGY FOR LSI (16K SRAM) - (NTT)
- AGING STUDIES OF Au-Ge-Ni CONTACTS (NEC)
- HIGH TEMPERATURE STABLE REFRACTORY (W-Ti-Si) SCHOTTKY CONTACTS - (Fujitsu)

### 8.4.1 Ge-GaAs Heterojunction Contact

An important lesson from interface science and technology is that lattice matched isoelectronic heterojunctions, e.g., GaAlAs/GaAs, do not exhibit Fermi level pinning at the interface when properly made. Indeed, this property has been the reason for the success of many recent high speed and optoelectronic devices. The Cornell University group has successfully applied the lattice matching concept by developing the Ge/GaAs heterojunction interface to make low resistance ohmic contacts. The reason for success is not entirely understood but is due in part to a low conduction band discontinuity at the Ge/GaAs interface (reported to range from 0.050 to 0.3 eV.), high As doping of the Ge and a metal/Ge barrier of only 0.5 eV (see Fig. 8-1). It appears capable of meeting required contact resistance criteria. Bases on Ge/GaAs phase diagrams this contact may not be stable above 725°C.

### 8.4.2 Graded Gap GaInAs/GaAs Contact

It is known that Fermi level pinning occurs in the conduction band at InAs surfaces. The consequence of this is that there is an electron accumulation layer at metal/n-InAs interfaces. This leads to an ideal ohmic contact because there is no barrier to electron flow (see Fig. 8-2a). Since Fermi level pinning produces A 0.8 eV Schottky barrier at metal/n-GaAs interfaces the question is "can InAs be used to eliminate this barrier?" The answer is "not directly." There are at least two reasons for this. First, it is thought that the band gap difference between InAs and GaAs is about 60-65% in the conduction band. This may produce a large band offset (see Fig. 8-2b) and thus would not have any advantage over a metal/GaAs interface. Second, there is a 7% lattice constant difference between InAs and GaAs. A heterojunction between the two would contain a large density of defects which are known to cause Fermi level pinning. However, this problem has been solved by a continuous grading in composition from GaAs to InAs. This produces a structure with no barrier to electron flow (see Fig. 8-2c) and a measured contact resistance of about 1 microohm $cm^2$.

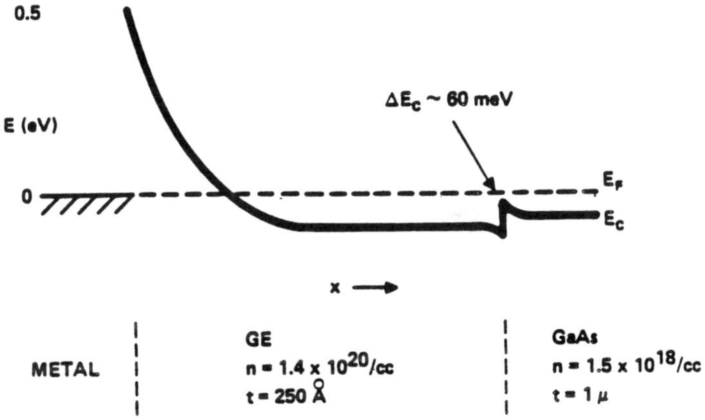


FIGURE 8-1. AFTER STALL, WOOD, BOARD AND EASTMAN ELECTRON LETT. 15 (24) 801-2 (1979).

Figure 8-2a.

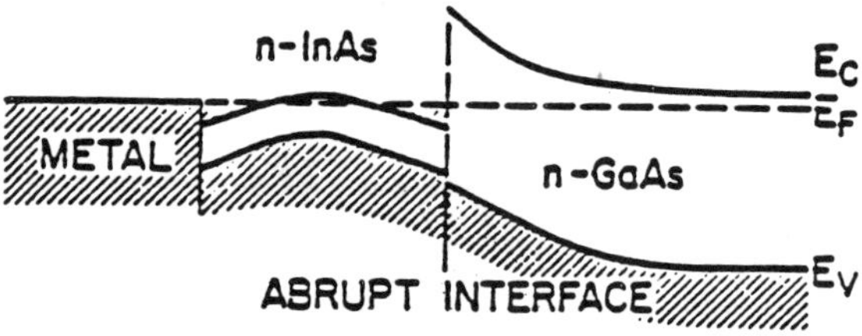


Figure 8-2b.

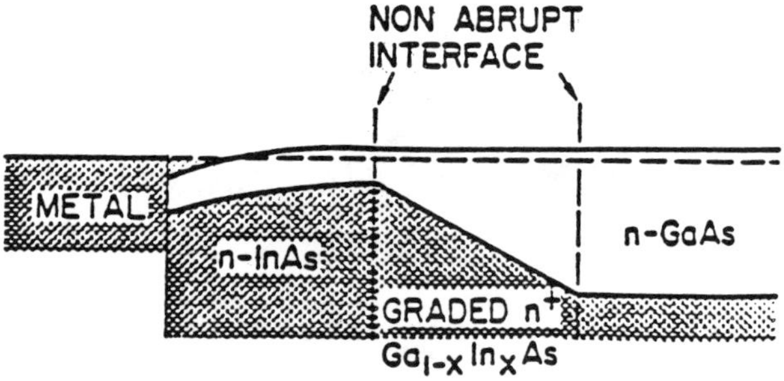


Figure 8-2c.

Graded Gap Contact.

### 8.4.3 Sn Doped GaAs by MBE

A barrier height of 0.8 eV requires a doping level of $10^{20}$ $cm^{-3}$ to produce a tunneling contact resistance of about 1 microohm $cm^2$. These values have been approached in layers grown by MBE and doped with Sn. A doping level of $6x10^{19}$ $cm^{-3}$ produced a non alloyed contact resistance of about 2 microohm $cm^2$. A disadvantage of this method is that for this doping level the Sn clusters on the surface during growth; this produces a rough morphology.

### 8.4.4 Si Doped GaAs by MBE

It has been recently found that non alloyed contacts to MBE grown GaAs doped with $10^{20}$ Si atoms $cm^{-3}$ have a resistance of 1.1 microohm $cm^2$. This is a surprising and important result since the measured electron concentration is only about $5x10^{18}$ $cm^{-3}$. Previously, it was observed that the electron concentration saturated at doping levels of about $10^{19}$ $cm^{-3}$ for nearly all n-type dopants and crystal growth methods. This had lead many workers to doubt that a low non alloyed contact resistance could be achieved by high doping. These new results should have a positive effect on continued research on high doping.

### 8.4.5 New Approaches to Rectifying Contacts

It is important to note the innovative work being done in the U.S. on Schottky barrier modification and multilayer majority carrier rectifiers. There is work on barrier height modification through near surface layer doping being done at Stanford University and the University of Illinois. Cornell has pioneered a diode structure known as planar doped barriers in which the rectification properties are controlled by doping and layer thickness using MBE. AT&T Bell Labs has studied a similar structure using compositionally graded layers of GaAlAs. The Japanese strategy seems to be more of a combination of trial and error, variations of current schemes, and adaptive engineering in this area.

## 8.5 Summary

The U.S. generally lags the Japanese in GaAs technology in both optoelectronic and high speed device applications. The reasons are complicated but include: 1) a lack of a perceived need for GaAs technology by major U.S computer companies. This relegates GaAs R&D to companies which rarely have large scale silicon system implementation experience; usually these are smaller or medium sized companies concerned with the fabrication of optoelectronic devices or devices for military applications, 2) a loss of expertise after a early U.S. R&D period. With respect to contact technology the U.S. and Japan are both using state-of-the-art in the use of Au-Ge-Ni ohmic and W-Ti-Si rectifying contacts. The U.S. has a considerable advantage in innovative concepts at this time. Action should be taken to maintain this advantage and assure early integration of these ideas into leading U.S. companies whose future may depend on GaAs LSI and VLSI.

## 8.6 References

Ge-GaAs Heterojunction

1. Stall, Wood, Board, and Eastman; Electron. Lett. 15(24) 800-1 (1979).

2. Stall, Wood, Board, Dandekar, Eastman and Devlin; J. Appl. Phys. 52(6) 4062-9(1981).

3. Ballingal, Stall, Wood, and Eastman; J. Appl. Phys. 52(6) 4098-103 (1981).

4. Metze, Stall, Wood and Eastman; Appl. Phys. Lett. 37(2) 165-7 (1980).

5. Katnani, Chiaradia, Sang and Bauer; J. Vac. Sci. Technol. B2(3) 471-5(1984).

InAs-GaAs Graded Heterojunction

6. Woodall, Jackson, Pettit, Freeouf, and Kirchner; J. Vac. Sci. Technol. 19(3) 626-7 (1981).

Sn Doping

7. Barnes and Cho, Appl. Phys. Lett. 33(7) 651-3(1978).

8. DiLorenzo, Niehaus, and Cho; J. Appl. Phys. 50(2) 951-4 (1979).

Si Doping

9. Kirchner, Jackson, and Woodall; To Be Published

10. Kirchner; Electronic Materials Conference, Santa Barbara, CA (1984).

11. Miller, Zehr, and Harris; J. Appl. Phys. 53(1) 744-8 (1982).

12. Casey, Panish, and Wolfstirn; J. Phys. Chem. Solids 32, 571-80-(1971).

Ohmic Contact Resistivity

13. Chang, Fang and Sze; Sol. St. Electron. 14, 541-50 (1971).

14. Schroder and Meier; IEEE Trans. Electron. Dev. ED-31(5) 637-47 (1984).

15. Braslan, J. Vac. Sci. Technol. 19, 803 (1981).

# 9. Brief Comments on the Development of Semiconductor Interface Studies in Japan

William E. Spicer, Stanford University

In the preceding section Jerry Woodall has given an overview of the present state of work on metal, semiconductor interfaces and related problems in Japan, the U.S., and elsewhere. This section offers some insights into the development of Japanese interface research. This differs from similar work in the United States and Europe in that it has grown quickly without the long and strong history of fundamental surface science work from which this country and Europe have profited.

In an earlier section of this report, "The Organization of the Japanese Effort on Non-Silicon Based Opto- and Micro-electrons," we briefly described the essence of the Japanese organization in this electronics. It may be of interest to see how semiconductor interface work grew within this framework.

Perhaps the first large effort which helped lay the foundations for Japan's present accelerating position was a "thrust" project funded in the mid 70's at Japanese universities by the Ministry of Education, Science and Culture (MESC). This effort was capped by the International Conference on Solid Thin Films and Surfaces held in Tokyo, July 5-8, 1978. The Proceedings of the conference were published as Vol. 86 of Surface Science, 1979. This conference was followed by a two-day post-conference symposium, "Basic and Applied Studies on Surfaces and Interfaces between Semiconductors and Other Materials Related to Electronic Devices" where some thirty leading foreign and Japanese scientists discussed critical problems.

This phase of university work ending in 1978 could be looked upon as part of a determined effort by the Japanese to bring themselves up to the state-of-the-art in surface and interface experimental techniques and overall knowledge relative to surface analysis and interface studies (particularly high vacuum technology) as well as other surface thin film studies. MBE was particularly and strongly emphasized. At about the same time, approximately $60M was committed by MESC for the building of the Photon Factory, a massive synchrotron radiation facility. On a smaller scale such facilities in the West have proved valuable in surface and interface studies as well as in other phases of science and technology important for 3-5 semiconductor interface studies. Thus, the possibilities of such studies were probably among the motivations for the investment in the Photon Factory, and it is now being used for such studies.

Other areas of interest for this study are also involved in the Photon Factory. One example is the use of synchrotron radiation for X-ray lithography. This is an area which, except for work at IBM, is largely ignored in this country. This is not to say that the Japanese seem convinced that X-ray lithography will be an important VLSI production tool, but they are determined to find out if this will be the case and to be in a leadership position if it is.

The Japanese are also increasing the amount of basic work they are doing. One example is the work using synchrotron radiation from the Photon Factory and photoemission or closely related techniques to study the fundamentals of surfaces and interfaces (for an example, see text) [1-4]. The fundamental work in this field is building up rapidly and growing at a much faster rate than anywhere else in the world.

Despite very rapid progress, the Japanese have a long way to go until they build the strong fundamental program which exists in the United States and in Europe. Thus, their practical work must depend more on an empirical approach than does work in this country despite outstanding contributions from individual workers. As can be seen from the organization chart of the Optoelectronics Joint Laboratory (Section 2.0), approximately one-sixth of the effort of this laboratory is devoted to interface work with well defined goals.

Self-aligned gates for GaAs MESFETS represent an important step in reaching an optimum technology for GaAs IC's. In this technology, the Schottky barrier gate is used to shield the volume under it from the ion implant; whereas, the areas adjacent to it are converted to highly conducting n+ material via the implant and subsequent anneal. By this approach it is possible to form a normally-off FET so that energy consumption can be minimized and packing density maximized. The critical problem was to find a gate metal which could withstand the high temperature anneals necessary to activate the implants, typically 750°C to 950°C depending on time of anneal [5], which can vary from 15 minutes to 6 seconds. A group at Fujitsu Laboratory tried TiW and, finding it unstable at the annealing temperatures, turned to TiW silicide which gave promising results [6]. The choice of TiW silicide was reported [6] to be based on the success of this material for high temperature in silicon technology [5,7]. Subsequently, W(x)Si was found to give quite a satisfactory gate [5,8] with optimum value of x being about 0.64. Ti(0.3)W(0.7)Si was found to make successful self-aligned gates, but W(x)Si was found more satisfactory because of the difficulty of controlling the Ti(0.3)W(0.7)Si composition as compared to that of W(x)Si. From their publications it is clear that the Fujitsu group has worked carefully and hard to find the optimum composition and annealing temperature for the gate. It is also clear that they had the device and manufacturing problems well in mind while doing this work so that products came out of the research laboratory suitable for introduction into manufacturing. It is also clear that this development is based on an almost purely empirical approach.

By this example one should not form the opinion that the Japanese are uniformly successful. While visiting a Japanese laboratory in 1983, I found considerable excitement about using a Si-Ge-boron alloy for a self-aligned gate. Later I was told that this work was abandoned because of lack of reproducibility. As in most research, the Japanese work on self-aligned gates seemed to be characterized by a number of groups working hard and with enthusiasm on a number of different approaches. The size of the efforts is an important consideration.

References:


1. K.L.I. Kobayashi, H. Daiman, and Y. Murata, Phys. Rev. Lett. 50, 1701 (1983).

2. K.L.I. Kobayashi, H. Daiman, and Y. Murata, Phys. Rev. Lett. 52, 1569 (1984).

3. K.L.I. Kobayashi, N. Watanabe, H. Nakashima, M. Kubota, H. Daiman, and Y. Murata, Phys. Rev. Lett. 52, 160 (1984).

4. M. Taniguchi, S. Suga, M. Seki, H. Sakamoto, H. Kanzaki, Y. Akahama, S. Endo, S. Terada, and S. Narita, Solid State Commun. 49, 867 (1984).

5. T. Ohnishi, Y. Yamaguchi, T. Inada, N. Yokoyama, and H. Nishi, IEEE Electron Device Letters, EDL-5, 403 (1984).

6. N. Yokoyama, T. Ohnishi, K. Odani, H. Onodera, and M. Abe, IEEE Transactions of Electron Devices, ED-29, 1541 (1982).

7. In visiting Japanese laboratories, the author has been struck by the close proximity of GaAs and Si work and by the movement of scientists and engineers form work on one technology to another. The development of this GaAs self-aligned gate technology may have benefited from such flexibility.

8. T. Ohnishi, N. Yokoyama, H. Onodera, S. Suzuki, and A. Shibatomi, Appl. Phys. Lett. 43, 600 (1983).

# 10. Equipment and Manufacturing Techniques

Robert Scace, National Bureau of Standards

I have been to Japan once a year starting in 1981 (twice in 1983), as well as on two earlier occasions. Visits to government and industrial laboratories were a part of all of these trips, for a total of 14 weeks in Japan. I have observed with interest the development of a few types of Japanese equipment, Tokuda Seishakusho reactive ion etcher, Takeda Riken test equipment, Anelva RF-excited plasma chemical vapor deposition (CVD) equipment, which may establish a time scale that can be applied to other cases.

The Tokuda etcher was on display at SEMICON/Japan in late 1977. Technically, it is distinguished from other designs in that the plasma (ionized reactive gas) is produced in a chamber separated from the one in which the etching of wafers takes place. This is said to reduce the influence of varying wafer loads on the plasma source, but adds the complication of conducting the plasma from one place to another. That is ordinarily difficult, because ionized gases tend to become de-ionized (and thus rendered less chemically active) by collisions with the walls of the apparatus. Apparently a special coating on the walls is used to control this effect. The etcher was initially developed by Toshiba. Tokuda is either a subsidiary of Toshiba or is at least partly owned by them.

The machine has gone through a series of design improvements. It was first shown at SEMICON/West in about 1981 or 1982, and is in some use in the U.S.

The Takeda Riken test equipment now appears in several models which are used for final electrical test of integrated circuits. The fastest model is a memory tester that operates at a 100-MHz data rate. It was announced at least five years ago and only a few have yet been sold (hearsay). Equipment from other makers is now approaching that speed. One suspects that the product was ahead of its time for wide application.

The original design was probably developed at Nippon Telegraph and Telephone Corporation (NTT) and conveyed to Takeda Riken, a corporate relative of Fujitsu, for commercialization. The original announcement stirred up great interest and did much to enhance the reputation of a hitherto little-known (in the U.S.) firm. Takeda Riken's equipment is shown regularly at SEMICON/West and is said by their representatives to be selling well in the U.S. It appears to be well engineered and built.

The Anelva RF plasma CVD equipment described in Tokyo at the SEMI Technology Symposium in 1983 (1) is apparently a commercial realization of an NTT development described at the same meeting in 1982 (2). Chemical vapor deposition is a process for thin film deposition which usually employs heat to drive the reactions that produce solid deposits from gaseous chemicals. By adding at least part of the required energy from a radio-frequency source, the process can be carried out at much lower temperatures. This has considerable technical advantage.

Only a part of the equipment needed to manufacture compound semiconductor devices is specialized for that purpose. Photolithography equipment, diffusion furnaces (used for annealing, seldom for diffusion), etching and wet chemistry stations, implanters, and thin film deposition equipment are all used with little or no change for processing either silicon or compound semiconductors. Processing atmospheres and reagents differ.

The following discussion of equipment therefore covers crystal-growing equipment, apparatus for epitaxial deposition of compound semiconductor materials, and focused ion beam equipment. Note that the chapter in this report by D.M. Collins also discusses some aspects of these kinds of equipment.

## 10.1 Crystal-Growing Equipment

Compound semiconductor crystals (of the III-V type of interest here) all have one or more relatively volatile constituent, which makes it necessary to grow crystals under pressure or in some other way that controls evaporation of the volatile portions. Bridgman (horizontal boat) methods will not be discussed here. This older technique produces non-circular wafers which present difficulties in large-volume processing, though they can have excellent physical properties.

Much attention has been given to adapting the Czochralski technique, used for the greater part of silicon production, to grow compound semiconductors. High-pressure furnaces made by Cambridge Instruments (England) are the older approach, and are widely used. They are made to grow GaAs ingots up to 4kg in weight.

A more recent approach employs conventional silicon crystal pullers, more or less modified. In either high or low pressure growth, a melt of boric oxide ($B_2O_3$) glass is floated on the GaAs melt. This melt slows the evaporation of arsenic sufficiently to allow crystals to be grown successfully. This is the liquid-encapsulated Czochralski (LEC) technique. Commerical LEC GaAs up to at least 75-mm diameter is available.

The structural perfection of GaAs is of great concern. Crystal defects have adverse effects on device performance. GaAs, in comparison with silicon, is extremely difficult to produce without dislocations and is also hard to process without introducing crystalline damage. Both U.S. and Japanese researchers are working on ways to grow good crystals and to keep them intact through the process of making devices. The level of Japanese effort is higher, and so is the quality of their grown crystals.

Improvements in crystal quality result from finding and correcting the causes of imperfections. One principal cause is mechanical stress caused by abrupt changes in temperature. Two major improvements in crystal growth have been pioneered by the Japanese. One is careful control of the temperature gradient in the growing crystal, by providing auxiliary heaters in the upper part of the apparatus. This approach appears to have been developed principally by NTT and is being applied to commercial materials by Sumitomo Metals. Computer control is reported to contribute to the effectiveness of the approach (3), with dislocation densities around 100 $cm^{-2}$ (compared with $10^4$ $cm^{-2}$ in typical conventional material).

The second approach is the use of magnetic fields to suppress convection currents in the molten GaAs. This technique was originated by SONY for silicon and has been applied at NTT (4) and at the Optoelectronics Joint Research Laboratory (OJRL) (5). The latter work employs a Cambridge MSR-6RA high-pressure puller fitted with a superconducting magnet. Extensive modifications to the puller were required. Computer control is employed. Dramatic improvements in crystal uniformity (reduced dislocation density, absence of striations) and electrical properties (undoped semi-insulating material) resulted.

Much of this work is in the development stage at present. Once the key variables are identified and controlled, it should be possible to engineer a new crystal puller design that can produce useful, well-controlled GaAs in commercial quantities. It is quite probable that Sumitomo is doing that now. Most large silicon manufacturers make their own pullers and regard them as an essential proprietary asset. It should be expected that advanced design GaAs pullers would also be regarded in this way. If this speculation is true, Japanese GaAs makers will be able to develop and hold a leading position.

Commercially available pullers of a newer generation would be expected to appear some years in the future and would probably not represent the state-of-the-art at that time.

Not all of the elements of what is required to achieve this result appear to exist simultaneously at present. The best OJRL results have been obtained on 2-in. diameter material. Up to 5-in. diameter crystals have been grown (6). Sumitomo's work is at 3-in. sizes (3). A large number of properties of the material (stray dopant levels, deep levels, microdefects, uniformity across the diameter and over the length of the ingot, stoichiometry) not mentioned earlier must be controlled. This is a challenging task that will take some years to accomplish. The Japanese are working harder at the job than anyone else, and have the best chance of success.

## 10.2 Epitaxial Deposition Equipment

For most device purposes, the material produced by the crystal growth process described above is a nearly insulating substrate, in or on which the active parts of devices will be formed. Epitaxial deposits, in which the crystal lattice is a continuous extension of that of the substrate, may have compositions that differ from the substrate composition. Aluminum or indium can substitute for part of the gallium in the GaAs, and antimony or phosphorus can replace part of the arsenic.

Such compositional differences are crucial to the functioning of laser diodes, high electron mobility transistors, and superlattice structures. Epitaxial deposits are required in silicon device processes, but only the electrical properties and film thickness are important control variables. The problem of epitaxy in compound semiconductors is much more complex.

Liquid-phase methods have been in use for years. In this process, wafers are put in contact with liquid material of the right composition and (in principle) some of the liquid freezes on the wafer surface to form an epitaxial layer. Nucleation of the film is difficult to control, as is thickness. These difficulties have led to the development of gas-phase processes.

Molecular-beam epitaxy (MBE) is done under clean, ultra-high vacuum conditions. The substrate and the materials to be deposited are each contained in internal, heated enclosures (ovens). The source ovens have shutters to interrupt the effusion of source material. Means are provided for cleaning the substrate surface (e.g., sputtering) to atomic levels of purity. Material of the desired composition is deposited on the substrate by manipulating the source temperatures and modulating the flow of material with the shutters. The process is complex and tedious, but is capable of exquisite control. In superlattice structure formation, the deposit is formed literally one layer of atoms at a time.

The apparatus is intricate and expensive (a half million dollars and up, depending on the auxiliary analytical tools that are included). It was first created as a joint Bell Labs-Riber (France) project, as a research tool. The maintenance required is substantial to keep the interior clean enough to allow the very low operating pressure to be obtained consistently. The prospect of using such equipment in a factory environment is not attractive.

A number of manufacturers build MBE equipment (Riber, Varian, Phi, Anelva). Only Anelva is Japanese. While details vary, all are conceptually equivalent. They all are research tools and none would qualify as production equipment.

A second gas-phase approach is metal-organic chemical vapor deposition (MOCVD). [There are alternative names: organometallic CVD (OMCVD), OM or MO vapor-phase epitaxy (OMVPE, MOVPE).] In MOCVD, gaseous compounds are metered into a chamber containing heated substrates where the gases react to form the epitaxial layer on the substrates and the reaction products are pumped away. The equipment resembles a diffusion furnace and is quite like the low-pressure CVD equipment widely used in silicon device manufacturing.

The equipment is reasonably inexpensive, and it is sufficiently undemanding from an engineering standpoint that to date most examples have been laboratory-built. Commercial versions are starting to appear (Cambridge, CVD Equipment, Biorad) in the U.S. The principal problems are in controlling the gaseous composition of the atmosphere well enough to allow growth of very thin layers with abrupt changes in composition, in the safety of the process, and in obtaining pure metal-organic reagents.

The reactants (trimethyl gallium, trimethyl aluminum, arsine, phosphine) are extremely toxic. One breath of phosphine at 2000 ppm concentration is fatal.

Meticulous equipment design, installation, operation, and maintenance are necessary for its safe use. While these are not unusual requirements in the semiconductor industry, meticulous care does not occur often enough in a laboratory setting for one to feel comfortable about the widespread use of home-built equipment using such hazardous materials.

The reactant purity problem is the chicken-and-egg issue usually encountered whenever new materials are needed for semiconductor purposes. The initial demand is so small that suppliers tend to ignore special needs for purity, or choose (for economic reasons) not to invest in whatever is needed to solve the problem. But without the right materials, the process never works well enough for volume demand to be generated. One suspects that the Japanese way of responding to customer needs may make this problem more soluble in Japan than elsewhere.

The performance of MOCVD in comparison with MBE can be gauged by the following quotations from Japanese authors:

> "The in-depth profile of an AlGaAs (60A)/GaAs(60A) superlattice obtained by Auger Electron Spectroscopy and the photoluminescence of the quantum wells indicated that the heterojunction abruptness and layer thickness, as well as the optical quality, were highly controllable during deposition.
>
> "The high Al-content MQW-DH laser consists of five $Al_{0.14}Ga_{0.86}As$ wells 150A thick separated by four $Al_{0.35}Ga_{0.65}As$ barriers 40A thick....."
> [MOCVD by Sony authors in 1983.] (7).

Later work by Sony authors extends the heterostructure dimensions down to 25A(8). A summary at the same meeting (9) compares MBE and MOCVD as having similar layer thickness capabilities, but with MBE having demonstrated the ability to fabricate superlattices of 100 layers. A Fujitsu author, in comparing the two methods, also cites MBE as superior for multilayer growth but states that MOCVD is better as regards etch-pit density and deep level concentration. He says, "If the controllability of thin films and multilayer growth of the MOCVD method catches up with that of the MBE method, the former will possibly replace the latter in the future" (10). Similar comments come from an Electrotechnical Laboratory (ETL) author (11).

One of the other panelists (Woodall) argues that, with respect to MBE performance, the U.S. and Japan are about equal but that Japan has the larger R&D effort. He also considers that the U.S. has a greater R&D effort in MOCVD. I have no reason to dispute this evaluation, and can add only the speculation that the strong focus by the Japanese on MBE work earlier may have caused them to neglect the potential of MOCVD, to their disadvantage.

## 10.3 Focused Ion Beam Equipment

GaAs surfaces are more sensitive to the environment during processing than silicon surfaces, which form a protective oxide spontaneously. (GaAs does not.) Surface defects thus formed affect device performance. For this rather global reason, Japanese workers are developing process techniques that will allow, in time, much GaAs device processing to be done in high vacuum (12). Some of these techniques could also be used in silicon processing.

One such tool is a focused ion beam apparatus, which is rather like a sophisticated ion implanter that only puts ions where they are wanted rather than uniformly over a whole wafer as today's implanters do. Two such instruments have been described recently, one at OJRL (possibly a Japan Electron Optics Laboratory development) and one by Hitachi. They differ in significant ways.

The OJRL machine has 100kV acceleration voltage, a liquid-metal source with an effective diameter of "several hundred" angstroms, and a beam spot size of about 0.1 µm. It uses an alloy source, and has a mass-analyzer in the column which allows quick changes from one ion species to another (13). Data are given for 160 keV Si implants (doubly ionized?) showing an effective implant resolution of roughly 0.2 µm at $10^{13}/cm^2$ doping. At high doses ($\sim 10^{15} cm^{-2}$) the ions spread to about 1 µm.

OJRL is linking this machine to an MBE equipment, which would give a facility for making multilayer structures without photolithography, thus omitting exposure of the surface to wet chemical operations (14).

The Hitachi machine, developed in collaboration with Toyohashi University, has a 30kV accelerating supply and produces an ion spot of 0.5 μm diameter. The source is an electron-beam heated oven which can heat its charge to 3000°C, allowing nearly any metal to be used. Sources must be changed to change ion species. The spot can be electrostatically deflected over a 2x2 mm area. No results of its use are reported (15).

The machine appears to be an early prototype without a mechanical stage for moving wafers under the scanning area. That step is to take about 3 years (from late 1983).

No similar equipment has been shown at any SEMICON show in the U.S. This is not necessarily indicative of a lack of U.S. development work. Normally U.S. manufacturers do not show equipment in early development, but wait until a more useful version is available.

## 10.4 Summary

All of the equipment described above is in a more or less early stage of development. MBE equipment is an exception, but as noted it is not well-suited for production use. MOCVD equipment is beginning to appear in the commercial market. Both a new generation of GaAs crystal pullers and focused ion beam equipment will be a few years in development before commercial versions can be expected.

Obviously, GaAs devices are being made now. The impact of these developments will be that (a) present-day devices can be made with better yields or improved characteristics, or (b) devices that are laboratory products today will then be manufacturable. The former will result from the use of all three classes of equipment; the latter will come from the use of advanced epitaxial processes.

The real significance of equipment developments lies in the impact of the products thus made possible. GaAs devices have unique capabilities or no one would go to the trouble of making them. Those capabilities lie principally in telecommunication and computers, and in military applications that exploit them. In some cases (optical communications and optical signal processing, for example) there is no other way to do the job.

It is therefore in the national interest of the U.S. to be aware of these developments and their implications for the future in commercial relations between the U.S. and Japan, as well as for future national security issues.

## 10.5 References


1. S. Matsuo, SEMI Technology Symposium, December 7-8, 1982.

2. T. Okada, SEMI Technology Symnposium, December 6-7, 1983.

3. Article in NIKKEI SANGYO SHIMBUN, February 2, 1983, p. 4.

4. Personal observation.

5. T. Fukuda, R+D Progress Report of Optoelectronics Joint Research Association, June 1984, pp. 3-12; K. Terashima and T. Fukuda, Abstract No. 326, 163rd Meeting of The Electrochemical Society; same authors, Oyo Buturi 53, pp. 42-47 (1984).

6. Brief report in Japan Semiconductor Technology News 3, pp. 21-22, June 1984

7. H. Kawai, K. Kajiwara, O. Matsuda, and K. Kaneko, paper EFM83-22, IEE Japan, August 1983.

8. H. Kawai, in Abstracts of the Third Symposium on Future Electron Devices, July 4-5, 1984.

9. M. Kawashima, ibid.

10. A. Shibatani, Oyo Buturi 53, pp. 33-37 (1984).

11. S. Kataoka, opening paper at 1st Symposium for New Generation Industrial Technology and 2nd Symposium for New Functional Device Technology, July 6-7, 1983.

12. I. Hayashi, Survey Reports on Materials for Opto Information Processing, Japan Electronics Industry Development Association, March, 1984.

13. I. Hayashi, Japan Science and Technology News 3, pp. 32-38, February 1984.

14. Y.S. Park, Office of Naval Research Far East Scientific Bulletin 9 (1) 84, pp. 152-157.

15. Technocrat 16 (12), pp. 29-30 (1983); Nikkei Electronics, October 24, 1983,.pp. 92-95.

# 11. Josephson Devices and Technology

Harry Kroger, Microelectronics and Computer Technology Corporation

## 11.1 Summary

The Josephson effect has been exploited in many ways since its discovery. In scientific areas, it has become the international voltage standard and is being used in such pure scientific investigations as monopole detectors by physicists. Such applications pose no competitive commercial threat to any country, and excellent work on standards is being performed at, among other places, NBS and MITI's Electrotechnical Laboratory (ETL). The Josephson effect is also potentially valuable in a host of nondigital applications such as magnetic anomoly detection, analog signal processing, parametric amplifiers, and analog-to-digital conversion. Structurally related devices, fabricated with the same technology, but which do not exploit the Josephson effect, have other interesting properties including being the detectors of millimeter to near infrared radiation which have the lowest known noise figures.

By far the greatest effort to exploit practical advantages of the Josephson effect has been in the digital field. The first and largest program was IBM's. This effort was remarkable in several ways. First, it was the most serious attempt to supplant semiconductor technology by an alternate integrated circuit technology. Second, the perceived advantages required the development of an entirely new hardware system. Indeed, it was system level performance advantages which were the driving force behind this development.

The IBM effort eventually attracted the interest of other U.S. computer companies and especially the Japanese. In an article in Science in 1982, an observer was quoted as saying that with the Josephson computer project, "IBM opened Pandora's box and out jumped the Japanese." In 1983, IBM closed down their Josephson computer development program because they perceived that it no longer offered significant system performance advantages compared to the anticipated advances in silicon technology.

The Japanese Josephson program has concentrated on the development of digital applications and is dominated by a government organization (MITI) and an organization which until recently was a "public" corporation (NTT, which went private in April of 1985). Only recently has there been some slight interest expressed in broadening the scope of the program to include other applications such as millimeter wave detection, microwave amplification and analog-to-digital conversion.

The Japanese digital Josephson program is now by far the largest in the world, and there are no apparent plans to curtail this development in spite of IBM's decision to drop its program. There are several reasons for the resoluteness of the Japanese. First, the Japanese program has been aimed, from its inception, towards the development of a superspeed computer, not a general purpose computer, as was the case for IBM. (The possibility of a different architecture is significant. For example, the IBM program demonstrated a technology whose level of integration was higher than that used in the Cray supercomputer, but not, according to IBM estimates, a level of integration necessary for a general purpose computer.) Second, most Japanese workers feel that there is no compelling reason for them to think that they cannot eventually succeed in their goals. In short, the Josephson researchers in Japan still believe in the potential superiority of superconducting computers. Third, in part because of its longer term goals, the Japanese program has evolved into one which is now exploring more advanced technologies than those which were seriously considered by IBM. (There has been a significant opportunity for them to learn from IBM and other efforts in this evolution.)

It should be commented that the IBM and Japanese attitudes are not necessarily diametrically opposed. For example, the rate of spending on any single one of the Japanese programs is susbstantially less than what was being spent at IBM. Also, one notes that the IBM effort was not cut back to zero, and that a program at a research level still exists.

The Electrotechnical Laboratory (ETL) program is aggressive and imaginative in all aspects, including material and fabrication technolgy and circuit design. The NTT program, at the Atsugi Electrical Communication

Laboratory (ECL), is marked by inventiveness in circuit design and contains the only significant exploration of Josephson packaging outside of IBM. NTT's Materials Laboratory, at the Ibaraki ECL, has been exploring new materials technologies. All the Japanese programs are staffed by imaginative and thorough workers.

The Japanese program began conservatively generally imitating the known IBM technology but, led by ETL, it has developed its own distinctive flair over the past three years. The ETL fabrication technolgy is clearly the most advanced reported anywhere. All of the MITI-supported companies (Fujitsu, Hitachi and NEC) now have in place at least a partial refractory materials technology and/or are planning to implement a completely refractory one immediately. Such technologies, especially the all-refractory ones, are now generally conceded to be the most extendable fabrication techniques.

The Japanese program is capable of long range success and should therefore be monitored carefully. I believe that an impressive and extendable superconducting computer technology will eventually be demonstrated if the funding for the programs continues. No lack of confidence among Japanese researchers regarding long range funding was apparent to American colleagues at the Applied Superconductivity Conference held in September, 1984. There are, however, some Japanese who do think that IBM made the correct decision in cancelling its Josephson program.

Monitoring the Japanese Josephson program should not be difficult if present Japanese attitudes persist because the Japanese researchers and their managements have consistently attempted to foster improved contact with American workers. Examples of this are: invitations to American workers to deliver invited papers at conferences in Japan (often accompanied by honorariums which cover at least the cost of air fare to Japan); an invitation from the Japanese Society for the Promotion of Science to the U.S. National Science Foundation for a U.S. delegation to meet with Japanese workers in Hawaii in 1983 (the NSF declined to financially support this workshop); and a recent invitation from Japanese trade associations to eight Americans to visit Tokyo for a workshop to discuss Josephson technology. This latter workshop took place in April 1984 and the trade associations, supported in part by MITI, provided approximately $12,000 to partially cover the travel expenses of the American delegation.

The major goal of the Japanese Josephson programs is the commercial development of a very high performance computer. Each program is a coordinated plan whose aim is development, not basic research. It therefore follows that any contribution of the Department of Commerce towards monitoring the Japanese programs should also involve the participation of U.S. computer companies. It is not obvious, however, that meaningful participation of many U.S. computer companies will be easy to obtain. Most U.S. computer companies adopted an ostrich strategy towards the IBM development program. Given the IBM decision, it is reasonable to suspect that most U.S. computer companies will continue their so-far successful strategy when it comes to the Japanese program. It may be possible to initially obtain cooperation from MCC and Hypres and eventually IBM in the monitoring process. The Commerce Department should provide leadership in obtaining wider cooperation for the monitoring and technical evaluation.

At the very least, the Commerce Department should position itself to be the recipient of future invitations for workshops, extend reciprocal invitations to the Japanese, and also support travel to such workshops for U.S. workers, including university professors and NBS researchers.

There is another means by which the DoC could financially support the aquisition of information from Japan (and other foreign countries). This is by way of contributions to help pay for foreign scientists to deliver talks at technical conferences in the United States. For example, the Applied Superconductivity Conference, Inc. (ASC) has only limited funds. While the Japanese trade associations have been generous in their support of travel of Americans to Japan (there is MITI support), it is difficult for a private, nonprofit organization such as the ASC to match these funds. The situation is aggravated by the fact that some foreign scientists (such as those who work for MITI) have limited travel funds. Some MITI travel restrictions do not seem rational. A trip to a conference cannot be used to visit a laboratory in the same country as the conference. The Japanese Ministry of Education, and not MITI, has supported travel of ETL scientists to visit laboratories around the world.

There are several scientific and technical publications which are published in Japanese which should be translated to English on a timely and regular basis. It would also be useful for the Commerce Department to encourage U.S. companies to provide instruction in written and spoken Japanese to their personnel. The utility of such programs transcends the Josephson program, of course. I am aware of one in-house Japanese language program which is being offered at the IBM Watson Research Center.

## 11.2 Introduction

### 11.2.1 History of Josephson Digital Research

The potential utility of industrial applications of Josephson devices was first seriously investigated at the IBM Watson Research Laboratory. A small program, which began in 1967, grew into a large one whose aim was to develop a high performance general purpose computer. An article in Science (November 4, 1983) stated that this program was spending at the rate of $20 million per year at the time it was terminated in September 1983. Reputedly this program was a "large" one for at least twelve years.

The activities at IBM began to attract responses from other industrial firms in the mid-to-late 1970's. Bell Telephone Laboratories, Sperry Research Center and later Sperry Univac in the United States initiated programs during this period. At the same time, larger programs were begun in Japan both by MITI and NTT. The first major public presentation of papers in English on Josephson digital technology by Japanese authors occurred at the 1980 Applied Superconductivity Conference held in Santa Fe, New Mexico. An interesting historical note from this conference was the fact that MITI's Electrotechnical Laboratory (ETL) presented their first papers in English which dealt with refractory super conductor Josephson tunnel junctions, an area of study which was not addressed with full vigor by IBM but which may prove to be of crucial importance.

The termination of the IBM Josephson computer development program (a research program still exists) has apparently not affected the Japanese Josephson digital programs. The Japanese programs represent by far the largest effort in the world on digital Josephson applications. Even before the termination of the IBM program, Juri Matisoo, who began the first investigation of Josephson computer applications at IBM, told me in 1981 that there were more workers in this field in Japan than in the United States. All of the ETL effort (with the exception of work on the Josephson voltage standard, which is pursued by a different group) and all of the NTT program (with the exception of a small effort on infrared detection) have as their obvious application Josephson digital systems. Most of the Japanese academic work also has application to digital electronics.

### 11.2.2 Non-Digital Applications of Superconducting Devices

The use of the word "superconducting" rather than "Josephson" is intentional. Passive superconducting delay lines and superconducting tunnel junctions, which may also show the Josephson effect and whose presence may be incidental or even an annoyance, are examples of such devices which are not "Josephson" devices. Their fabrication may, however, involve the same techniques useful for Josephson devices, and "true" Josephson devices can sometimes be part of the same circuit or system.

There are several applications of superconducting electronics which are far less controversial than the computer applications. In many of these areas, workers in the United States hold an essentially unchallenged lead. Since there has been no significant Japanese activity, they will be discussed only briefly. It should be noted that at a recent workshop on Josephson electronics held in Japan to which American researchers were invited, Professor Ko Hara of the University of Tokyo stated that the Americans were wise to have distributed their efforts among several fields of application, and that Japanese workers would be wise to emulate this.

#### 11.2.2.1 Analog Signal Processing

This field, which is being pursued by MIT Lincoln Laboratories or a start-up company which it has spawned, is an outgrowth of SAW (surface acoustic wave) signal processing. Using the higher bandwidth of superconducting delay lines (and their essentially lossless and dispersionless properties) enables these devices to process information at extremely high effective rates, sometimes exceeding that of Cray supercomputers by orders of magnitude. Some of these devices are purely passive; others also incorporate Josephson devices. A recent review has been given by Ralston (1984).

#### 11.2.2.2 Microwave Amplifiers

The ac Josephson effect can be used to produce parameteric amplifiers since this nonlinear phenomenon operates at high frequencies and can be modelled accurately. Recent interesting experimental results have been described by Silver (1984).

#### 11.2.2.3 Instruments

The dc Josephson effect can be used to produce very fast samplers as described by Faris and Hamilton. These purely electronic (non-optical) systems can resolve signals with about a 2 ps resolution which have only several microamps of currrent. These devices are also useful for diagnosing Josephson digital systems and are also being aggressively studied by ETL and NTT. It is reputed that instruments based on such phenomena are being developed by Hypres, Inc., a start-up company led by Sadeg Faris. Faris, who made major contributions to the IBM Josephson computer program between 1975 and 1980, and who worked in the Physical Sciences Laboratory at Yorktown Heights from 1980 to 1983, left IBM in 1983 to found Hypres. There is no known comparable entrepreneurial activity in Japan.

#### 11.2.2.4 Millimeter and Microwave Detectors and Mixers

So-called "S-I-S" (superconducting-insulator-superconductor) detectors are the most sensitive known detectors of high frequency microwave to far-infrared radiation. These devices make use of the extreme nonlinearity of superconducting tunnel junctions' current-voltage characteristics, but the simultaneous presence of the Josephson effects is an annoyance at best. (However, an ideal S-I-S detector has low subgap leakage current, which is highly desireable for the most common Josephson logic gates, so that there is a common technology thread woven between both of these applications.) These devices can approach the quantum limit in detection sensitivity (noise limited by the arrival rate of discrete photons), which is usually achieved only by visible or shorter wavelength detectors. Impressive work in the United States has been demonstrated by Paul Richard's group at Berkeley, Tony Kerr's group at NASA's Goddard Institute of Space Sciences, and by Mike Tinkham's group at Harvard for far infrared radiation. No comparable work is known in Japan.

#### 11.2.2.5 Signal Processing and Analog-To-Digital Conversion

Some of these applications are also dircectly useful in digital computers or similar systems. We differentiate between these applications and the strictly "digital" applications on the basis that these can be profitably used in non-mainframe computers, most often in military processors. Hamilton, Kautz and Lloyd of NBS (1983) have demonstrated a 6-bit A/D converter that operates with a 0.5 ns conversion time, and Hamilton and Lloyd (1982) have demonstrated a binary counter that operates at a rate greater than 100 GHz with power dissipation of only 0.1 uW (1982). Ted Van Duzer, at Berkeley, and Silver, at TRW, have also studied digital signal processing and A/D conversion. While there is not as wide a range of effort in Japan, ETL's recent work with multipliers (Section V.A) falls within this application area.

#### 11.2.2.6 Narrow Applications

There are several specific applications of Josephson "SQUID" (superconducting quantum interference devices) which have been developed primarily in the United States, and have not been vigorously pursued in Japan. These fields include magnetotel lurics studied by John Clarke's group at Berkeley and biomagnetism studied at NYU and more recently at JPL (among other centers).

11.2.2.7 Infrared Detectors

The NTT Ibaraki Electrical Communication Laboratory (ECL) has pursued what is apparently the only experimental program in the world investigating the properties of lead-barium-bismuth oxide (described in Section 11.7) as infrared detectors, with excellent results. The presumed application is in fiber optic communication system repeaters, but a casual appraisal of this work invites its consideration as an interesting material in "non-equilibrium," "three-terminal," digital switches (see Appendix II B). This is the one nondigital area in which Japanese researchers have been far more active than American researchers.

11.2.2.8 The Josephson Voltage Standard

The ac Josephson effect is the current method of defining the volt according to NBS and most standards laboratories around the world. Excellent work is being pursued both at NBS and MITI's ETL, among other places, to improve this standard. This work poses no competitive commercial threat and will therefore not be mentioned again.

11.2.3 Outline of the Remainder of the Paper

The remainder of the paper will contentrate on digital applications of Josephson devices since this is the major thrust of the Japanese programs, and is also, according to some estimates in the United States, the most significant potential commercial market for this technology. Since the main Japanese programs are project oriented, this report will concentrate on program goals rather than science.

Section 11.3 is a guide to the appendices which contain descriptions of the Josephson technology and the reasons why it is attractive for high performance computing systems. More importantly, Section 11.3 also explains the basic technical assumptions on which the rest of the report is based. Section 11.4 describes the problems with this technology which were uncovered by IBM and which formed the basis for the techno-economic decision to drop their development program. This section is included because it forms an important part of the background to any discussion on the potential threat of the Japanese development programs.

An overview of the Japanese Josephson programs is presented in Section 11.5. The MITI programs (ETL, Fujitsu, Hitachi and NEC) are described in Section 11.6. The NTT program is discussed in Section 11.7, and other Japanese activities in Section 11.8. A comparison between the Japanese and U.S. Josephson programs is made in Section 11.9.

Recommendations to the Commerce Department in Section 11.10 conclude the report. This section stresses the importance of personal contacts between researchers in different countries (or in the same country) as the most effective means to exchange information. Illustrations from my personal experiences which support this contention are mentioned throughout the report.

## 11.3 The Josephson Technology and Basic Assumptions of This Report

The basic assumption of this report is that the Japanese technologists working on Josephson computers are very capable. Therefore, one should assume that they can succeed in their goals, and that they constitute a threat to the U.S. computer industry.

This report does not attempt to convince anyone of the correctness of the IBM decision to cut back on their Josephson program, or of the wisdom of the Japanese to continue with theirs. But it is within the scope of this report to support the reasonableness of the threat posed by the Japanese programs. To this end, a basic description of the Josephson technology is presented in Appendix I. Appendix I.A is a brief introduction to the Josephson digital technology. Approaches to superconductive elec tronics, which were apparently neither considered nor seriously investigated by IBM, are partially discussed in Appendix I.B. This section also corrects some major and minor misconceptions that compare semiconductor and superconductor technologies. The anticipated superior systems level performance of Josephson computers was the major reason for interest in them. Appendix II contains a short discussion of presently understood advantages and disadvantages of superconductive computers. Several minor corrections to common wisdom are mentioned.

## 11.4 The IBM Decision

In spite of the fact that the IBM Josephson program had great success in producing logic chips, the development program was terminated in September, 1983. The following reasons were given:

1. A problem with the design of the cache memory was encountered. The IBM goal was to fabricate a 4 K cache memory; this was the mimimum size which was acceptable for future systems which employed Josephson technology and was on the critical path for a significant system demonstration of the technology. (Several IBM researchers told me that a 1 K memory was possible to produce, but that this would not satisfy their future system needs for a general purpose, high performance computer.) The problem was not that they felt that a 4 K memory could never be fabricated, but that, given the known spread in fabrication related parameters, the then known circuit design would not have acceptable margins.

2. A redesign of the circuit was possible so that acceptable margins would be obtained, but, given the complexity of this task, it would require an additional two years. This delay was of major significance in the decision. An additional two years spent in development would mean that the system level performance would be only marginally better than silicon technology, perhaps only two or two and a half times the speed. It was felt that the introduction of a radically new technology would have to offer a greater advantage. A decade earlier, the Josephson computer was supposed to offer an order of magnitude of improved performance.

3. There was also concern that the problems in fabricating a 4 K memory would mean that larger memories, which would eventually be necessary in order to stay competitive with semiconductor technology, would be even more difficult to fabricate. This implied that the technology was not easily

extendable. Theoretical analysis also showed that, while the performance of the memory would scale well to 1 μm design rules from the 2.5 μm rules being used, the performance would not scale well below 1 μm.

Given the IBM goals and the rate of spending on the program, it is understandable why the decision was made. On the other hand, the reasons do not constitute a proof of impossibility of success with Josephson computers in general, or even with the architecture which IBM was considering. While I am aware that IBM did devote effort to the study of other memory schemes such as vortex devices (private communication, H. H. Zappe) it is impossible to consider all options in a finite period of time. There are also those who felt that the level of integration already demonstrated would be easily adequate for special purpose supercomputers. It is also possible that greatly improved fabrication techniques could have a substantial impact on margins.

## 11.5 The Japanese Josephson Programs

There are several distinct Japanese programs in Josephson electronics. Three of these are industrial programs: the MITI program, which is part of the Super-Speed Scientific Program (a national goal); the NTT program, which is independent of MITI; and small independent programs such as Mitsubishi's. Mitsubishi was originally supported by MITI, but failed to win a follow-on contract after the first round of support which ended about three years ago. Finally, there are also a number of university research programs supported by the Ministry of Education.

While there is no obvious coordination between these programs, there is at least an accidental advantage in NTT exclusively tackling the Josephson packaging problem, which was already solved in principle by IBM, leaving ETL to explore more virgin territory. The packaging technology is by no means trivial, and it is expensive and time consuming. But NTT has devoted sufficient effort to it to surpass some of the performance achieved by IBM. Some Americans maintain that there is a history of a flow of vital technology from NTT to NEC, Hitachi, and Fujitsu. If true, this could explain why none of these other programs has mounted any significant known packaging program; the technology could be supplied when and if needed in the future.

Japanese academic workers appear to be pursuing their own interests, with no obvious lead being taken from the other programs. There is some interaction between the MITI-supported companies (Fujitsu, Hitachi, and NEC) but their meetings, in so far as they relate to the MITI program, take place on neutral ground. Each supported company is free to pursue its own distinct technological approaches, being judged on how well their own approach works.

These programs do not exist in total isolation. For example, the Ministry of Education has supported the travel of ETL (MITI) scientists in visiting Josephson laboratories around the world as well as direct support for the university programs. Travel funds from MITI are apparently limited and for restricted purposes. There are also regular professional and/or trade association meetings at which technical information can be exchanged. The trade associations are supported by MITI and by private companies.

## 11.6 The MITI Programs

Dr. Hisao Hayakawa heads the MITI Josephson program at MITI's ETL (Electrotechnical Laboratory) in Ibaraki. In this capacity, Hayakawa wears two hats. He heads the Cryoelectronics Section of the Fundamental Science Division and reports to the Division Supervisor. The Cryoelectronics Section is the premier research group on Josephson technology in Japan. He is also the head of the Special Section on Josephson Computer Technology and reports to the head of ETL. The responsibiltiy of the latter job is to coordinate interdivisional work on the Josephson project within ETL. In this capacity, Hayakawa can call upon resources from the Electronic Device Section for semiconductor processing technology, the Materials Division for materials and materials processing technology, and the Computer Science Division for architecture studies. Hayakawa also is the chief contract administrator for the MITI industrial programs. MITI is now supporting Fujitsu (19 professionals), Hitachi (12 professionals), and NEC (9 professionals).

Hayakawa told the U.S. delegation to the Tokyo workshop in April that the approximate division of resources on the device aspects of the Super-Speed Scientific Computer Program is:

| Technology | Josephson: 1/3 | GaAs: 1/3 | HEMT: 1/3 |
| --- | --- | --- | --- |
| Companies: | Fujitsu<br>NEC<br>Hitachi | NEC<br>Hitachi<br>Toshiba<br>Mitsubishi<br>Oki | Fujitsu<br>Oki |

The original division of resources for devices were:

Josephson: 1/2 GaAs: 1/4 HEMT: 1/4

There are also software/architectural aspects to the Super-Speed Program. These originally were about 20% of the total effort; I received no update on this during my recent visit to Japan. Hayakawa referred to this as "parallel processing." He said that the goal was 10 GFLOPS.

Hayakawa described the original (simplified) time scale of the "Super-Scientific" Computer Project as follows :

```
           Devices              Systems
  1980---------------1985------------1989
```

and then presented his recommended revision of this plan as:

```
       Devices           ?       Systems    ?

1980---------------------1987--------------1989
```

simply because none of the devices technologies would be advanced enough for a commitment to be made by 1985. Clearly he feels that this is especially so for Josephson devices, which he feels have the greatest potential.

I sensed that Hayakawa was concerned about the future of Josephson technology not only in Japan but elsewhere. However, he made reference to his long range plans and the necessity of each group deciding independently whether further work was warrented. He expressed concern about the effect of the dissolution of groups which have contributed to the Josephson effort such as at IBM and Sperry. I did not sense that he was immediately concerned about his group or that he could not lead MITI to eventual success. His research program is planned as if he was confident of long range funding.

At my most recent meeting with Hayakawa at the Applied Superconductivity Conference in San Diego in September, he indicated that there was some movement within MITI to give greater support to long-range projects such as the Josephson technology and less to GaAs, which is approaching the status of a commercially viable product. He said that he supported the decoupling of Josephson research from the Superspeed Computer Project.

Any company which MITI supports is also expected to financially contribute to the program. The Japanese government owns title to any patents developed during MITI-supported research. The government in turn can license the patents to any company it chooses.

Technology transfer is also a problem in Japan. On my first visit to ETL two years ago, Hayakawa told me that, in order to overcome resistance to all-refractory Josephson technology, he was having each of the MITI-supported companies send one or more engineers to ETL for an extended assignment to work on all-refractory technology. This assignment was also necessary in order to develop more collaboration between the companies, Hayakawa thought. They tend to act too much like competitors. Individuals from each of the companies have just completed a year's assignment at ETL. NEC and Hitachi both have at least a partial refractory technology in place (lower electrode is niobium or niobium nitride), and a Fujitsu engineer told me in San Diego that they have begun to work on an all-refractory technology.

### 11.6.1 The ETL Program

From its beginning, the ETL program has always been concerned with the manufacturability of Josephson circuits. Hayakawa steered the program towards all-refractory device technology and away from IBM's lead-alloy technology. (His reasons were to assure reproducibility and not simply to improve recyclability - a radical thought to most workers at that time.) The early use of niobium nitride, rather than niobium, as the refractory material was dictated only in part because its oxide forms a superior tunneling barrier than does niobium. ETL recognized early that the larger superconducting energy gap of niobium nitride would provide a greater capability to drive other gates and therefore improve margins in complex circuits.

ETL was an early practitioner of "full wafer processing" which is noted in Appendix I.B as being capable of producing more reproducible devices. The junction defining step at ETL is reactive ion etching which, of course, would not be possible with lead-alloy materials. The higher processing temperatures, which the use of niobium nitride permits, also allow ETL to use other modern semiconductor fabrication techniques including annealing of resistors and a wider choice in planarization methods.

The circuit designs developed at ETL have been continually motivated to insure reproducibility from both a theoretical point of view (wide margins) and from a practical point of view (manufacturability). One example of this was to eliminate resistors from logic gates so that only the properties of Josephson junctions would determine the gates' thresholds and margins. A practical example of the ETL philosophy is illustrated by Hayakawa's statement that various materials and processes must be replaced in ETL's current fabrication technology. To understand the significance of this statement, I should point out that none of his competitors ever admitted as much and that he is currently sitting on top of the "best" Josephson technology for LSI. This is certainly vastly superior, as far as extendability is concerned, to IBM's last effort in "edge junctions" which Hayakawa stated he would avoid two years ago - another seemingly radical decision at the time.

The current ETL approach is to imitate advanced silicon technology fabrication techniques, but with perhaps even greater emphasis on perfect planarization than is practiced in the semiconductor industry (Kosaka, Shoji, Aoyagi, Shinoki, Tahara, Ohigashi, Nakagawa, Takada, and Hayakawa, 1984). The emphasis on planarization is generally important for improving the manufacturability of memory circuits. These require control lines which overlay the interferometers, but becomes crucial if one wishes to move towards 3-D circuits in which one layer of circuitry is deposited on top of previously formed layers. (See Appendix I.) This is part of Hayakawa's theme to "let superconductors do what superconductors do best." The attraction to a 3-D circiut is that no single crystal layer is needed to fabricate circuits and that heat dissipation is never likely to be a problem.

The motivation behind the continued emphasis on direct coupled logic at ETL, rather than upon interferometers, was to increase the packing density of Josephson gates. Note that this is one of the fundamental requirements discussed in Appendix II for any hardware tecnnology which is applicable to high performance computers.

Hayakawa also anticipated, more than two years ago, some of the difficulty which IBM might experience in designing and manufacturing dense memories, mentioning area and margins to me. One ETL approach has been to take small steps towards designing denser, more manufacturable memories. At the recent Applied Superconductivity Conference, Hayakawa's invited talk included an update on the recent results obtained using the word-organized memory originally described by Kurosawa et al. (Kurosawa, Yagi, Nakagawa, and Hayakawa, 1983), which required only x and y address lines. (It did not require the additional diagonal address in the bit organized IBM design.) This resulted in simpler decoding and address circuits. A critical path through the 1 K lead-alloy memory had an access time of 1.2 ns. The circuit had minimum dimensions of 5 μm with a 2.5 μm minimum dimension. The access time would be subnanosecond. A 2 K destructive read-out memory was also described by Hayakawa. All memory elements were successfully addressed and the access time was 700 picoseconds.

ETL is also making progress with logic and arithemetic functions. Using a unit cell of two OR and one AND gates, they have demonstrated an 8-bit adder with a logic delay of 400 picoseconds and with a power consumption of about 200 microwatts. ETL has also operated a 4 X 4 bit multiplier with a propagation delay of 1 nanosecond. The former result is "almost an order of magnitude faster than GaAs," and the latter result "five times faster than GaAs." Privately, Hayakawa told me that such results and technological opportunities convince him that, "Josephson is an easier technology to exploit than semiconductors." He made the point that these results were obtained with "two orders of magnitude fewer workers than those who work on GaAs," which in turn has far fewer workers than does silicon.

Two years ago, Hayakawa told me that he had a small group, "maybe two or three," working on vortex devices and non-equilbrium devices (see Appendix I.B for descriptions of such devices), but that he was not free to talk about these results yet. He did admit that he had high hopes for this research for the "second generation" Josephson computer. On my most recent visit, we were treated to the first ETL papers on vortex structures. They represent the beginning of the methodical building of a repertoire of useful and carefully categorized vortex devices. Some of this work has been recently published (Sakai, Akoh, Yagi and Hayakawa, 1984, and Sakai, Akoh, and Hayakawa, 1984). These are beautiful papers because they use elegant but simple device structures to measure very basic properties of Josephson transmission lines. They also hint at practical applications with the statement that their model shows that a memory storage line could be placed within a 5 μm diameter circle. No overall design for a memory chip was even hinted at, however.

No paper or talk in English on nonequilibrium devices has appeared. However, I have read an abstract of a paper written in Japanese on nonequilibrium superconductivity which was published in a special issue on Josephson computer technology of the Bulletin of the Electrotechnical Laboratory, vol. 48, no. 4 (1984). (Akoh and Kajimura, 1984). The abstract seems to indicate that this paper deals with the experimental physics of nonequilibium quasiparticle injection, and was probably not a paper on an improved device.

I am unaware of any ETL paper on the architecture of Josephson computers, with perhaps the exception of two papers on two-phase clocks. They find, "the system provides higher speed of operation (142%) and smaller size...than those of the ac [IBM type] power supply system." (Okada, Hamazaki, and Sogawa, 1984, and Sogawa, Nakagawa, Hamazaki, Takada, Okada, and Hayakawa, 1984). (I tend to classify these papers more as "circuit papers" rather than "architecture papers," but this impression may arise because I now have access to only the abstracts.) However, this topic may be what Hayakawa meant when he told me, two years ago, that he had people assigned from the computer science lab at ETL to work on Josephson computers. At that time, he also indicated that two-phase clocks might be better than single phase, but he eventually hoped to get rid of latching logic altogether. The results of these papers are very significant because they directly impact the system level performance of a Josephson computer. (Improved gate speed would not have the same impact.) The architecture topic could eventually prove to be one of the most important in Josephson computer technology.

### 11.6.2 The Fujitsu Program

Fujitsu is the largest group supported by MITI. They have 19 professionals working on the program who act as their own technicians. The Fujitsu group moved last year from Kawasaki to Atsugi to new, less cramped facilities which they have put to good use. The U.S. trade press reported that Fujitsu was not enthusiastic about continuing on with their Josephson program after IBM dropped out. This attitiude was confirmed at a recent visit by Dr. Kaneyuki Kurokowa, Director of the Atsugi Laboratory. Fujitsu perceives itself to be leading the world in HEMT (high electron mobility transistor) and that it is a significantly smaller company than either Hitachi or NEC, its chief competitors. They view that their plate is already full in attempting to keep ahead of the pack in HEMT without also pursuing Josephson which, according to Kurokowa, is even further in the future.

When I visited Fujitsu two years ago, I detected a real concern that Josephson must demonstrate its absolute superiority over HEMT in a relatively short time frame in order for the project to have a long range future. This attitude was unique at Fujitsu as far as I could determine. I also felt that this attitude might be short-sighted in that one could imagine different areas of preference for both HEMT and Josephson.

At the recent Josephson electronics workshop in Japan, Fujitsu presented papers on reliability of lead-alloy circuits and their memory circuits to the entire U.S. delegation when they visited their Atsugi labs. These papers on fabrication dealt exclusively with lead-alloy technology; it was clear from a complete tour of their laboratories that this was the only fabrication technology which was then being pursued. More recently, at the Applied Superconductivity Conference in San Diego, a Fujitsu researcher told me that they had switched to all refractory technology and that they hoped to present papers on this work at the next conference.

Their lead-alloy fabrication technology made use of a number of very clever procedures. Undoubtedly they solved the problem of recycling lead-alloy circuits between cryogenic and room temperatures. The trick they used was to deposit the lower lead-alloy electrode film on a niobium film which stabilized the lead-alloy. It is clear that they chose to abandon this particular technology for reasons which go beyond those positive reasons for preferring an all refractory technology. They were still using the infamous lead-bismuth alloy counter electrode which permits the fabrication of excellent devices, which, unfortunately, change with time when stored at room temperature. The use of the counter-electrode material was abandoned three years ago by IBM, but this fact was never published.

At the workshop in Tokyo, Dr. S. Hasuo, who is directly in charge of Fujitsu's Josephson project, presented a detailed analysis of an older 64-bit Fujitsu memory and how this analysis has affected their design of a 16-K bit DRO memory. This work was excellent both for its thoroughness and how it encompassed both theoretical and practical considerations. Hasuo's talk was the most interesting one I have heard on the difficulties of designing memories.

### 11.6.3. The NEC Program

Mr. H. Abe heads an enthusiastic and dynamic group of researchers at NEC. Abe, who spent time with Ted Van Duzer's group at Berkeley several years ago, recently told Van Duzer that NEC would continue its Josephson program in spite of the IBM decision even if MITI did not continue to support the project. On a visit to the U.S. about a year ago, he also told Larry Smith of Lincoln Laboratory that NEC would continue its work stating that good plans should be followed through as long as one cannot see why they are wrong.

At the workshop in Tokyo, Abe described the basic structures, logic circuits, memory and peripheral circuits, and the high speed performance of the NEC program. The standard NEC process is still lead-alloy, but I detected real enthusiasm for moving to all-refractory technology. Until recently, all of the NEC fabrication technology was basically a copy of the IBM lead-alloy technology. More recently, they have begun to work on an all niobium full-wafer process which defines the active area of devices by reactive ion etching.

Part of Abe's talk described simulations on switching speed, power and area consumption of logic circuits for circuits fabicated with 5 micron and 1 micron design rules. Abe assumed a standard cell design which consisted of two OR and 1 AND cicuits, and that these cells must drive other cells which are on average 20 cells away. He concluded that for 1K gate chips:

| Design rule: | 5 micron | 1 micron |
|---|---|---|
| Gate delay (ps) | 33 | 14 |
| Gate area (square microns) | 110 | 22 |
| Chip size ( mm/side) | 3.4 | 0.94 |
| Total chip power (mW) | 7.5 | 1.5 |

When Ted Fulton of Bell Telephone Laboratories visited NEC two years ago, he gave them very high marks for quality fabrication technology and for enthusiasm. NEC has continued an aggressive program for a relatively small group. This group is probably not large enough by itself to cope with the full range of problems which discouraged IBM, but if they rapidly institute an all-refractory technology, they may find that they have spent developmental resources as wisely as anyone in the field.

### 11.6.4 The Hitachi Program

Dr. Ushio Kawabe, who heads Hitachi's Josephson project, described the evolution of their lead-alloy technology at the workshop in Tokyo in April. He gave due credit to IBM and showed how Hitachi had extended it. This work will have no long range impact on the future of Josephson technology, other than as a means for Hitachi to implement novel circuit designs. Sadeg Faris, who visited Hitachi Central Research Laboratory was impressed with their circuit designs. During a break in the workshop program, I learned in a private conversation that Hitachi has already begun work on Nb/NbN/oxide/lead-alloy devices. This is the first step towards the ETL Nb/NbN/oxide/NbN/Nb technology, but is not a sufficiently large step to reap the full benefits of all-refractory technology.

Kuroda told Joe Logue, who headed the IBM Josephson program at the time of its closing and who visited Hitachi the week before the workshop, that Hitachi had designed a memory chip with a 600 ps access time. Kuroda did admit, however, that he had no detailed analysis to show that Hitachi could actually fabricate such a chip consistently with its known circuit margins. I would not have expected such a complete analysis yet from such a small group, even though I have a very good opinion of their work.

Hitachi is spending some effort towards the development of "three terminal", non-Josephson active devices. This is an agressive attitude for a relatively small program. They also have developed elegant test apparatus to measure properties of chips using fiber optics, closed circuit TV, and a high resolution sampler. These observations support their own contention that they are committed to superconducting electronics for the long run.

## 11.7 The NTT Program

NTT has the largest Josephson program in Japan (and in the world). There are 25 professionals working on fabrication and circuit design at their new and very impressive Electrical Communications Laboratory (ECL) at Atsugi. This group moved, a year ago, from the central ECL facility at Musashino. (The Atsugi Laboratory also includes most of NTT's most aggressive semiconductor fabrication and crystal growth facilities.) There is a group of 7 professionals still at the Musashino Laboratory who work on vortex devices, and two groups at the Ibaraki Laboratory who work on related superconducting materials development and materials processing. There are a total of 13 professionals in the Ibaraki groups.

Two years ago, NTT told me they had 40 professionals working on Josephson technology; the above figures, therefore, represent a minor expansion. However, I am aware of rumors that the number of workers who were actually engaged in Josephson computer development at NTT was far in excess of what they would publically admit. "In excess of 100," was quoted to me, but I have no independent means of corroborating this assertion.

Dr. Akira Ishida, who headed the Josephson work first at Musashino and later at Atsugi, has been transferred back to Musashino to work on planning NTT research. Dr. Takatsuga Hattanda now has the title of Chief of the Superconducting Device Section of the Functional Device Development Division at Atsugi. The Atsugi lab has an impressive fabrication facility. They have refined lead-alloy technolgy to a high art form. Their fabrication studies have apparently been concentated on 1K memory chips. As reported in the literature, they are using a larger cell than IBM which can increase margins. A critical path through the memory has been operated, but it is not completely clear if fully functional chips had been tested when we visited Atsugi in April. (The "full" operation is a matter of some controversy; an abstract handed out in Astugi did seem to say that the memory was fully operational. In response to private questions following Hayakawa's plenary session talk at the ASC in September, I thought that Hayakawa said that NTT had, in fact, operated all elements in a memory array. On the other hand, some IBM researchers, who were asking the questions, thought that he said that not all elements had been successfully operated; the repeated questions made too many uses of English negatives. At the very least NTT has probably demonstrated the most successful memory circuit of any Josephson research group, and this was achieved with the difficult lead-alloy technology.)

Professor Malcolm Beasley of Stanford, who worked at the Ibaraki ECL last summer, told me that the Atsugi lead-alloy pilot line had been closed down and replaced with a refractory technology.

There were two very different reactions to this work from the U.S. delegation. On one hand, some were unimpressed because it was just a copy of old IBM technology. On the other hand, Sadeg Faris and I were very impressed. Faris has maintained for some time that this work is beyond what IBM ever demonstrated with memory. Nevertheless, this work has lead NTT to ask most of the questions which IBM asked itself when they made the decision to stop their development.

The NTT design was not an exact duplicate of the IBM design. Being more conservative, it is likely to provide a convenient stepping stone on the path to a Josephson memory. Finally, one observes that NTT did concentrate its resources on memory, not logic, and therefore attacked the critical problem which IBM found to exist in circuit fabrication.

The Atsugi group also contains some of the best Josephson logic circuit designers to be found anywhere. Careful analyses were presented of very original gates including improvements to their wide-margin OR and AND gates, when the U.S. delegation visited the Atsugi ECL.

The NTT sampling and packaging technology was described by Hayuo Yoshikiyo when the U.S. delegation visited the Atsugi laboratory in April. They handed around for inspection their package parts, which at first look like a direct copy of IBM's silicon chip-board-card technology. Faris and I were very impressed. Faris's reasons are compelling: even the micropins-into-mercury in etched holes in silicon had very impressive performance; the rise time was less then 50 ps; and no lumped capacitor was used to prevent ringing. This technology was described by Aoki, Tazoh, and Yoshikiyo (1984) at the recent ASC. This performance is better than what was ever publically demonstrated by IBM and better than anything of which Faris was aware when he was an IBM employee. The measured cross-talk (which they had studied extensively) was also better than that reported by IBM. This was very careful and thorough work. However, the NTT pins were spaced farther apart than IBM's, so a direct comparison is difficult to make. This is just another illustration of NTT successfully tackling an intermediate, but more accessible, goal to use as a stepping stone in their program. Faris was also impressed with their sampler which they used to measure the properties of the package.

A seven-man group which is working on refractory materials and device processing at Ibaraki is headed by M. Igarashi. They have performed good work over the past three to four years on the deposition and preparation of lower electrode NbN and A-15 superconductors for oxidation in order to form tunnel barriers. At Ibaraki, they presented papers on new methods of chemically and physically cleaning NbN electrodes with fluorocarbon plasmas. They are well equipped with surface analysis equipment, and have used it in ways which have paid dividends to them in improving device properties. They did not describe any of their A-15 work other than to mention that Nb-Ge, Nb-Sn, and Nb-Al were all being studied. They made no reference to refractory counterelectrodes, but it difficult to imagine that they haven't tried them.

Mac Beasley of Stanford was invited to spend the summer at Ibaraki to work on the physics of tunnel barriers. Beasley has told me that he was surprised to realize that NTT had guessed that his professional interests had recently switched from the superconducting electrodes to the barrier itself. (Beasley also received an invitation from Hayakawa to spend the spring semester at ETL, but he had to decline because he couldn't get free of responsibilities at Stanford.) I have just received a preprint of an Ibaraki ECL paper on niobium-niobium junctions in which metal florides are used as low capacitance barriers. Beasley was a co-author.

The Ibaraki ECL is also doing very interesting work with an unusual superconductor, Pb-Ba-Bi-oxide. This material has a transistion temperature above 10K, and was first prepared at DuPont. It is transparent (with a greenish color) when prepared as a single crystal grown from a flux. NTT also prepares it by sputtering from a composite target followed by annealing in oxygen to adjust the stoichiometry. The sputtered material is polycrystalline and each grain boundary acts like a Josephson junction. A thin film of the material is therefore a series-parallel arrangement of a random array of tunnel junctions.

They are using this material in the form of a constriction as an infrared detector. Radiation from an optical fiber placed above the constriction is absorbed by the material. They told us that the PBBO was the most sensitive known infrared detector in the 1-4 micron region. Its sensitivity is derived only in part from the low temperature of operation. The material reflects only about 5% of the incident radiation. The detection is accomplished by monitoring the current or voltage of the device when it is biased at the sum of the superconducting energy gaps of the grains which comprise the constriction. The superconducting energy gap is suppressed because of the excess quasiparticles which are generated by the weakly absorbed radiation. A large gap suppression occurs because of the low density of states in this material. (I have not yet had as opportunity to study their papers but the explanation is believable; the material is transparent which indicates that the electronic density in the normal state must be low.) They have measured the response to modulated radiation up to 400MHz, with no fall-off in sensitivity.

This material also invites obvious consideration as the "base" electrode in 3-terminal, non-equilibrium QUITERON type structures (see Appendix 1.B for a description of such devices) because of the apparent ease in suppressing the energy gap.

No concrete information was available on the activities of the group working on vortex devices at Musashino.

## 11.8 Other Japanese Programs

### 11.8.1 The Mitsubishi Program

Mitsubishi has a small program and recently published a paper on digital applications. At the workshop in Tokyo, K. Hamanaka described work on a microstrip structure for SIS mixers at 30 GHz, a planar DC SQUID for IF applications, and experiments on nonequilibrium superconductivity. The size of this group is estimated at six professionals. This small group is attempting to survey the state of the art and the expected state of the art in superconducting microelectronics. From my limited knowledge, I would rate it as probably an excellent group on a man-for-man basis, but it apparently has neither the equipment nor the manpower to engage in anything but lead-alloy technology, and certainly not in any serious development of Josephson LSI. The probable management purposes of this group is to scout the surrounding territory in order to shout an early warning signal if a serious threat in Josephson technology is mounted by any competitor.

### 11.8.2 University Research

Much of the university Josephson research in Japan is theoretical, or at least conducted by groups which have greater theoretical competence than experimental facilities. I heard from a Japanese visitor, about two years ago, that the Ministry of Education was sponsoring many Josephson groups in Japanese universities. On the basis of published work, it is clear that most of these programs are not better equipped than most U.S. universities. In fact, I suspect that several U.S. university programs are much better equipped than their Japanese counterparts. To the extent that this is true, it is either specifically due to DoD support for Josephson electronics or "center-wide" support for microelectronics which transcends interest in Josephson devices. Advances in materials and processing of Josephson circuits are more likely to flow from U.S. universities than from Japanese universities in the foreseeable future.

There is a great deal of theoretical work on phase-mode circuits in Japanese university programs. It is improbable that most of this will have any near term pay-off in practical terms. On the other hand, several of my collegues were very impressed with the work of Nakajima, of Tohoku University, who described a theoretical design of a Josephson phase-mode microprocessor (Nakajima, Oya, and Sawada, 1983). Nakajima recently spent a year with VanDuzer's group at Berkeley. High quality work on the physics of Josephson devices has been pursued by Professor Hara's group at the University of Tokyo (Nakanishi and Hara, 1984, for example).

## 11.9 Comparison of the Japanese and American Josephson Programs

### 11.9.1 Strengths of the Japanese Programs

#### 11.9.1.1 Participation by Several Industrial Laboratories

Joe Logue, who was the most recent manager of the IBM program, said publically three years ago that the introduction of the Josephson technology would require an industry-wide effort, and that no single company, even IBM, could bring it about by itself. The total effort in Japan is closer to being industry-wide than the U.S. effort. Besides IBM, only Sperry and Bell Labs, later Sperry Univac, had any digital Josephson program in the U.S., but these latter programs were relatively small and are currently nonexistent (as digital, development programs), whereas ETL, Fujitsu, Hitachi, Mitsubishi, NEC, and NTT are all involved in Japan.

The wider participation permits more approaches to be considered by different groups. As the Japanese and non-IBM programs in the U.S. began, there simultaneously arose the first serious efforts on full-wafer processing and deposited barriers (which may still prove to be part of the new "standard" technology in Japan) on both sides of the Pacific. The Japanese programs eventually reach some consensus after an evolution of the best ideas. I suggest that since this consensus is reached by industrial laboratories, the likelihood that the consensus is a healthy one is increased. (Questions such as manufacturability are more likely to be considered realistically.)

Even within the MITI program, divergent approaches are encouraged. Fujitsu, Hitachi, and NEC each have designs of logic gates and memory cells which are unique; the same is true for their choices of materials and fabrication techniques. Hayakawa told me that he only encourages generic approaches among the MITI-sponsored companies, such as the use of refractory materials, but that he would not attempt to dictate the choice of a particular logic gate, for example.

There appears to be a more robust exchange of information among the participants in Japan than there was in the United States. Obviously, ETL must know a great deal about what Fujitsu, Hitachi, and NEC are doing. But beyond this, I believe that there are more technical meetings on the subject in Japan than in the U.S. I do not believe that there is any exchange of proprietary information until patent applications are filed, but rather that there are just more meetings. Some of these are professional society meetings; others are sponsored by trade associations. Until 1979, there was no technical meeting devoted exclusively to digital Josephson electronics in the United States. The "Josephson Digital Workshop" has been held only every other year since 1979.

### 11.9.1.2 Total Size of the Japanese Programs

In 1982, I was told by Professor Hara, of the University of Tokyo, that there were more than 200 workers in Josephson digital electronics in Japan. I assume that this estimate included university researchers and technicians. Nevertheless, this was a larger number of workers than in the U.S. (according to Juri Matisoo of IBM, who was in a position to know the size of the IBM program).

The greater number of researchers is an obvious advantage over the U.S., but the total number of workers in Japan may not yet be great enough to bring a Josephson computer to the marketplace.

### 11.9.1.3 Goals of the Japanese Programs

The production of a superspeed scientific computer is a national goal in Japan, and the Josephson program at MITI has so far been tied to that goal. Since neither MITI nor NTT is immediately pursuing a general purpose computer, they are able to evaluate advanced and risky technologies in a different light than did IBM. This is probably an atmosphere which is more conducive to introducing a nonsemiconductor technology, such as Josephson or superconducting integrated circuits.

There are still pressures to achieve results and these pressures are probably a healthy stimulus for progress. (See comments above in descriptions of the MITI and Fujitsu programs.) However, I suspect that any technology which will be used for the super-speed computer will eventually be reevaluated for use in general purpose computers. When I asked Kurokawa, who is a critic of the Josephson program, why Fujitsu was interested in any supercomputer work, he said, "Today's supercomputer can become tommorrow's general purpose computer."

### 11.9.1.4 Talent of the Japanese Workers

The quality of Japanese research is first rate by any measure. All the work is thorough and clever; the productivity of average Japanese researchers engaged in Josephson work is extremely high. The ETL program, especially, is highly original and practically oriented (see earlier section). Several other American visitors have noted the enthusiasm of the Japanese workers. The Japanese Josephson researchers are justly proud of their accomplishments. Hayakawa commented privately that the results which he described during his invited talk at the Applied Superconductivity Conference (which were significantly better than what had been attained by GaAs technology) were achieved with "fewer than 1% as many workers." The "can-do" attitude of Japanese Josephson workers is an important factor in my high overall opinion of what they may yet accomplish.

11.9.1.5 Technical Position of the Japanese Programs

Many of the technical accomplishments of the Japanese Josephson programs are not only impressive, but they seem much more extendable than the technology developed by IBM. Included in this category are the all-refractory fabrication technology and the full-wafer processing of ETL, and the simplified memory addressing also introduced by ETL. One should also include the demonstration of a two-phase clock, the greater imitation of silicon fabrication methods to improve planarization, and the greater early emphasis in developing micron-sized features in the fabrication process. Several of the Japanese programs claim to have developed circuit designs which seem to have better margins and occupy a smaller area than those developed at IBM; some of these results I have only heard about, and I have not had an opportunity to study them all in English. But if generally true, then this also assists the Japanese programs to be in a better position to make more rapid advances towards a Josephson computer than was IBM. Much of the early Japanese Josephson work was imitative of the then-known IBM work, and that was a necessary phase since IBM had the only program for many years. ETL was the first Japanese program to strike out on its own using original ideas. To the extent that some of the other programs might now follow the ETL lead, these programs will progress faster for two reasons. First, the ETL technology is more extendable than IBM's. (Beyond the merits of the specific approaches chosen by ETL, everyone had an opportunity to learn from IBM.) Second, the other programs will have more complete access to the ETL technology than they did to IBM's. An obvious example is the lead-bismuth alloy counterelectrode first publically described by IBM in 1978. While this counterelectrode material makes excellent devices, the devices change their characteristics during storage at room temperature. I believe that IBM dropped the use of this alloy more than three years ago, but it was still being studied by Fujitsu this year. I am unaware of any IBM publication which described the problems in using this material.

Some of the so-called "imitative" work has been much more than that. The packaging work at NTT has gone beyond IBM's, according to Faris. Even if this work does not represent a fundamental breakthrough but simply good engineering, the results could have a major impact on the confidence with which one can predict the future system performance of Josephson computers.

11.9.1.6 Pursuit of Information from the USA

The Japanese have gone to more effort and expense in learning about Josephson research in the United States than have most American companies in learning about research in Japan. There are several methods which they have used:

a. Study at American universities. Three of the staff of NEC's program have spent time performing research at U.S. universities or have enrolled as graduate students. Japanese workers have been fully supported by their companies at Ted Van Duzer's group at Berkeley. (I am not aware of specific arrangements at other institutions, but I would be suprised if these were different.)

b. Invitations to American professors to do research in Japan. Mac Beasley, of Stanford, spent this past summer at NTT's Ibaraki ECL; he also received an invitation to spend the spring semester at ETL but his obligations at Stanford did not permit him to accept this offer.

c. Invitations to American researchers to visit Japan. Ted Fulton of Bell Labs and I were invited to speak at the Cryogenics Engineering Conference in Kobe in 1982. Although we didn't expect it, honorariums were presented to us from the cryogenics trade association which almost paid for our airfare. We were also invited to speak in Tokyo at a meeting of the Japan Electronic Industry Development Association's group on Josephson technology where other honorariums were received. I would be suprised to find that these cases were unique.

On both trips to Japan, I was invited to visit all of their laboratories, although there wasn't time for me to accept all of the invitations. Talks were presented by some workers at each place, and I always was given a very open tour of their laboratory facilities. I felt that my Japanese hosts were more open than, say, Sperry and IBM or BTL generally were to each other. These were "win-win" situations. I felt that I gained far more valuable information than I revealed, but perhaps my hosts had the same feeling. Neither side described any proprietary information. Both sides, if asked about incomplete work or ideas for which patents had not yet been filed, politely stated that they could not yet talk about such topics.

d. Invitations to American groups for workshops. I am aware of two workshops on Josephson technology which were suggested by the Japanese. The Japan Society for the Promotion of Science (which I understand to be the rough equivalent of our National Science Foundation) suggested, in 1983, that a workshop on Josephson technology be held in Hawaii during the summer of 1984. The suggestion did not receive support of the NSF, so it was not held.

Early this year, the Japan Electronics Industry Development Association invited eight Americans to a workshop on Josephson electronics in Tokyo. About $12,000 was provided to help pay for the airfare for the members of the American delegation. I believe that both MITI and private industry are supporters of this trade association. The stated purpose of the meeting was to foster improved communication and even collaboration with U.S. workers. Our hosts were also anxious to get a reading on the thoughts of other U.S. researchers on IBM's decision to close their Josephson program. The stated purpose of the meeting is quite believable, however. Greater exchange of information is consistent with all previous desires of the Japanese. Hayakawa told several of us that "high officials in MITI" were interested in colloboration with U.S. workers. The Japanese apparently view research in a much less competitive way than they do manufacturing technology. I suggested to Hayakawa that some U.S. businesses might be afraid to collaborate on research because they viewed the Japanese capability to manufacture with so much respect. Invitations to all the Americans to visit the industrial laboratories were, of course, part of this trip.

e. Visits by Japanese scientists to the U.S. Japanese scientists seem to visit this country more often than Americans visit Japan. Besides attending conferences, they will often request to visit U.S. laboratories. They view such visits as a simple reciprocal exchange for visits which we have had to their laboratories. I wholeheartedly agree with their sense of etiquette on this point. I again feel that these are "win-win" situations in which no one gains any significant advantage over the other in the exchange of information.

f. Japanese scientists and engineers learn English. Their knowledge of English is what makes the above activities possible.

As an aside on this issue, it should be noted that it is a laborious task for even Japanese skilled in English to write papers in English. This is one reason that some Japanese prefer to publish in Japanese. It is interesting to note, however, that the papers submitted by the Japanese workers to the recent Applied Superconductivity Conference used English as did most papers submitted by Americans.

### 11.9.2 Strengths of the American Program

#### 11.9.2.1 Wide Range of Application Interests

The Japanese recognize that there is a wider range of interest in various applications of superconducting electronics in the United States. Generally, this makes for a healthier advance of technology. If one application falls into temporary disfavor (as it has with the IBM decision) then all progress in the technology is not brought to a halt. There can also be cross fertilization between different applications; a specific example is mentioned below under military applications (see Topic 3 in this Section). Finally, success in any one application area can make acceptance of what seems to be a very exotic technology easier in other areas simply because psychological barriers have been overcome.

#### 11.9.2.2 American University Programs

The university research programs in the United States appear to me to be of more importance than the research programs in Japan, but this is an opinion which is difficult to substantiate since I have not been an industrial worker in Japan. My opinion is based on the following considerations: first, there has arisen an unusually friendly and cooperative spirit in the U.S. between academic and industrial scientists. Second, there is the content of the specific programs themselves.

For example, Ted Van Duzer's program at Berkeley is the only academic research program in the U.S. which is truly in the mainstream of Josephson digital electronics or in a closely related field. There is, perhaps, only one other mainstream digital Josephson program in the world -- W. Jutzi's at the University of Karlsruhe in Germany. (This is not meant to put down the interesting work which is being done elsewhere in the U.S., Japan, Europe, or Russia; but these other programs are either not specifically concerned with digital applications, or are only theoretical, or are aimed at a generation ahead of what industrial engineers are designing, or they do not have the appropriate equipment to fabricate modern devices and circuits.) Van Duzer's program has developed a wide variety of theoretical and practical circuits, including A/D as well as strictly digital applications. His group has often "filled in the gaps" in explaining the potential significance of alternate approaches rejected by IBM in the early days of the program. Lead-alloy and, more recently, niobium technology have been developed at Berkeley. Van Duzer also coauthored the only text which deals extensively with Josephson digital electronics. Two other programs which, among others, have greatly benefited the U.S. industrial research in superconducting electronics are Beasley's program at Stanford and Bob Buhrman's at Cornell. The Stanford group has pioneered the use of refractory materials in Josephson tunnel junctions, and the Cornell group has explored the use of novel submicron fabrication techniques to produce Josephson devices and the physics of "nonequilibrium" devices.

As valuable as these and other academic research programs are (some that are not mentioned have greatly advanced knowledge in nondigital Josephson fields or have produced graduate students who have become very prolific industrial contributors), they should not be viewed as a substitute for industrial research and development. There are important questions which are appropriate only for industrial labs. These include manufacturing methods which might require almost yearly updates in equipment, extensive CAD tools, and exhaustive analyses of projected margins and yields of various circuit designs. Current advances in silicon technology are being purchased only with the concerted effort of large teams of workers. There is no reason to suppose that Josephson LSI and VLSI can find a much easier road to travel. The full demonstration of Josephson digital technology can therefore be expected to be carried out only in industry.

#### 11.9.2.3 DoD Programs

I am not aware of any activity in superconducting electronics on the part of the Japanese military. On the other hand, there has been considerable interest, sponsorship, and scientific contributions by the U.S. DoD in the field of superconducting electronics. Most of the application areas mentioned above, including that of digital applications, have been greatly advanced by this interest.

The military applications of Josephson devices can act as a cross fertilization mechanism to advance purely commercial ones. One important example is worth special mention. For more than a decade, the Navy has been interested in investigating refractory materials in order to improve the reliability of Josephson devices. A conversation which Marty Nisenoff, of the Naval Research Laboratory, had with me during the early days of Sperry's involvement with Josephson devices (about 1976) was very influential in making Sperry rethink the advantages of an all refractory technology compared to the partial refractory (niobium-lead) technology which Sperry was then using. As a follow up to this, mainly through technology transfer, but also through direct contract support, NRL was very influential in inducing Sperry to establish a niobium nitride technology. The primary interest of the Navy was simply to produce devices whose superconducting transition temperature was high enough to permit the use of closed cycle refrigerators. The ultimate potential importance to superconducting digital microelectronics was appreciated at Sperry later. (It was also appreciated by Hayakawa and still later at BTL.) But it was the military interest which first spurred Sperry into action. The original perscription of how to deposite niobium nitride and much of the remaining U.S. industrial competence in niobium nitride technology still flows from NRL.

As with the case of the academic programs, military applications should not be considered a substitute for commercial interests. The Japanese program is meant to eventually capture commercial markets; the military applications of superconducting electronics can sometimes be orthogonal to commercial interests, in spite of the above example. But often there is the possibility - even the probability - of commercial applications being coherently associated with military ones. For example, DoD might not easily be able to afford to purchase certain hardware because it is too expensive at that time; on the other hand, if it (or some closely related parts) were being developed as commercial products, then the price might be acceptable.

DoD has contributed significant support to DoC through support of some very outstanding research at NBS. Most of this has been supported by the Office of Naval Research in such areas as analog-to-digital conversion and counting (mentioned in Appendix I). These applications are likely to have their first application in military electronics, but they also have significant potential commercial utility.

#### 11.9.2.4 Entrepreneurial Activity

Current wisdom dictates that entrepreneurial enterprises are the way to succeed in high technology. There are two recent start-ups in the United States. Sadeg Faris founded Hypres, Inc. He had a brilliant career both as a contributor to the IBM Josephson program and as a constructive critic. He is reputed to be developing superconducting instrumentation, but is still publically advocating superconducting technology as the only presently known viable approach to the ultimate performance in supercomputers. Hypres has made outstanding technical progress during the past year, but it remains for Faris to demonstrate his first product and to prove that a small enterprise has relevance to superconducting digital electronics.

More recently, a company called Micrilor has been founded which may pursue superconducting electronics. Its founders are John Cafarella and Stan Reible, who were previously employed at Lincoln Laboratories. They are generally in the signal processing business, which for them includes both SAW (surface acoustic wave) devices and analog superconducting signal processors. The founders are experienced in both technologies.

There is no known comparable activity in Japan.

## 11.10 Recommendations to the Department of Commerce

### 11.10.1 General Recommendations

#### 11.10.1.1 Rapid translation of certain journals

Some journals, which are not now regularly translated from Japanese, should be rapidly translated. Most of these will also be useful to workers in fields which are not specifically related to superconducting electronics. This would be a very valuable function which the Department of Commerce could perfrom. Some of these journals are:

a. Japan Applied Physics Society (suggested by Hayakawa and Kim)

b. Japan Transistor Society (suggested by Hayakawa)

c. Tokyo Device Meeting (Mentioned to me as expected first publication of papers in my two trips to Japan; this appears much later in the USA as bound and carefully edited volumes.)

d. Any proceedings of the Cryogenics Section of the Japan Electronic Industry Development Association and/or the Cryogenic Association of Japan. (My two trips to Japan were supported by both organizations. I don't know if proceedings or notes are commonly distributed; if they are distributed they may be an example of the type of limited distribution publications described by Professor Y. Kim.)

e. Bulletin of the Electrotechnical Laboratory.

### 11.10.1.2 Encourage American Scientists To Learn Japanese

Direct conversation is the best way to learn scientific and technical progress. The ability to carry on technical conversations with the Japanese in their own language would be a significant way to improve this method of technology transfer. It would also be a friendly gesture on the part of those American technologists who learn their language. I suspect that many technical meetings in Japan held in Japanese would not be off limits to foreign visitors. One should remember that it is an extra burden for Japanese scientists to prepare talks in English and to ask and answer questions in a foreign language. The capability of many U.S. scientists and engineers to read Japanese would eliminate the necessity of waiting for translations and would permit reading those publications which aren't translated.

The Department of Commerce should encourage American scientists, engineers, and businessmen to learn Japanese. Publicity on the value of learning the language is one way to provide encouragement. I am aware of a Japanese language course which is being offered at the IBM Watson Research Laboratory. Perhaps the Commerce Department could stimulate other companies to offer similar courses through publicity or by even recommending tax credits for those high-tech industries which offer them.

### 11.10.1.3 Recommendations on Josephson Technology

Josephson technology differs from all other microelectronics technologies under development in Japan in that an entirely new system is under development. The most significant results are likely to come from the NTT and MITI programs. The progress of these programs should therefore be monitored very carefully in the United States. But the significance of future results should be intrepeted in a techno-economic sense, rather than in purely technical one. As examples of what is meant by "techno-economic", the following kinds of questions should be answered if the Japanese programs are to continue to show technical progress:

1. Can Japanese Josephson technology create a superior performance supercomputer?

2. Will the likely market for such a supercomputer be large, given its anticipated performance?

3. Can the technology also be applied to a high-performance, general purpose computer?

4. If the Japanese <u>can</u> produce a better supercomputer or general purpose computer, <u>will</u> they?

Monitoring the MITI and NTT programs should therefore involve the participation of U.S. computer firms. As long as the potential answer to any of the above questions is "yes," the Department of Commerce should clearly be involved with the monitoring along with the DoD and the DoE.

I believe that there is an important role which the Department of Commerce could play in this monitoring process. Many U.S. computer companies are facing such difficult competition from IBM and the Japanese that there are few resources left to address questions which involve only the long range future. On the other hand, the Japanese have the advantage of government support for investigating risky, long-term development programs. It is not clear to me that direct support of comparable programs by the Department of Commerce is either necessary or desireable. Beyond merely collecting information, DoC should occasionally provide educational forums, especially if significant progress is made in Japan, by holding special conferences and workshops to which workers from several areas of employment are invited. Besides representatives of the U.S. computer industry, there are many other resources within the United States, including those of the government, such as NBS, DoD, and DoE, who employ individuals who are knowledgeable about facets of this area.

It is not obvious, however, that meaningful participation of many U.S. computer companies will be easy to obtain. Most other U.S. computer companies adopted an ostrich strategy towards the IBM Josephson development program. Given the IBM decision to close down their program, it is reasonable to expect that most companies will continue their successful strategy when it comes to the Japanese programs. It might be possible to obtain initial cooperation from MCC and Hypres, for example, and eventually IBM, in the monitoring process. The Commerce Department should provide leadership in obtaining wider cooperation for the monitoring process and, perhaps, education to other U.S. computer companies from NBS, among other resourses.

Holding special seminars and workshops may be the best means available to eventually interest U.S. computer companies. It is important that invitations to such events go both to senior executives of the corporations (both those involved with general as well as technical management) and to influential scientists and engineers within those corporations. One workshop already in existence, the "Josephson Digital Workshop," is held every odd year in the early fall. Although experts in non-Josephson technologies are invited to speak and interact with the main contingent of Josephson workers at every meeting, this meeting is too technical for many purposes. But it is deserving of DoC support as are the more general purpose meetings recommended above. The Japanese themselves are the most important source of information on their own program and on Josephson technology in general. The Department of Commerce should attempt to open itself as much as possible to, what has been so far, freely flowing information. Japanese researchers and their managements have consistently attempted to foster improved contacts with American scientists. Examples of this are: invitations to Americans to deliver papers at conferences in Japan (often accompanied by honorariums which cover the cost of air fare to Japan); an invitation from the Japanese Society for the Promotion of Science to the U.S. National Science Foundation for a U.S. delegation to meet with Japanese workers in Hawaii in 1983 (the NSF declined to financially support this workshop); and a recent invitation from Japanese trade associations to eight Americans to visit Tokyo for a workshop to discuss Josephson technology. This latter workshop took place in Tokyo in April 1984, and the trade associations, supported in part by MITI, provided approximately $12,000 to partially cover the travel expenses of the American delegation.

At the very least, the Commerce Department should position itself to be the recipient of future invitations for workshops, extend reciprocal invitations to the Japanese when appropriate (perhaps on a regular basis), and also support travel to such workshops for U.S. personnel, including university professors and NBS researchers.

There is one other means by which DoC could financially support the aquisition of information from Japan (and other foreign countries). This is in the area of contributions to help pay for foreign scientists to deliver talks at technical conferences in the United States.

# 11. APPENDIX I. JOSEPHSON DIGITAL TECHNOLOGY

## A. BASIC DEVICE AND CIRCUIT PROPERTIES OF JOSEPHSON JUNCTIONS

### 1. PHYSICAL AND TECHNOLOGICAL BACKGROUND

The Josephson junction, in its most common form, consists of two superconducting films separated by an insulating layer some tens of angstroms thick. Conduction current can be transferred across the insulating barrier by two channels: one, the normal tunneling of single electrons from one electrode to the other, can be modeled as a nonlinear resistor. The second is the Josephson tunneling of superconducting pairs of electrons. A Josephson junction can carry a DC current smaller than its so-called critical current with no voltage drop. If a larger current is forced across the junction, it must flow through the nonlinear resistor, and a voltage appears across the junction. The Josephson current then oscillates at a frequency proportional to the voltage. These oscillations are largely shorted out through the junction capacitance in common Josephson devices.

The circuit model that incorporates these features has proven to be extremely accurate and valuable in predicting the behavior of individual devices and complicated circuits. The fundamental physics of the Josephson effect is on a firm theoretical footing. Real and nonideal device properties are manifested only in the form of the single-electron nonlinear resistance, and in the magnitude of the Josephson critical current relative to that resistance.

A Josephson junction is, therefore, a two-terminal device. Useful gates with three or more terminals are constructed of junctions and other circuit elements such as inductors and resistors. Magnetically controlled gates are called interferometers. These gates are the exact duals of FETS, and therefore have some of the properties of "true" three-terminal devices (Davidson and Beasley 1979). A "duality" exists if there is an exact analogy between devices if inductance is replaced by negative capacitance, resistance by conductance, and voltage by current. This duality is useful to invent interferometer circuits from FET circuits and vice versa, but the duality is only qualitative. Interferometer circuits have low gain, and this poses one of the major difficulties in designing a Josephson computer.

The interferometers consist of two or more junctions connected in two or more loops, and are current-biased in the zero-voltage state. A control current is coupled to the interferometer loop by a mutual inductance, and causes the net current flowing through one of the junctions to exceed its critical current, thereby switching the interferometer to the voltage state and diverting the bias current to control downstream gates. An advantage of magnetically coupled gates is series fan-out capability, limited only by the maximum acceptable timing delay between the switching of the first gate and the last gate on any single control line. A major disadvantage is that the inductors occupy most of the area of the gate, with the result that these gates tend to be area consumptive.

Another family of gates is characterized by directly coupled control currents. These gates have to rely soley on parallel fan-out, which is maximized by biasing the gates as close to the critical current as manufacturing and circuit tolerances will allow. The result is that smaller control currents are needed for switching, and that larger output currents are available as control currents for downstream gates. These gates do not require large inductors, and therefore occupy less area than interferometers.

The memory function is achieved by storing persistent, circulating currents in superconducting loops. "Persistent" is used here in the literal sense. The currents do not decay, and will last as long as the films remain at the operating temperature and no substantial external fields are applied. Individual junctions or gates are used for both writing and sensing.

Signals propagate between circuits on superconducting transmission lines located over a superconducting ground plane. These lines are nearly lossless and dispersionless for frequencies up to about $10^{12}$ Hz, while the superconducting ground plane confines the electric and magnetic fields so that crosstalk between adjacent transmission lines is negligible, even for separations of 2.5 μm. The lines are usually terminated in matched loads, and driven by gates of the same impedance level. This is an important advantage of Josephson circuits.

Compared to transistor-based circuits, conventional Josephson circuits have some major inconveniences: low gain, no convenient inverter, and latching gates. The lack of an inverter may cause difficulty for logic designers but it should be noted that some potentially high performance CMOS circuits of the self-timed variety also effectively cannot employ inverters except in setting the clock. The latching nature of the conventional Josephson gates also increases their sensitivity to noise and power disturbance. Circuit designers have made extensive use of the simulations to overcome these obstacles, and the accuracy of the simulations has contributed greatly to the development of practical working circuits.

## 2. REASONS FOR INTEREST IN SUPERCONDUCTING DIGITAL TECHNOLOGY

The major force behind the development of Josephson technology is its potential application in ultra-high performance computers. The following attributes of the technology are particularly important for this application:

a. High Speed. Individual junctions switch from the zero-voltage to the nonzero-voltage state in a time determined by the product of the junction capacitance and the shunt resistance. This time is typically on the order of 10 picoseconds. Several Japanese programs (ETL and Hitachi) have reported switching delays of less than 10 ps for unloaded gates, and switching times of less than 20 ps for fully loaded gates when using 2.5 μm minimum dimensions. This delay is expected to scale roughly linearly with the linewidth. However, we emphasize that the gate delays are not expected to be the limiting factor in overall system performance of a computer. In fact, in comparing Josephson vs. semiconductor technologies for computer applications, gate delay may be only a secondary issue, because the overall speed will probably be determined by the packaging (see item c, below).

b. Low power. The performance of silicon technology may be limited by the ability to cool circuits. This is not the case for Josephson technology. Common Josephson devices operate at typical voltage levels of approximately 2 millivolts, and typical imepdance levels of 10 ohms. For the 2.5 μm design rule devices, typical power dissipation is on the order of 2 microwatts per gate. Tens of thousands of such gates per square centimeter can be effectively cooled by immersion in liquid helium at 4.2 degrees K. Furthermore, "superfluid" helium below 2.2 degrees K could provide more orders of magnitude of cooling capacity if it were needed for still denser circuits. Liquid helium refrigeration technology is mature and reliable. Closed-cycle refrigerators are routinely used in large-scale applications for cooling superconducting magnets in accelerators. The IRAS infrared astronomy satellite was maintained at 2 degrees K with a year's supply of superfluid helium. Rather than being an obstacle to the practical realization of this technology, cryogenics will more likely prove to be its essential advantage.

c. The package. The "package" contributes approximately two-thirds of the total delay in typical high-performance machines (van der Hoeven 1983). These delays include time-of-flight delays (proportional to the physical length of the signal lines), charging times for lines in semiconductor technology, and package loading delays (due to large inductances in getting on and off chip and packaging parts). Josephson technology has three major effects on the packaging technology. First, the proven lack of a thermal bottleneck allows for a dense, three-dimensional package which semiconductor technologies may not rival. Second, the availability of superconducting transmission lines permits packaging with only one level of wiring in each of the x and y directions, thus simplifiying the construction of the actual interconnection. Third, the lack of necessity of single crystal material together with the above properties could make possible both three dimensional wafers (multiple layers of active devices) and wafer scale integration without concern for lossy long-line interconnections.

B. ALTERNATE APPROACHES TO SUPERCONDUCTING ELECTRONICS

There are a number of approaches to superconducting electronics which were not a part of the IBM program and are therefore not as well known to those who are not Jospehson researchers. Some of these approaches such as "full wafer processing," have almost become part of a new "standard" approach;

others have just begun to be seriously investigated. A brief discussion is made of some of these in this section so that certain current thrusts of the Japanese programs can be placed in perspective, and so that some of the alternate approaches can be more fully appreciated. This discussion is not exhaustive; I am constrained from mentioning some of them.

1. FULL WAFER PROCESSING

Full wafer processing is the formation of a Josephson junction over the entire wafer before any processing (or, at least, before any significant processing) is performed on the wafer; this strategy allows for greater cleanliness and, therefore, improved uniformity of critical current density and a greater control of device area, which is proportional to the critical current. The critical current is the single most important device parameter and the one which is most difficult to control.

The first report of full wafer processing was by the Sperry group (Kroger, Smith and Jillie, 1981; Smith, Jillie and Kroger, 1983). The Sperry group used anodization of the upper electrode of the full-wafer superconductor-barrier-superconductor "trilayer" to mutually isolate the individual Josephson devices. The next group, of which I am aware, using a full wafer process apparently for some time, was ETL (private communication, H. Hayakawa, 1982); a number of ETL papers appeared in the interim (see, for example, Shoji, Kosaka, Shinoki, Aoyagi, and Hayakawa, 1983). The ETL group used reactive ion etching (RIE) of the upper electrode to isolate the devices. Mike Gurvitch of Bell Telephone Laboratories (BTL) used a combination of RIE and anodization to produce very high quality nibium devices which used oxidized aluminum barriers (Gurvitch, Washington, and Huggins, 1983). Independent of these activities, Faris, then with IBM, advocated a full-wafer process, which he called a "pancake" device.

The application of the full-wafer processed devices to integrated circuits was reported by Jillie et al. (Jillie, Smith, Kroger, Currier, Payer, Potter, and Shaw, 1983). Subsequent to the closing of the Sperry Research Center, publications on the use of this method to process integrated circuits have originated exclusively from ETL (Kosaka, Shoji, Aoyagi, Shinoki,

Nakagawa, Takada, and Hayakawa, 1983; Kosaka, Shoji, Aoyagi, Shinoki, Tahara, Ohigashi, Nakagawa, Takada, and Hayakawa, 1984), although I have reason to believe that other groups are also using it.

All of the full-wafer processes require, in practice, the use of only refractory superconducting materials because only such materials can be anodized or because only such materials (and their barriers) can withstand the temperatures imposed by RIE. While no report in the literature has been made of uniformity of full-wafer-processed devices which rivals the uniformity of the devices fabricated by IBM, it should be very clearly noted that neither the Sperry, ETL, nor BTL work was produced by anything more than a laboratory scale apparatus; the IBM results, on the other hand (which have not been reported in the literature), were obtained from a very disciplined pilot line. None of the practitioners of full-wafer processing has any doubt, however, that it cannot be used to fabricate more uniform devices than any alternate means. Future efforts to improve Josephson device processing in both the United States and Japan will almost certainly involve full wafer processing. I believe that the desirability of full wafer processing is one of the reasons why Hayakawa perfers all-refractory device technology.

## 2. VORTEX DEVICES

"Vortex" devices are "long" Josephson transmission lines. These nonlinear transmission lines support "solitons" in the sense of applied mathematics, which have particle-like properties (in the sense of physics). The particles are "fluxons" (quantized flux quanta) whose velocity is contolled by the bias current density applied to the active transmission line. Their particle-antiparticle behavior (including creation and annihilation of particle antiparticle pairs) permits the construction of logic and memory devices.

While the properties of vortex devices have been extensively explored theoretically, not much has been published on practical Josephson vortex devices. A memory address circuit was described at the 1980 ASC (Rajeevakumar,1981). Arnett, of IBM, described the difficulties of scaling vortex memory devices to smaller dimensions in a talk at the 1983 Josephson Digital Workshop. Unfortunately, this work has not been published. At the very least, this talk described some practical difficulties which would be encountered if an attempt were made to make vortex devices small enough so that they could be used in a reasonably high density memory of conventional architecture. At the most, the paper might have described fundamental limitations to the practical utility of such devices in a dense memory. Several attendees felt that Arnett only conclusively covered the former point, although he seemed to be convinced that he had covered the latter point as well. The lack of publication of this important and interesting talk is doubly unfortunate because it occurred at the conference when IBM publically announced that its Josephson development program had been terminated. Many attendees were, therefore, not in the mood to listen as carefully as they normally would have been.

The recent ETL paper was a more modest thrust into the field (Akoh, Sakai, Yagi, and Hayakawa, 1984). It described the fundamental properties of a Josephson vortex line from a very basic point of view, which emphasized the comparison between theory and experiment and methods of obtaining measurements using Josephson samplers. While device application were discussed, no assumption was made that a vortex memory should be constrained to fit into the usual design of memory circuits. In Section V.A of the main text, it was noted that two papers which appeared in English in Japanese publications mentioned memory applications without proposing the full structure of a memory circuit.

3. NONEQUILIBRIUM DEVICES

"Nonequilibrium" devices utilize phenomena other than the Josephson effect in order to achieve useful logic functions. (Some of these proposed devices also show the Josephson effect, but this is usually incidental, or perhaps even an annoyance, to the intended operation of the device.) Nonequilibrium superconductive devices have a long history beginning with Giaever's proposal for a two tunnel junction series-arranged device in 1969 (Giaever, 1969) to Gray's superconducting transistor (Gray, 1978) and more recently to the Quiteron proposed by Faris (Faris 1982; Faris, Raider, Gallagher, and Drake, 1983). Other proposals have been raised by Rogers and Buhrman (1983) and even a structure which requires only one tunneling barrier, but which uses a semiconductor "collector," has been proposed by Frank, Brady, and Davidson

(1984). Several other proposals, both published and privately disclosed, are known to me. The Quiteron has received the most publicity and is currently the most controversial. In general, the motivation for the Quiteron is not controversial: to provide a "true" three terminal device which has greater power and current gain than a Josephson gate, thereby improving the margins in superconducting logic circuits, without increasing the speed-power product above that of Josephson devices, and to also provide a natural inverter. Some dc characteristics of Quiterons are promising, but there has not yet been a wide acceptance of the speed and useful power gain of these devices. There is also some controversy regarding the mechanism responsible for the operation of these devices. To oversimplify, perhaps, the pessimists maintain that it is "only" heating; the optimists either maintain that it is some nonthermal nonequilibrium phenomenon, or, if it is just heating, point out that thermal cooling can occur very rapidly in a properly designed device. A review of this field has recently been presented by Gallagher (1984), which is relatively negative on the Quiteron, per se, but not necessarily on "three terminal" superconducting devices in general. No publication on this subject in English has appeared by any of the major Japanese Josephson programs. However, in a recent special issue of the Bulletin of the Electrotechnical Laboratory, Vol. 48, No. 4 (1984), on Josephson computer technology, a paper on nonequilibrium superconductivity appeared in Japanese (Akoh and Kajimura, 1984). I now have access only to the English abstract which stresses inhomogeneous gap states and apparently is not primarily concerned with practical devices. This untranslated and as yet not understood paper is a prime example of the deficiency of an American scientist's ability to read Japanese and of the lack of rapid availability of English translations of some important Japanese technical publications.

## 4. PHASE MODE CIRCUITS

All of the common Josephson circuits described above are ones in which gates switch and latch to a finite voltage under the circumstances of a logic "1" state. These gates make primary use of the "dc Josephson effect", which describes the circumstances of a device remaining (or not remaining) in the zero voltage state. There is another category of Josephson digital circuits which are not well recognized by workers outside of the Josephson field because they have as yet not been in the mainstream of Josephson device development. These are "phase-mode" circuits which rely on the "ac Josephson effect" and/or on switching flux quanta from one superconducting loop to another. The basic ideas are not recent: circuits of this type have a long history among both American and Russian workers (Fulton, Dynes, and Anderson, 1973; Hurrell, and Silver, 1978; and Likharev, 1977, for example). These circuits require that the Josephson devices make only transitory departures from the zero voltage state (for the order of a picosecond or so) and then return to the zero voltage state. The power dissipation of these circuits is orders of magnitude lower than for the conventional Josephson circuits.

It follows that the commonly quoted speed-power product of "conventional" (IBM) Josephson gates is not nearly as low as what can ultimately be expected from Josephson electronics. This observation is not meant to herald the imminent arrival of a new and wonderous technology, but simply to point out the difficulty of predicting the ultimate performance limits of any technology by using only the most well-known data which has been published.

Two extremely interesting papers have been published on this subject within the past several years. Nakajima et al. (Nakajima, Oya, and Sawada, 1983), of Tohoku University, described a theoretical design of a microprocessor based upon phase mode circuits, and, in an even more important paper, Hamilton and Lloyd (1982), of NBS, described the experimental operation of a binary counter that operated at a rate greater than 100 GHz while dissipating only 0.1 microwatt of power. There have been a number of proposals to use phase mode cicuits for the more conventional logic operations, but apparently there is currently no serious attempt to reduce these ideas to practice.

## 11. APPENDIX II.

### PROPERTIES REQUIRED OF ANY HARDWARE TECHNOLOGY FOR HIGH PERFORMANCE COMPUTERS

There are seven properties which detemine the system level performance of any hardware technology in high performance computers. These are:

1. Speed of loaded logic gate delay. (with fan-out, not ring oscillator delay.)

2. Speed of interconnection. (A lossless, dispersionless transmission line is ideal; a lossy line which is not too long is acceptable; special driver circuits which might be required to send signals along lossy interconnections which are really resistance/capacitance networks could impact not only the speed of signal transmission, but also the useful density of logic gates because of excess power consumption.)

3. Capability to interconnect devices in three-dimensions or in "two and a half-dimensions". (By "2 1/2 dimensions" is meant two-dimensional packaging on circuit boards and an efficient backplane which interconnects the boards so that they are in close proximity in the third dimension.)

4. Useful volume density of logic gates. (Could be limited by heat removal or by the volume consumed by power to the gates, for example.)

5. Volume density of memory.

6. Manufacturability. (The manufacturing tolerances in an LSI or a VLSI format could prohibit the implimentation of the fastest known circuits, for example.)

7. Suitability of fit of hardware to architecture. (This has not been explicitedly discussed in the literature. It was recently invoked implicitly by Gheewala (1984) in down grading the expected system performance of Josephson computers by assuming that latching logic with a single phase power supply was used with all the prohibitions peculiar to the IBM technology.) The properties of Josephson devices make them by general, but not universal, consensus clear winners on the first three requirements: loaded gate delays of less than 20 ps with 2.5 um technology (this is only a slight advantage to Josephson compared with HEMT, but as we observed in Appendix I, this is not expected to be a major factor in system performance as long as the gates are "fast enough"). Only a superconducting interconnection has loss less and dispersionless transmission of signals, and only the Josephson technology has demonstrated the capability of dense packing of circuit boards, first by IBM and more recently by NTT.

It is difficult to call the expected winner on the fourth property, the useful volume density of logic gates. On one hand, the area density of silicon logic gates is clearly superior to that of the relatively immature Josephson devices. On the other hand, it is clear that the useful volume density of Josephson gates could easily exceed that of already demonstrated high speed silicon. This follows from a relatively simple extrapolation of the IBM results on "2 1/2-D" packaging (Ketchen et al., 1981). This argument should not be viewed as conclusive, however, since the Josephson results depend upon immersion cooling (in a liquid helium bath) and no significant demonstration of immersion cooling of silicon has been reported.

Immersion cooling could greatly improve the silicon results. It should be noted that the ability to remove the required heat (ratio of heat produced by logic switches to the capability of heat removal by immersion cooling) is not a sensitive function of operating temperature. That is, while the heat produced by Josephson gates is much less than produced by semiconductor gates, the capability of nonsuperfluid liquid helium to remove heat is significantly less than higher boiling temperature liquids. Thus, it is not clear that the required heat removal from standard Josephson gates with nonsuperfluid helium will be significantly different from that of semiconductor devices.

A precise judgment would require far more careful study than is presently useful, but a "tilt" towards Josephson devices on this question is possible because of practical and theoretical considerations. First, only for Josephson devices has the capability of sufficient heat removal from dense volume packing actually been demonstrated, and there remains the problems of supplying the input power to semiconductor chips in a 3-D packaging scheme without robbing volume which is needed for heat removal and interconnection. Second, the Josephson results appear to be far more extendable. The required heat removal for Josephson devices was judged to be comparable to immersion cooled semiconductor devices for only standard Josephson gates and under the assumption that the liquid helium bath was not superfluid. Nonstandard gates, "phase mode" circuits (which were discussed in Appendix I), for example, have orders of magnitude of lower power dissipation than the common Josephson gates. Superfluid helium, by virtue of its properties as a thermal superconductor, can remove heat at a rate of about an order of magnitude greater than normal helium. The fifth and sixth requirements, volume density of memory and manufacturability, involve the problems which caused IBM to revise its optimism on Josephson computers. As noted in Section III of the main text, it was the delay anticipated in arriving at a suitable design for high area density of memory which was a significant factor in the IBM decision. In part, this problem also involved manufacturability since, if all devices had identical properties, then the memory design could have been fabricated. Several approaches mentioned above are being exploited by ETL and other programs to alleviate or eliminate this problem. An unusual approach is to simply abandon the idea of storing information as flux quanta in loops, although only one such concrete proposal has been published (Bradley and Van Duzer, 1984). The most radical ideas, which ETL is apparently studying, are the construction of 3-D chips (multiple layers of circuits on the same chip), and wafer scale integration to increase the volume density of Josephson cashe memory, and investigation of the possible utility of vortex devices.

The question of manufacturability of the basic IBM type of circuit and device design has been criticized for some time by some knowledgeable workers, some of whom predicted the eventual termination of the IBM program as long as five years ago for nearly the reasons eventually given by IBM. Some of these same critics nevertheless believe that superconducting computers will still become the wave of the future. Several approaches to improving manufacturability have been noted above, including the use of higher gain non-Josephson devices, and improved materials and fabrication processes. I am also aware of alternate approaches which have not been publically described, but which I judge to be extremely promising.

The presently demonstrated level of memory density and manufacturability are areas in which Josephson technology significantly lags behind semiconductor fabrication capability. Whether the density of Josephson memory will ever rival or surpass that of semiconductor technology is difficult to predict at present. It should be noted, however, that it is not impossible that it could. Assessing the probability that it will is fruitless, at present, without knowing the results of designs and experiments which are currently being planned or undertaken in various laboratories around the world.

In conclusion the relative performance of different hardware technologies in high performance computers is a complex system question. The most unique advantage of Josephson technology is in the superconductive interconnection which makes use of lossless and dispersionless transmission lines which have low cross talk. It is impossible to predict, with absolute certainty, the optimum hardware technology of the future; it may well depend upon the architecture of the computer. But it would be wrong to dismiss the Japanese Josephson programs as having no commercial relevance simply because IBM dropped its Josephson project. The Japanese programs do pose a technological and commercial threat to the position of U.S. computer technology.

## 11. REFERENCES

H. Akoh and K. Kajimura, Bull.Electrotech.Lab., Vol.4, 364 (1984).

H. Akoh, S. Sakai, A. Yagi, and H. Hayakawa, (paper presented at the 1984 Applied Superconductivity Conference; to be published in the IEEE Trans. Magn., 1985).

K. Aoki, Y. Tazoh, and H. Yoshikiyo, (paper presented at the 1984 Applied Superconductivity Conference; to be published in the IEEE Trans. Magn., 1985).

P. Bradley and T. Van Duzer, (paper presented at the 1984 Applied Superconductivity Conference; to be published in the IEEE Trans. Magn., 1985).

A. Davidson and M. R. Beasley, IEEE J. Solid State Circuits, Vol. SC-14, 758 (1979).

S. M. Faris, U.S. Patent 4,334,158.

S. M. Faris, S. I. Raider, W. J. Gallagher, and R. E. Drake, IEEE Trans. Magn., Vol. MAG-19, 1293 (1983).

D. J. Frank, M. J. Brady, and A. Davidson, (paper presented at the 1984 Applied Superconductivity Conference; to be published in the IEEE Trans. Magn., 1985).

T. A. Fulton, R. C. Dynes, and P. W. Anderson, Proc.IEEE, Vol.61, 28 (1973).

W. J. Gallagher, (Invited paper presented at the 1984 Applied Superconductivity Conference, to be published in the IEEE Trans. Magn., 1985).

T. R. Gheewala, paper presented at the 1984 IEEE Computer Design Conference, October, 1984.

I. Giaever, in Tunneling Phenomena in Solids, edited by E.Burstein and S. Lundqvist, Plenum Press, New York (1969), p.271.

K. E. Gray, Appl. Phys. Lett., Vol.32, 392 (1978).

M. Gurvitch, M. A. Washington, and H. A. Huggins, Appl.Phys.Lett., Vol.42, 472 (1983).

C. A. Hamilton and Frances L. Lloyd, IEEE Elec. Dev. Lett., Vol. EDL-3, 335 (1982).

Hamilton, Kautz, and Lloyd, 1983:
C. A. Hamilton and Frances L. Lloyd, IEEE Trans. Magn., Vol.MAG-19, 1259 (1983).
R. L. Kautz and Frances L. Lloyd, IEEE Trans. Magn., Vol. MAG-19, 1186 (1983).

J. P. Hurrell and A. H. Silver, in Future Trends in Superconductive Electronics, edited by B. S. Deaver, Jr., C. M. Falco, J. H. Harris, and S. A. Wolf, American Institute of Physics, New York (1978), p.437.

D. W. Jillie, L. N. Smith, H. Kroger, L. W. Currier, R. L. Payer, C. N. Potter, and D. M. Shaw, IEEE J. Solid-State Circuits, Vol.SC-18, 173 (1983).

M. B. Ketchen et al., IEEE Elec.Dev.Lett., Vol.EDL-2, 265 (1981).

S. Kosaka, A. Shoji, M. Aoyagi, F. Shinoki, H. Nakagawa, S. Takada, and H. Hayakawa, Appl.Phys.Lett., Vol.43, 213 (1983).

S. Kosaka, A. Shoji, M. Aoyagi, F. Shinoki, S. Tahara, H. Ohigashi, S. Takada, and H. Hayakawa, 1984; (Invited paper presented at the 1984 Applied Superconductivity Conference, to be published in the IEEE Trans. Magn., 1985).

H. Kroger, L. N. Smith and D. W. Jillie, Appl.Phys.Lett., Vol.39, 285 (1981).

I. Kurosawa, A. Yagi, H. Nakagawa, and H. Hayakawa, Appl. Phys. Lett., Vol 43, 1067 (1983).

K. K. Likharev, IEEE Trans. Magn., Vol.MAG-13, 242 (1977).

K. Nakajima, G. Oya, and Y. Sawada, IEEE Trans. Magn., Vol.MAG-19, 1201 (1983).

Y. Okada, Y. Hamazaki, and E. Sogawa, Bull. Electrotech. Lab., Vol.4, 319 (1984).

T. V. Rajeevakumar, IEEE Trans. Magn., Vol.MAG-17, 591 (1981).

E. Sogawa, H. Nakagawa, Y. Hamazaki, S. Takada, Y. Okada, and H. Hayakawa, Bull. Electrotech. Lab., Vol.4, 326 (1984).

R. W. Ralston, 1984; (Invited paper presented at the 1984 Applied Superconductivity Conference, to be published in the IEEE Trans. Magn. (1985)).

S. Sakai, H. Akoh, and H. Hayakawa, Japan.J.Appl.Phys., Vol.23, L610 (1984).

S. Sakai, H. Akoh, A. Yagi, and H. Hayakawa, Extended Abstracts of the 16th International Conference on Solid State Devices and Materials, Kobe, 631 (1984).

A. Shoji, S. Kosaka, F. Shinoki, M. Aoyagi, and H. Hayakawa, IEEE Trans. Magn., Vol.MAG-19, 827 (1983).

A. D. Smith, R. D. Sandell, J. F. Burch, and A. H. Silver, 1984 (Invited paper presented at the 1984 Applied Superconductivity Conference, to be published in the IEEE Trans. Magn., 1985).

L. N. Smith, H. Kroger, and D. W. Jillie, IEEE Trans.Magn., Vol. MAG-19, 787 (1983).

G. Uehara, M. Nakanishi, and K. Hara, Jpn.J.Appl.Phys., Vol.23, L321 (1984).

B. J. van der Hoeven, Invited paper at the 1982 Applied Superconductivity Conference (not printed).

# 12. III-V Compound Semiconductor Insulated Gate Field Effect Transistors

H.H. Wieder, University of California at San Diego

## 12.1 Introduction

The projected applications of intermetallic semiconducting III-V compounds for high speed digital and analog field-effect transistors (FET's) require that a judicious choice be made among the available binary, ternary and quaternary alloys of these compounds. The synthesis and material technology of such alloys are more complicated than those of the elemental semiconductors. Preservation of their stoichiometry, reduction of residual impurities, control of their homogeneity, impurity diffusion and ion implanting require basic information and data which is available only to a limited extent. The bandstructure parameters, the physical and chemical properties of the surfaces and the interfaces between them and dielectric or metal overlayers impose severe constraints on their technological utilization. They also provide hitherto unavailable options for the implementation of high frequency and microwave devices which may not be matched by the ubiquitous silicon-based FET and integrated circuit technology.

The surface and interface properties of III-V compound semiconductors are among the most important material parameters which define and constrain the available technological options for making FET. Among these are the surface barrier $\phi_B$ of Schottky barrier gate metal-semiconductor-field-effect transistors (MESFETs), the surface Fermi level $E_F^*$ and the surface or interface state density $N_{SS}$ of dielectrically-insulated gate metal-insulator semiconductor field-effect transistors (MISFETs). Of particular interest is the extent of displacement of the surface potential $\Psi_s$ produced by $V_g$, the voltage applied to the gate of such transistors and the feasibility of driving the surfaces of such compounds into inversion, depletion, and through flatband, into accumulation.

The potential advantages of a III-V compound semiconductor insulated gate field effect transistor and integrated circuit technology intended to emulate the ubiquitous metal-oxide-silicon (MOS) technology have been recognized for some time. However, serious attempts to make such devices began only in the 1970's. The impetus for these investigations is the expected improvement in the gain-bandwidth product of metal-insulator-semiconductor field effect transistors (MISFET) available in III-V semiconductors whose low field and peak electron velocities are substantially greater than those of Si. The evolution of such a MISFET technology requires a thorough understanding and technological control of the interfaces between these semiconductors and their native oxides which evolve upon contact with the ambient environment and of syntehtic dielectric layers deposited on their surfaces. In many instances the ~ 20 Å thick native oxide cannot be removed and such structures have an interfacial native oxide layer between the semiconduotor and a synthetic dielectric layer. A comprehensive model of such interfaces is not yet at hand; however, sufficient information has been accumulated during the past few years on their physical and chemical properties than similarities and differences between them and $SiO_2$-Si interfaces are evident. First order correlations can also be established between the electrical properties of such interfaces and the corresponding properties of MISFET structures.

An arbitrary classification of insulating dielectric layers used for III-V compound MISFET can be made in terms of their preparation and their macroscopic properties. Homomorphic dielectric layers are those produced by subjecting various III-V compound semiconductors to electrochemical anodization, thermal or plasma oxidation or ion bombardment processes. Such layers are often compositionally inhomogeneous, they contain variable quantities of the atomic constituents of the respective III-V compounds, oxygen and impurities such as C introduced during their growth. Their crystalline phase, order and morphology change as a function of preparation. Native oxides are neither stoichiometric nor spatially homogeneous and are much too conductive to qualify as gate dielectric layers adequate for MISFET.

Heteromorphic or synthetic dielectric insulating layers are usually deposited by means of chemical vapor phase transport reactions, pyrolisis or sputtering, are usually amorphous, are more nearly homogeneous and their dielectric-semiconductor interfaces are more nearly abrupt than those of homomorphic layers.

An evaluation of the electronic properites of metal-insulator-semiconductor (MIS) two-terminal capacitors employing homomorphic and hetero-morphic layers can be made to first order within the context of the theoretical framework developed for the interpretation of experimental measurements made on the silicon-silicon oxide system.

The large electron to hole mobility ratio in III-V compounds precludes the use of complementary n-channel and p-channel structures such as those employed in the Si technology. It is only n-doped semiconductors that are suitable for such applications. However, both depletion mode and enhancement mode MISFET are feasible. A depletion mode MISFET (D-MISFET) is "normally on" i.e., a large source-drain current appears for zero gate voltage and this current can be reduced by the depletion of electrons from the conducting channel by an applied negative gate voltage, $V_g$. An enhancement MISFET can be made by forcing the channel to be, with $V_g = 0$, thin enough so as to be totally depleted. In that case a positive $V_g$ is required to decrease the channel depletion and thus to produce a source-drain current. Alternatively, with a MISFET made of a p-type semiconductor with n-type source and drain contacts, no current flows between them because they are reverse biased, unless a large enough positive $V_g$ is applied to the gate to cause surface inversion of a thin conducting n-type region between them.

In any case, it is clear that a MISFET, unlike a Schottky barrier gate field effect transistor (MESFET) will not draw gate current even though the applied gate voltage is such as to force the surface into, or above, its flatband value.

MISFET allow the application of large gate voltages of either polarity; these are limited primarily by the breakdown strength of the gate dielectric layer. Large $V_g$ values contribute to the large dynamic range of MISFET and simplify the task of circuit designers. An insulated gate also provides a margin of safety; switching transients which might affect, adversely, the gate current of MESFET, have a small or negligible effect on MISFET. The insulator prevents such current flow. Furthermore, metal-semiconductor interdiffusion and degradation of the electrical properties of the interface inherent in MESFET are also prevented by the gate insulating layers. It is evident that the considerations enumerated above apply not only for discrete devices but also for integrated circuits (IC's). A considerable effort has been expended in Japan as well as in the U.S. on developing such a technology based primarily on GaAs and InP because substrates of these binary semiconducting materials are available in the form of semi-insulating (SI) wafers.

## 12.2 GaAs-Insulator Interfaces and MOS Structures

Homomorphic and heteromorphic dielectric layers grown on or deposited on the surfaces of n and p-type GaAs have been the most extensively investigated III-V compound MIS structures with the expectation that a suitable dielectric interface will be found to make a GaAs MISFET technology feasible. Homomorphic layers have been made by anodic oxidation, microwave or RF plasma oxidation or by means of magnetically confined plasma-beam oxidation. A large number of investigations were made on anodically oxidized dielectric layers using an aqueous solution of glycol as well as other electrolytes. To first order, the C-V curves measured on all homomorphic MIS capacitors exhibit a strong frequency dispersion whose characteristics resemble those of Si MOS structures only superficially. Methods of analysis, usually applied to Si MOS structures using 1 MHz sinusoidal potentials to obtain C-V curves in order to derive their surface state densities must be used with caution; in GaAs MIS structures fast surface states may respond to frequencies in excess of 10 MHz.

Hetermorphic MIS structures on GaAs have been made by a variety of methods which include the pyrolisis of silane, chemical vapor deposition of silicon nitride and of silicon oxynitride layers, and growth of aluminum oxide by the pyrolisis of aluminum isopropylate. In an attempt to reproduce the interfacial properties of the silicon-silicon oxide system a wide variety of methods have been tried; none have yielded the hoped-for characteristics. A review and assessment of the properties of GaAs MIS structures and of the corresponding MISFET technology has been published in Thin Solid Films[1,2].

In Japan, research on the oxidation of GaAs surfaces and on the properties of the resultant oxide layers was intended to provide at least two specific functions: the passivation of the surfaces of discrete devices and ICs and suitable gate insulators for MISFET. Such work and various facets pertaining to it, including the surface physics and chemistry of homomorphic and heteromorphic insulators and fabrication of prototype devices, was carried out in university research laboratories as well as in industrial research centers.

Sugano and Mori[3] described apparatus and techniques used to oxidize GaAs and $GaAs_xP_{1-x}$ in a high frequency oxygen plasma at a pressure of 0.1 to 1 Torr. They found the oxidation rate to be strongly dependent on the plasma characteristics. The oxides thus produced were chemically stable although they had relatively low resistivities of $10^8$ to $10^{10}$ ohm-cm. This work, performed at the University of Tokyo, was supported by a grant-in-aid from the Ministry of Education.

The anodization of GaAs by liquid procedures was carried forward in Japan, initially by Hasegawa[4] and his associates at the University of Hokkaido. Following his return from the University of Newcastle-on-Tyne in the U.K. where, working with Hartnagel and his collaborators they found good electrical and dielectric properties of oxide layers grown in a solution of glycol and water, sufficient to demonstrate elementary aspects of field effect transistors and speculated on improving the GaAs-oxide interface in order to develop an appropriate MISFET technology. However, Sawada and Hasegawa[5] found subsequently that the capacitance-voltage (C-V) characteristics of GaAs-oxide MOS structures were anomalous. They attributed this to a high density of interface states present in n-type GaAs within 0.8 eV below the conduction band edge. This limited the feasible excursion of the surface potential (from its equilibrium position for $V_g = 0$) towards the conduction band edge. They presumed that no such impediments are present on p-type GaAs MOS structures. This work was supported by the Ministry of Education and was also performed at Hokkaido University, Sapporo. Further work, performed by Hasegawa and Suzuki[6], provided details of the electrochemical anodization under dark and illuminated conditions. Other investigations of the process of anodic oxidation of GaAs and GaP and the properties of their corresponding MOS devices were made by Ikoma et al[7] at the University of Tokyo. They found that using various electrolytes, the oxide on p-GaAs grows in a constant field while on n-type GaAS the oxide grows under a constant resistivity condition. Oxide films produced in this manner were found to have a high breakdown strength of ~ 7 x $10^6$ V/cm and a resistivity of the order of $10^{14}$ ohm-cm.

In a paper presented at the 7th International Vacuum Congress in Vienna in 1977, Hasegawa and Sawada[8] correctly identified the displacement of the surface potential in anodized n and p-type GaAs MOS structures to be limited to the lower half of the fundamental bandgap, i.e., between the valence band edge and midgap. However, they also stated that n-GaAs exhibits normal surface inversion while p-type GaAs exhibits normal accumulation. These last two statements are erroneous; subsequently obtained evidence, primarily in the U.S., demonstrates that neither accumulation nor inversion can be obtained in GaAs MOS or MIS structures.

Searching for an improvement in the electrical properties of wet chemically anodized GaAs, Tokuda et al[9] proposed that such oxide layers be sugjected to annealing at about 600°C which lowers their effective resistivity from about 5 x $10^{14}$ ohm-cm to ~ 2 x $10^8$ ohm-cm. By a reanodization at a higher potential they managed to increase the resistivity of these annealed oxide layers to ~ 6 x $10^{12}$ ohm-cm. This work was performed at the University of Tokyo with partial support of the Ministry of Education.

Investigations on the potential of anodically oxidized surface layers on GaAs for use as an encapsulator intended for post-ion implantation annealing at temperatures up to 700°C was investigated by Yokomizo and Ikoma[10] at the University of Tokyo. The results appeared promising but not necessarily better than synthetic dielectric layers deposited on GaAs for the same purpose. The potential application of the glycol-water wet anodization process to gallium phosphide was investigated by Hasegawa and Sakai[11]. They found that such GaP MOS structures have a very long charge retention and that the oxide may serve as a mediator for ion diffusion. This work was performed at Hokkaido University as part of a continuing grant from the Ministry of Education.

Investigations on the properties of homomorphic insulating layers grown on GaAs were also performed in Japanese Industrial Laboratories. At the Nippon Telegraph and Telephone Corporation Shinoda and Yamaguchi[12] investigated the preparation and properties of oxide layers grown by first bombarding GaAs surfaces with positive ions producing a Ga-rich oxide layer. Thereafter these layers were anodized by a 13.6 MHz RF plasma anodization process. They found that such oxide layers have a low pin-hole density and high breakdown strength of ~ $3 \times 10^6$ V/cm and exhibited, as well, less hysteresis in their C-V characteristics.

Chemical vapor phase transport procedures were also used by Shinoda and Kobayashi[13] of the Nippon Telephone and Telegraph, Musashino Electrical Communication Laboratory to grow $Al_2O_3$ layers on GaAs using aluminum tri-isopropoxide as a source material. The deposited layers were presumably non-uniform following ion beam bombardment; however, no electrical data was presented.

A comparison of oxide layers grown on GaAs by thermal, anodic and plasma oxidation was made by Watanabe et al[14] of Hokkaido University using Auger electron spectroscopy and secondary ion mass spectroscopy. They found that oxides grown by thermal oxidation consisted only of $Ga_2O_3$ and $As_2O_3$. By contrast, oxides formed by means of anodic and plasma oxidation contained both $Ga_2O_3$ and $As_2O_3$. The former had a large pile-up of arsenic at the substrate-oxide interface while the latter had a much smaller concentration of As at the interface.

A different technique for the synthesis of hetermorphic $Al_2O_3$ layers on GaAs was used by Yokoyama et al[15] of Hiroshima University. They employed a photochemical reaction between aluminum and oxygen molecular beams excited by ultraviolet light to produce such insulating layers. However, the capacitance-voltage data obtained on such films indicated the presence of the same anomalous frequency and voltage-dependent dispersion of the capacitance found on other heteromorphic or homomorphic layers.

By 1977 it had appeared to many of us in the U.S. that there are serious impediments to improving the dielectric-GaAs interface so as to make a

GaAs MISFET technology feasible. In Japan the work continued towards this end and continued, furthermore, well after the series of papers published in the U.S. suggested that the problem is related to the pinning of the surface Fermi level $E_F^*$ of GaAs surfaces. The results obtained on both n-type and p-type MIS structures by Meiners[16] on both homomorphic and heteromorphic dielectric layers, which have confirmed the earlier measurements of Hasegawa and Sawada[8], indicate that the position of the surface Fermi level, $E_F^*$ for $V_g = 0$ calculated from the frequency dispersion of C-V and G-V measurements[17] is between 0.8 and 0.9 eV below the conduction band minimum (CBM) of n-type and between 0.6 and 0.7 eV above the valence band maximum (VBM) of p-type GaAs, in good agreement with the XPS measurements of Fermi level pinning obtained on cleaved (100)-oriented GaAs surfaces[18,19]. The surface density $N_{ss}$ dependence on energy is approximately U shaped. The surface potential excursion is limited to - 0.45 eV in the lower half of the band gap with a minimum, $N_{ss} = 2 \times 10^{12}/cm^2$-eV, reaching values in excess of $N_{ss} = 10^{13}/cm^2$-eV at the boundaries of this zone. Neither flatband nor steady-state surface inversion was achieved with either n-type or p-type MOS capacitors with electric fields of the order of $10^6$V/cm. The surface states appear to be quite slow, and thus tend to affect the low frequency response of MIS structures far more than their high frequency response.

## 12.3 GaAs MISFET

At the Fujitsu Laboratories, Mimura et al[20] investigated the potential application of plasma-grown homomorphic gate dielectric layers for MISFET. By controlling the etching depth of the channel between the $n^+$ source and drain electrodes they were able to produce either depletion mode MISFET or by reducing the channel depth below its full depletion value in a MISFET they were able to obtain enhancement-mode operation. The microwave response characteristics of such enhancement-mode devices were reported to yield 3 dB unilateral gain at 10 GHz. At the University of Tokyo Tokuda et al[21] using 1.5 μm long gates deposited on $n = 1.8 \times 10^{17}/cm^3$ GaAs with an anodically grown gate insulating layer found a maximum oscillation frequency, $f_{max} = 22$ GHz and a noise figure of 6.1 dB. They measured a higher transconductance, $g_m$, at 50 Hz as well as in the μ-wave region than at intermediate frequencies and attributed this to a high density of fast surface states. While the increase in $g_m$ in the microwave region can be attributed to the inability of these states to charge and discharge at a high rate, the higher $g_m$ in the low frequency region is an artifact. In a subsequent paper describing work performed at the University of Tokyo (Institute of Industrial Science) Ikoma et al[22] inferred from C-V measurements and DLTS measurements that the surface Fermi level of GaAs MIS structures changes only within a narrow range of the bandgap at room temperature. They suggested that highly localized electronic states might be distributed spatially within the oxide of the anodized layers as well as within the bandgap of GaAs; fast states might have a density of the order of $10^{11}$ $cm^{-2}$ $eV^{-1}$ while slow state densities might be of the order of (or higher than) $10^{12}$ $cm^{-2}$ $eV^{-1}$ located within about 10 Å of the oxide-GaAs interface. By assuming that in MISFET the difference in $g_m$ between its DC value and that obtained at 1 GHz is due to interface states they estimated these to be between $3 \times 10^{12}$ and $10^{13}$ $cm^{-2}$ $eV^{-1}$.

A paper by Mimura et al[23] of the Fujitsu Laboratories published in 1979 described an evaluation of the native oxide-GaAs interface covered by a plasma-grown oxide layer. The properties of GaAs MISFET operated under pulsed gate excitation in the deep depletion regime were investigated. They proposed that a high density of electron trapping centers is present near the GaAs-oxide interface under the gate with injection and subsequent trapping of electrons in this oxide under an applied gate voltage pulse. This causes a fast flatband voltage shift. Backtunneling from these traps produces a reduction in density of conduction electrons induced at the interface, at the initial phase while the latter portion of this process is presumed to consist of detrapping from a 0.60 eV Poole-Frenkel center in the bulk of the oxide. This work was supported by the Ministry of International Trade and Industry.

In a paper presented at the 37th Annual Device Research Conference at the University of Colorado, Boulder (1979) Mimura et al[24] described the use of enhancement type MISFETs made of plasma anodized GaAs using conventional photolithographic technology to fabricate devices with 1.5 µm gate length in n 13 stage ring oscillator with a speed power product of 26 fJ per gate and with a propagation delay of 380ps. However, they observed significant hysteresis in the low frequency characteristics of the inverters used to make this ring oscillator; they attributed this feature to tunneling of electrons into the native oxide under the gate. This work was also supported by MITI.

In 1978 Mimura and his associates[25] published a review paper on the microwave properties of GaAs MISFETs made by means of the above cited techniques, including an analysis of the equivalent circuit of these devices, performance criteria, and a comparison with GaAs MESFETs. They found that depletion mode devices with gate lengths of the order of 2 µm used as a class A amplifier produced 0.4 W output power at 6.5 GHz. These MISFETs had a maximum frequency of oscillation of 22 GHz which is 10% larger than the best analogous GaAs MESFET and an intrinsic current gain cutoff frequency of 4.5 GHz which is some 22% higher than that of analogous MESFET; these favorable properties were attributed to the smaller gate parasitic capacitances of the former compared to the latter. Enhancement mode MISFET demonstrated useful unilateral power gain in the 2-8 GHz frequency range and a maximum frequency of oscillation of 22 GHz.

The evolution of a planar GaAs MISFET integrated logic circuit technology based, primarily, on the Fujitsu developed and promoted gate insulators produced by plasma oxidation was described by Yokoyama et al[26]. Selective multiple ion implantation into semi-insulating GaAs was used to produce deep depletion enhancement and depletion mode MISFETS with 1.2 µm gate lengths. A propagation delay of 72ps and a power-delay product of 139 fJ was achieved with a 27 stage ring oscillator. They suggested that MISFET might be potentially superior to MESFET, primarily because of their higher cutoff frequency compared to MESFET. This paper is interesting not only on account of its subject matter and contents but also because of relevant material not included in it. It must have been clear to the authors that no DC or bias control can be obtained with such GaAs MISFETs and that, in consequence, these devices and integrated circuits would not be useful for conventional logic circuit applications. Although dynamic logic circuits have been demonstrated in Si they require more complex circuitry and higher power consumption than conventional logic circuits and therefore have not been adopted. There is no indication that such circuit configurations were contemplated. One is left with the assumption that in Japan there must have been still some hope of solving the lack of DC response and adequate stability of such MISFET. This, however, was not borne out by subsequent events.

The present day situation with respect to GaAs MISFET and integrated circuits, worldwide, is not a favorable one. The performance and the properties of GaAs MISFET reflect those of MIS structures. Although the high frequency and microwave properties of MESFET and MISFET of the same doping density and geometrical aspect ratio are nearly the same, their low frequency and dc properties are quite different. The lack of dc bias control and the inability to invert the surfaces of GaAs MISFET precludes their use in other than dynamic mode circuit applications. Decreasing the gate length and source to drain spacing in order to improve the electron transit time can produce a transfer of electrons from the Γ to the L vally of the GaAs conduction band with a consequent reduction in electron velocity. In high electric fields the saturated electron velocity of GaAs is nearly the same as that of Si. To maintain a high electron mobility in GaAs the potential applied between the source and drain of either MESFET or MISFET would have to be reduced below the threshold for electron transfer which corresponds to an electric field $E = 3.6 \times 10^3$ V/cm. This restricts the dynamic operational range of narrow gate high speed, high frequency transistors. Unless and until Fermi level pinning at dielectric-GaAs interfaces can be lifted there is little hope of a viable GaAs MIS technology. Research on obtaining a better understanding of such interfaces is underway in Japan and elsewhere. However, the driving impulse for technological utilization has attenuated considerably and most of the principals involved in industrial research laboratories in Japan have turned to other more promising approaches. As we shall see, research on homomorphic and heteromorphic gate dielectric layers using other III-V compound semiconductors continues. In Japan the principal focus appears to reside in universities, in particular, at the University of Tokyo.

## 12.4 Dielectric-InP Interfaces

The properties of thermally grown oxides in dry oxygen on InP were investigated in the U.S. by Wager and Wilmsen[27]. They found that the oxides, composed of 70% to 75% $In_2O_3$ and 25% to 30% $P_2O_5$, grow very slowly at temperatures below 340°C and rapidly above this temperature and that the composition of such oxides differs from that of the thin (~ 60 A) native oxides. Wilmsen[28] was the first to investigate the properties of MIS capacitors made with electrochemically anodized insulating layers and found that the surface potential can be changed by an applied gate voltage in contrast with MIS capacitors made with sputter-deposited $SiO_2$ which had their Fermi level pinned in accumulation probably by a high density of fixed positive charge trapped in the oxide as well as by the generation of a high density of surface donor centers on the InP substrate. Although the initial impetus for concentrating on the favorable properties of InP came from the U.S., demonstrating its applications for MISFET, research in Japan, after a slow start, continued to emphasize both the fundamental aspects of InP-oxide and synthetic dielectric interfaces as well as that of various relevant MISFET structures and properties.

In the U.S., Meiners[29] found that MIS structures made of low temperature chemical vapor phase-deposited (CVD)$SiO_2$ on n-InP can be accumulated, that 1 MHz is a sufficiently high frequency to which no surface states can respond, that the Fermi level is pinned to within 0.2 eV of the CBM and that the interface state density is an order of magnitude smaller near midgap than that of GaAs. Essentially the same results were obtained with electrochemically anodized or plasma oxidized surfaces. MIS structures made in a similar manner with p-type InP have a strongly depleted surface; $E_F^* = 1.12$ ev in good agreement with that determined by Spicer[30] on (110)-oriented surfaces and by Waldrop et al[31] on (100)-oriented surfaces, $0.9 < E_F^* < 1.2$ eV.

In Japan, Kawakami and Okamura[32] of Nippon Telegraph and Telephone Public Corporation used chemical vapor deposited $Al_2O_3$ as gate insulators to investigate the properties of inversion layers made on p-type InP. Yamamoto and Uemura[33] of the same organization investigated the interfacial properties of anodic oxide-InP grown from a tartaric acid, ammonium hydroxide-glycol solution. They found an interface state density, near midgap of the order of $10^{11}$ $cm^{-2}$ $eV^{-1}$ rising rapidly near the conduction band edge to more than $10^{13}$ $cm^{-2}$ $eV^{-1}$. Yamaguchi[34], also affiliated with NTT, of their Ibaraki Electrical Communication Laboratory, produced thin insulating layers on InP by the direct thermal reaction of InP with ammonia gas at temperatures between 530 and 560°C. The dielectric layers were found to be primarily phosphorus nitride and indium nitride and had a rather low resistivity of $10^{11}$ to $10^{12}$ ohm-cm. Thermal oxidation of InP and the resultant properties of the oxide layers were investigated by Yamaguchi and Ando[35] of NTT. They found that the oxide layer is composed of polycrystalline $InPO_4$ and that above 620°C they form lower oxides such as $In_4O_2$ and $In_2O$ due to the evaporation of phosphorus. The activation rate for oxidation is 2.06 eV attributed to the oxygen diffusion through the oxide and the resistivity of such layers is low, typically of the order of $10^8$ to $10^9$ ohm-cm. In order to improve the characteristics of the oxide Yamaguchi[36] proposed a modified thermal oxidation process using phosphorus pentoxide vapor at temperatures between 400 and 500°C. This improved oxide layer resistivities to $10^{11}$ or $10^{12}$ ohm-cm. However the C-V data appeared to be anomalous and the advantages, if any, of this process are highly questionable.

The extensive investigations on plasma anodization of GaAs begun at the University of Tokyo by Sugano and his associates were continued subsequently with specific emphasis on developing a suitable MISFET gate dielectric layer. These will be discussed in the following section. A specific emphasis on dielectric-InP interfaces was presented by Hirayama et al[37] in their paper concerned with plasma anodization of Al-InP, i.e., anodization in an oxygen plasma of a vacuum deposited Al layer on InP. The characteristics of the $Al_2O_3$ layers produced in this manner were found to be dependent on the end-point of the anodization process. C-V measurements made at both room temperature and 77°K suggested that the minimum of the interface state density was $\sim 4 \times 10^{11} cm^{-2} eV^{-1}$ while the average value for nearly the entire bandgap range is $\sim 10^{12} cm^{-2} eV^{-1}$. Electron transport through the $Al_2O_3$ layer was interpreted in terms of the Poole-Frenkel effect; its dynamic dielectric constant was estimated to be ~4.8 and its resistivity to be between $10^{10}$ and $10^{12}$ ohm-cm.

## 12.5 InP MISFET

The favorable dielectric-InP interface properties, compared to those of GaAs make depletion-mode, accumulation-mode, and inversion mode MISFET feasible. Depletion and accumulation mode MISFET can be made by direct, selected area $n^+$ ion implantation into SI InP of the Ohmic source and drain contacts and of the conducting channel in the depletion mode devices which require an additional shallow implantation of donors. Alternatively, epitaxial growth techniques can be used to process devices. D-MISFET were found to have microwave power gain, noise figure and maximum power added efficiency comparable to those of GaAs D-MESFET. Unlike GaAs the transconductance and gain of such MISFET is essentially frequency independent and their operating points can be controlled by means of dc bias. InP enhancement-type MISFET (E-MISFET) can be made either in the form of accumulation-mode transistors (A-MISFET) or inversion-mode transistors (I-MISFET). The latter require p-type substrates and blocking source and drain electrodes. For integrated circuit applications they appear to offer, at this time, fewer advantages in comparison with A-MISFET which can be incorporated into planar monolithic integrated circuits by means of relatively simple procedures. Most of the early pioneering work on such MISFET was performed in the U.S. However, a substantial amount of effort has been and is being spent in Japan to develop a viable InP mISFET technology and Japanese researchers had made important contributions to this end.

Enhancement-type InP MISFETs were made by Hirayama et al [43] using plasma anodized aluminum oxide layers deposited on p-type InP or on semi-insulating (100)-oriented InP. The former were, clearly, I-MISFET while the latter are accumulation mode MISFET. Si ion implantation was used to form the source and drain electrodes and photolithographic lift-off techniques were used to make the discrete devices. An effective electron mobility in the channel of $8 \times 10^2$ $cm^2/V\text{-sec}$ was obtained with a threshold voltage, $V_T = 0.95$ V. C-V measurements suggested that the Fermi level is not pinned within the bandgap.

Further investigations on MISFETs mode by the same procedure were performed by Hirayama et al[44] at the University of Tokyo. Devices made by means of the techniques he described earlier were investigated, at room temperature and at 80°K. They found an increase in the effective electron mobility and a decrease in the threshold voltage $V_t$ of the MISFETs at 80°K compared to their room temperature values attributed to a decrease in the density of interface states. At 300°K they found that the DC drain current drifts for a few minutes and thereafter becomes stable (for a few hours). We shall return subsequently to this drift/instability problem because it constitutes one of the hurdles that must be overcome if a digital MISFET integrated circuit technology is to become viable. Details of the research on InP surfaces and on MISFETs performed under the direction of professor Sugano at the University of Tokyo are contained in their annual research reviews (in English). At Hokkaido University, Professor Hasegawa and his associates have continued their investigations on anodically grown oxide MIS structures and on InP MISFETs[45]. Sawada and Hasegawa[46] obtained field effect mobilities of $1.5 \times 10^3$ to $3 \times 10^3$ $cm^2$/v-sec on enhancement-type MISFET mode with a double layer gate insulator: an initial anodically grown native oxide layer followed by the anodization of a vacuum-deposited Al layer. They claimed that this type of device configuration increased the effective channel mobility and decreased the instability and drift at DC of the drain current. Furthermore, they suggested that in contrast with the suggestions made by Okamura and Kobayashi[47], that the presence of the native oxide at the synthetic dielectric InP interface is the cause of the instability, the specific incorporation of the native anodized oxide layer does not affect adversely the electronic properties of the MISFET and may, in fact, improve their performance.

The properties of enhancement-type MISFET made with $Al_2O_3$ gate insulating layers deposited by chemical vapor phase transport procedures were investigated at the Nippon Telegraph and Telephone Corporation in Tokyo by Kawakami and Okamura[48]. These were made by the use of sulfur-diffused n-type source and drain contacts on Si InP. They found that these devices had a source-drain capacitance two orders of magnitude smaller than those of inversion mode devices made on p-type InP of the same dimensions. It is, however, noteworthy that only curve-tracer data was presented as evidence for improvement in the MISFET characteristics. The advantages of such a reduction of the capacitance have not been explored in terms of the high frequency properties or gain bandwidth product of MISFET.

The contact resistance of the source and drain contacts of InP MISFET is an important parameter which determines the transconductance and gain-bandwidth product of such transistors. At the Tokyo Laboratories of NTT,

Yamaguchi et al[49] investigated the nature of contacts made to Si-implanted $n^+$ junction on Si InP. They found that Au-Ni/Au-Ge-Ni overlayers yielded a specific contact resistance $\rho_c = 10^{-5}$ ohm-cm using a transmission line method of analysis for a Si implantation dose of 2 x $10^{14}/cm^2$ at $10^2$ or 2 x $10^2$ keV and an alloying cycle of 400°C for 7 sec.

At the Basic Research Laboratories of the Nippon Electric Co. in Kawasaki, Ohata et al[50] made accumulation mode MISFET using chemical vapor phase transport procedures to deposit $SiO_2$ gate insulating layers on SiInP. They also made a variety of measurements on two-terminal MIS capacitors with this type of dielectric layers on n-type InP. Their results are rather peculiar in that they find a minimum interface state density near flatband increasing towards midgap, in contrast with evidence obtained by others, worldwide. They reported a field-effect mobility of 1.2 x $10^3$ $cm^2$/V-sec for 1.5 μm long and 28 μm wide gates with ~ 0.09 μm thick dielectric layers, a transconductance of 64 mS/mm and a 7 dB maximum power gain at 4 GHz with a cutoff frequency of 9 GHz of such MISFET.

Itoh and Ohata[51] of the NEC Corporation's Kawasaki Laboratories found that self-aligned InP A-MISFET devices with high transconductances and microwave response are feasible. For this purpose, chemical vapor deposition was used to provide a $SiO_2$ self-aligned mask for the vacuum deposition of Al gates onto ~ 650 Å thick $SiO_2$ gate insulating layers. These led to a channel length of 0.8 μm, gate length of 1 μm and width of 280 μm. A typical gm = 17 mS/mm, a minimum noise figure of 1.87 dB with an associated gain of 10 dB was obtained at 4 GHz with power outputs of 1/7 W/mm and 1.0 W/mm at 6.5 and 11.5 GHz, respectively. The maximum power added efficiency at 6.5 GHz was 43.5%. However, these devices were also found to have long term drift in their DC characteristics. A comparison with two-dimensional electron gas field effect transistors (TEGFET) at room temperature might favor the A-MISFET.

Matsui et al[52] at the University of Tokyo investigated the properties of A-MISFET using using a double layer gate insulator: an oxide layer grown by plasma anodization followed by a plasma anodized $Al_2O_3$ layer. Such A-MISFET were found to have electron field effect mobilities typically 2.1 x $10^3$ to 2.6 x $10^3$ $cm^2$/V-sec and threshold voltages of 0.3 to 0.9 V.

Their most significant finding is that such a structure has a sharp reduction of the DC drain current instability from 5 µs to ~ 5 x $10^4$ s of ±4% compared to that of conventional A-MISFET. Further investigations of such two-layer plasma-oxidized structures were made by Fuyuki et al[53] at the University of Kyoto. The structure and formation of these layers were investigated by Auger spectroscopic methods. In the first stage a native oxide of InP grows to thicknes of ~ 0.05 µm within about 20 min. in a magnetically confined plasma. Then Al sputtered from inner discharge electrodes deposits on this native oxide taking the form of aluminum oxide. They obtained in this manner a typical resistivity of $10^{11}$ ohm-$cm^2$ and breakdown strength of 1.5 x $10^6$ V/cm. The interface state density, measured in the vicinity of the Fermi level is between 2.5 and 5 x $10^{12}$ $cm^{-2}$ $eV^{-1}$.

The uniformity of MISFET characteristics over a wafer and between wafers as well as their reproducibility under ostensibly similar conditions has not been investigated, as yet, insufficient detail. Nonuniformity and lack of reproducibility may be attributed to uncontrolled variations in the deposition of the gate insulating layers, the predeposition surface preparation of the semiconductor, and spatial fluctuation in the properties of the semi-insulating substrates. Steady progress is being made in overcoming these difficulties and there is little doubt that these issues can be reduced to acceptable levels. More serious, however, is the experimentally observed long term drift in MISFET characteristics, in particular, the drift in drain current with applied steady state drain voltage and gate voltage. It is assumed that the experimentally observed hysteresis in C-V curves of MIS structures of the same materials is closely related to the same phenomena which gives rise to the drift and instabilities of MISFET. Fritzsche[54] interpreted the change in channel conductivity with time of I-MISFET in terms of electron tunneling into traps distributed through the native oxide. Okamura and Kobayashi[55] have analyzed these phenomena in terms of a model which contains one trapping level in the synthetic dielectric $Al_2O_3$ layer and a second trapping level located in the native oxide. Tunneling into the former is considered the primary transport mechanism with a long time constant and this trap, which is located below the Fermi level, is presumed to be insensitive to temperature. The trap in the oxide is above the Fermi level; it is presumed to be temperature dependent and elimination of the native oxide should, therefore, reduce the contribution of these traps. These assumptions appeared to be confirmed by their experimental data; using HCl to etch and remove the native oxide prior to the deposition of the synthetic dielectric reduced the instability. However, their subsequently published data[56] obtained by first oxidizing the InP surface and then depositing the $Al_2O_3$ gate dielectric layer also led to a reduction in drift. It may be assumed that a procedure that reduces the surface and interfacial disorder also reduces MISFET instability.

Lile et al.[57,58] described their investigations made on A-MISFET using a variety of surface preparation techniques, including HCl etching and oxidation; they concluded that these procedures do not improve the drift. Furthermore, they found a considerably greater drift in MISFET using $Al_2O_3$ dielectric layers than those with $SiO_2$ layers. The drain current, after an initial steady-state period varies exponentially with reciprocal temperature and logarithmically with time. They concluded that these phenomena can be represented by a model which consists of a thermally activated tunneling process; however, the physical origin of the states involved remain undetermined.

Matsui et al.[52] have reduced, as stated earlier, the drift of A-MISFET to less than ± 4% for periods from 5s to 5 x $10^4$s by growing an ~ 0.01 μm thick native oxide by plasma anodization, thick enough to prevent tunneling into the superposed $Al_2O_3$ layer which was also grown by plasma anodization. This suggests that at least some options are becoming available for controlling drift. In fact some A-MISFET with $SiO_2$ gate insulators selected from a batch exhibit no drift whatsoever and it is not yet possible to correlate this with specific fabrication procedures. The solution of these problems might be considered the most important impediment to progress towards an integrated circuit technology employing complementary D-MISFET and A-MISFET.

## 12.6 References


1. "Special issue on Semiconducting III-V compound MIS Structures" Thin Solid Films, 56, No. 1/2 (1979), H. H. Wieder and C. W. Wilmsen, ed.

2. "Second Special issue on Semiconducting III-V Compound MIS Structures" Thin Solid Films, 103, No. 1/2 (1983), H. H. Wieder and C. W. Wilmsen, ed.

3. T. Sugano and Y. Mori, J. Electrochem. Soc. 121, 113 (1974).

4. H. Hasegawa et al, Appl. Phys. Lett. 26, 567 (1975).

5. T. Sawada and H. Hasegawa, Electron. Lett. 12, 471 (1976).

6. H. Hasegawa and T. Suzuki, Japan J. Appl. Phys. 15, 2489 (1976).

7. T. Ikoma et al, Proc. 8th Conf. Solid State Dev. Tokyo (1976) and Japan J. Appl. Phys. Suppl. 16-1, 475 (1977).

8. H. Hasegawa and T. Sawada, Proc. 7th Internat. Vac. Congress 1977 and 3rd Conf. Solid Surfaces, Vienna, P.O. Box 300/A-1082 Vienna, Austria, pp. 549-552.

9. Tokuda et al, Electron. Lett. 14, 163 (1978).

10. H. Yokomizo and T. Ikoma, Japan J. Appl. Phys. 17, 1685 (1978).

11. H. Hasegawa and T. Sakai, J. Appl. Phys. 49, 4459 (1978).

12. Y. Shinoda and M. Yamaguchi, Appl. Phys. Lett. 34, 485 (1979).

13. Y. Shinoda and T. Kobayashi, Japan. J. Appl. Phys. 19, L299 (1980).

14. K. Watanabe et al, Thin Solid Films, 56, 63 (1979).

15. S. Yokoyama et al, Thin Solid Films 56, 81 (1979).

16. L. G. Meiners, J. Vac. Sci. Technol. 15, 1402 (1978).

17. L. G. Meiners, Colorado State Univ. Rept. No. SF19 (1979).

18. I. Lindau et al, J. Vac. Sci. Technol. 15, 1337 (1978).

19. W. E. Spicer et al, Phys. Rev. Lett. 44, 420 (1980).

20. T. Mimura et al, Proc. 9th Conf. Solid State Dev. Tokyo 1977; Japan. J. Appl. Phys. 17-1, 153 (1978).

21. H. Tokuda et al, Electron. Lett. 13, 761 (1977).

22. T. Ikoma et al, Proc. 10th Conf. Solid State Dev. Tokyo 1978) Japan J. Appl. Phys. 18-1, 131 (1979).

23. T. Mimura et al, Appl. Phys. Lett. 34, 642 (1979).

24. T. Mimura et al, paper MP-4, 37th Ann. Dev. Research Conf. Univ. of Colorado, Boulder (1979).

25. T. Mimura et al, IEEE Trans. Electron. Dev. ED-25, 573 (1978).

26. N. Yokoyama et al, IEEE Trans. Electron. Dev. ED-27, 1124 (1980).

27. J. F. Wager and C. W. Wilmsen, J. Appl. Phys. 51, 812 (1980).

28. C. W. Wilmsen, Crit. Rev. Solid St. Sci. 5, 313 (1975).

29. L. G. Meiners, Thin Solid Films, 56, 201 (1979)

30. W. E. Spicer et al, J. Vac. Sci. Technol. 16, 1422 (1979).

31. J. R. Waldrop et al, Appl. Phys. Lett. 42, 454 (1983).

32. T. Kawakami and M. Okamura, Electron. Lett. 15, 502 (1979).

33. A. Yamamoto and C. Uemura, Electron. Lett. 18, 64 (1982).

34. M. Yamaguchi, Japan J. Appl. Phys. 19, L401 (1980).

35. M. Yamaguchi, and K. Ando, J. Appl. Phys. 51, 5007 (1980).

36. M. Yamaguchi, J. Appl. Phys. 52, 4885 (1981).

37. Y. Hirayama et al, J. Electron. Mater. 11, 1011 (1982).

38. L. Messick, D. L. Lile, and A. R. Clawson, Appl. Phys. Lett. 32, 494 (1978).

39. L. Messick, Solid-State Electron. 22, 71 (1979).

40. D. L. Lile and D. A. Collins, Thin Solid Films 56, 225 (1979).

41. L. Messick, Solid-State Electron. 23, 51 (1980).

42. L. G. Meiners, D. L. Lile, and D. A. Collins, Electron. Lett. 15, 578 (1979).

43. Y. Hirayama et al, Inst. Phys. Conf. Ser. No. 63, Internat. Conf. on GaAs and Related Compounds Oiso, Japan (1981), p. 431.

44. Y. Hirayama et al, Appl. Phys. Lett. 40, 712 (1982).

45. T. Sawada et al, Proc. 13th Conf. Solid State Dev. 1981 Tokyo, Japan J. Appl. Phys. Suppl. 21-1, 397 (1982).

46. T. Sawada and H. Hasegawa, Electron. Lett. 18, 742 (1982).

47. M. Okamura and T. Kobayashi, Japan. J. Appl. Phys. 19, 2143 (1980).

48. T. Kawakami and M. Okamura, Electron. Lett. 15, 743 (1979).

49. E. Yamaguchi et al, Solid State Electron. 24, 263 (1981).

50. K. Ohata et al, Inst. Phys. Conf. Ser. No. 63, Internat. Conf. on GaAs and Related Compounds, Oiso, Japan, 1981, p.353.

51. T. Itoh and K. Ohata, IEEE Trans. Electron. Dev. ED-30, 811 (1983).

52. M. Matsui et al, IEEE Electron dev. Lett. EDL-4, 308 (1983).

53. T. Fuyuki et al, Japan. J. Appl. Phys. 22, 1574 (1983).

54. D. Fritzsche, Inst. Phys. Conf. Ser. No. 50, 258 (1980).

55. M. Okamura and T. Kobayashi, Japan, J. Appl. Phys. 19, 2151 (1980).

56. T. Kobayashi, M. Okamura, E. Yamaguchi, Y. Shinoda and Y. Hirota, J. Appl. Phys. 52, 6434 (1981).

57. D. Lile, M. Taylor and L. Meiners, Japa. J. Appl. Phys. 22, Supl. 22-1, 389 (1983).

58. D. L. Lile and M. J. Taylor, J. Appl. Phys. 54, 260 (1983).

# Glossary

| | |
|---|---|
| APD | avalanche photodetector |
| BH | buried heterostructure |
| CVD | chemical vapor deposition |
| DH | double heterostructure |
| ECL | electrical communication laboratory |
| ETL | electrotechnical laboratory |
| FET | field effect transistor |
| HEMT | high electron mobility transistor |
| IC | integrated circuit |
| IMR | integrated multichannel repeater |
| INS | information network system |
| LAN | local area networks |
| LED | light emitting diode |
| LPE | liquid phase epitaxy |
| LSI | large scale integrated (circuit) |
| MBE | molecular beam epitaxy |
| MEC | Ministry of Education and Culture |
| MITI | Ministry of International Trade and Industry |
| MESFET | metal semiconductor field effect transistor |
| MISFET | metal insulator semiconductor field effect transistor |
| MIS. | metal insulator semiconductor (capacitor) |
| MOS | metal oxide semiconductor (capacitor |
| MOVPE | metal organic vapor phase epitaxy |
| MOCVD | metal organic chemical vapor phase deposition |
| MODFET | modulation-doped field effect transistor |
| MQW-DH | multiquantum-well double heterostructure |

| | |
|---|---|
| NTT | Nippon Telegraph and Telephone |
| OEIC | optoelectronic integrated circuit |
| OJRL | optoelectronic joint research laboratory |
| OMVPE | organometallic vapor phase epitaxy |
| QW | quantum well |
| RAM | random access memory |
| ROM | read only memory |
| SAM | separate avalanche and multiplication (regions) |
| SCH | separate confinement heterostructure |
| VLSI | very large scale integrated (circuit) |
| VPE | vapor phase epitaxy |

# Appendix: Literature Support to Opto- and Microelectronics Panel*

In addition to technical assessments, an important objective of the JTECH Program is the identification of relevant and timely Japanese S&T source material for the technical areas addressed by the panels. Thus, panel members are requested to evaluate, from their technical perspective, the usefulness of the various types of source materials provided. Panel members are also encouraged to give recommendations on Japanese source material in their technical areas which they feel should be acquired and dissiminated to the U.S. technical community on a continuing basis.

In this appendix, we summarize:

- the JTECH approach in providing literature/translation support to the panels, and a description of the general types of material provided,
- open-source Japanese literature available in the opto- & microelectronics area, and
- literature provided to the opto- & microelectronics panel by the JTECH staff.

## A.1 Approach

The following approach has been followed in providing panel members with pertinent source material:

1. Initially, at the panel "kickoff" meeting, the JTECH staff provides general source material for panel members' review. This material is mainly in English and usually of an overview nature. Also presented and discussed is background information on the various types of Japanese publications, technical society and working group meetings, organizations, etc. which are

---

* This appendix written by the JTECH Program Staff; the staff member responsible for literature and translation support is Dr. Y. Kim.

potential sources of information for particular technical topics.

2. Each panel member then ascertains particular source material (including papers in Japanese) needed for their technical topic.

3. The JTECH staff then collects, evaluates, and translates those requests by the panel members.

(Steps 2 and 3 are iterative, with several interactions taking place during the panel assessments together with trips to Japan by Y. Kim to collect information.)

4. Toward the end of the panel life, the panel members, together with the JTECH staff, assess the source materials found to be most useful for the technical area. Also, panel members are encouraged to offer recommendations related to more effective methods of acquisition/translation/dissemination for the U.S. technical community.

Under the JTECH literature support approach, two factors are emphasized:

A. Timely, Pertinent Source Material. In addition to providing panelists with readily available "open" material (such as excerpts from Japanese technical journals and periodicals), the emphasis of the approach is in providing very recent "semi-open" sources of Japanese research accomplishments and plans (such as results of Japanese ad hoc seminars and working groups). (These different categories of source material are discussed in the next section.)

B. Selectivity. To be effective under practical time and budget constraints, it is, of course, important that some selectivity be exercised in the source material provided. In providing the panelists with translated material, the following "selective" approach is used: (1) The titles of papers from a Japanese document are translated and distributed to panelists; (2) for seemingly interesting titles, a panelist requests that certain abstracts be translated (this step is usually done by telephone); and, (3) based on the abstracts, the full translation of the most relevant papers is performed. (In applying this procedure for the Opto- & Microelectronics Panel, only about 10% of the total Japanese source material collected required full translation, as indicated later in Section A.5.)

## A.2 Categories of Source Material

In the JTECH approach for providing literature support to panelists, we consider the Japanese S&T source material in terms of three categories. These categories are defined below, and the material provided to the computer science panel in each category is listed in the next section.

### A. General Material (mostly in English)

These are materials initially provided to assist panel members in identifying specific source material needed for their particular area of expertise. This material is typically shallow in technical depth, but provides a general program background. Included are popular overview articles, previous technology assessment studies, and translations of Japanese Government publications on overall programs, research objectives, participating organizations, funding levels, etc.

### B. Open Material

This category consists "openly available", regularly published technical publications. A list of the technical journals pertinent to computer science is given in the next Section.

### C. Semi-Open or Closely-Held Material

The emphasis under the JTECH program is in providing panelists with this category of source material (described below), which is of a more timely benefit, as illustrated in Figure A-1.

Japanese academic societies hold biannual or annual meetings, and participants are provided with the abstracts of the meeting. Many manuscripts of these abstracts are hand-written; they are not reproduced in regular journal publications. Frequently these abstracts contain very current and useful information, but they have to be accessed through privately-held material. In this sense, the meeting abstracts should be regarded as semi-open.

In the Japanese high tech community, frequently ad hoc seminars and conferences are organized to pool and disseminate the state-of-the-art research information. Such meetings are attended usually by the invitees only, and the conference proceedings are printed in Japanese with most manuscripts being handwritten to allow entry of the latest information. Some of the more formal technical information contained in these proceedings eventually work their way through the appropriate professional journals published in English, but with a typical delay of six to eighteen months.

Japanese industries participate in the national R&D projects under a "Research Association" type of arrangement. Technical progress made in these projects is disseminated first in the technical committees composed of member companies, and refined (edited) technical papers reach public domain at a much later time. In highly competitive R&D, timely access to raw data is important. (The ongoing national projects in the opto & microelectronics area are: Supercomputer, Optoelectronics, Future Electronics Devices, and 5G Computers.)

Most Japanese high-technology industries maintain in-house R&D centers that are staffed with leading technical expertise. Many companies publish periodical reports that contain some useful technical information. Of course, commercially sensitive technical information is closely-held as proprietary.

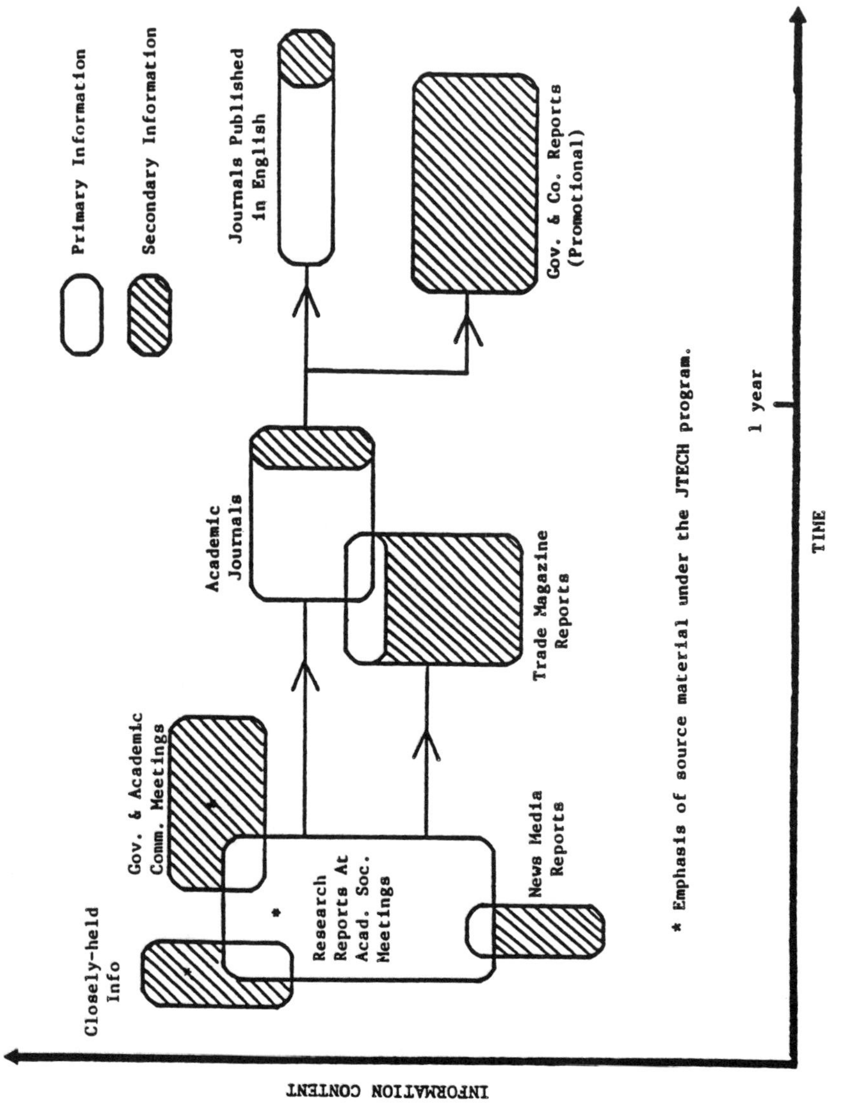


* Emphasis of source material under the JTECH program.

Figure A-1. Flow of Technical Information from Japan.

A.3 Open Sources of Japanese Literature in Opto & Microelectronics

1. Society Publications

A. Journal of Applied Physics
Abstracts of Society Meetings

B. Journal of Electronics & Communication Society
Abstracts of Society Meetings

C. Japan Metallurgical Society

D. Japan Society of Precision Engineering
Journal of..

E. Journal of Japan Society of Crystallography
Journal of...

F. Society for Instrument & Control Engineering (SICE)
Publication: Transactions of...

G. Japan Electrical Society
Publication: Journal of...

H. Japan Information Processing Society
Publication: Journal of...

I. Japan Machinery Society
Publication: Journal of ...

2. Technical Magazines

1. Electronic Materials
2. Materials Science
3. Surfaces

4. Surface Sciences
5. Solid State Physics
6. Semiconductor Research
7. Vacuum
8. Metallurgical Materials
9. Metals
10. Machinery Technology
11. Metal Surfaces Technology
12. Anticorrosion Technology
13. Iron & Steel
14. Sciences
15. Optics
16. Polymers

3. Research Reports

(A) National Projects
  (1) Future Electron Devices
  (2) Supercomputer
  (3) Optoelectronics
  (4) Fifth Generation Computer Systems (5G)

(B) Laboratory Research Reports
  (1) ETL (Electrotechnical Laboratory) Reports
  (2) ETL Research Reports
  (3) OEJRL (Optoelectronics Joint Research Laboratory) Reports
  (4) Electrocommunication Application Laboratory Reports
  (5) NTT Electrical Communications Laboratory Reports

(C) In-house Industry Laboratory Reports

Most Japanese industrial laboratories publish periodical research reports. Some examples are: Hitachi Review, NEC Technical Reports, Toshiba Review, Sony Central Research Laboratory Review.

4. Popular Magazines & Promotional Publications

(1) Nikkei Mechanical
* (2) Nikkei Electronics
(3) Computer White Paper, JIPDEC
(4) INS (Information Network System)
(5) Frontier Technologies
(6) Electronics Industries Monthly

5. Japanese Science & Technology Literature Available in the United States

The Library of Congress has an impressive holding of Japanese S&T literature, including almost all of the open-source material mentioned above. Most periodicals appear to arrive with a four to six month delay. The cataloging of Japanese S&T literature is done mostly in terms of romanization of Japanese titles. Therefore, it is rather difficult to make use of these materials unless the researcher is familiar with the Japanese language.

## A.4 Literature Provided to JTECH Opto & Microelectronics Panel

### A. General

1. George Mu, "Statement for the record submitted to the Subcommittee on Science, Research and Technology" (March 7, 1984) Congressional Hearings on the Availability of Japanese Science and Technology Information.

2. Yoon Soo Park, "III-V Compound Semiconductor Research at the Optoelectronics Joint Research Laboratory," Scientific Bulletin, ONR Far East, Tokyo. January-March, 1984, Volume 9, Number 1. pp. 152-157.

3. Toshio Takai (EIAJ), "Setting the Record Straight on

Superconductors," Journal of Japanese Trade & Industry, (MITI) No. 4, 1983.

4. Yoshitatsu Tsutsumi, "Moving Ahead--at the Speed of Light: The Japanese Optoelectronic Industry," Journal of Japanese Trade & Industry, Number 4, 1984. pp. 28-30.

5. "Assessment Research on Japanese Superconducting Electronics Technology," Advanced Materials Technology. February, 1984.

6. "Prospects on Market Demands in Electronic Industries," JEIDA, July 1983.

7. "Trends in Electronic Technology Reflected in U.S. Patents," Electronics Industries Monthly, March 1984, Volume 26, Number 3. pp. 31-40.

B. Open

1. Izuo Hayashi (OEJRL), "Development of Optoelectronic IC's in Japan," Japan Science and Technology News, Volume 3, Number 1, February 1984. pp. 32-38.

2. T. Itoh and others, "Implantepitaxy by Means of Silicon PI-MBE." n.p., n.d.

3. Yoon Soo Park, "Semiconductor Lasers and Crystal Growth Technology Report From a Topical Meeting of the Japan Society of Applied Physics (JSAP)," Park. ONR/Far East Scientific Bulletin, Volume 9, Number 2, 1984. pp. 79-85.

4. DJIT: Digest of Japanese Industry & Technology, Number 198, 1984. pp. 31-35.

5. "Discovery of New Superconducting Phenomenon Under Ultrahigh Pressure," Science & Technology in Japan (Jul.-Sept., 1984) pp. 33.

6. "Explanation on Patents Related to IC," Electronics Industries Monthly, March 1984, Volume 26, Number 3. pp. 41-43.

7. "Outlook of Policies Related to Information," Electronics Industries Monthly, March 1984, Volume 26, Number 3. pp. 2-10.

8. Recent Technical Papers Published in Japanese on Semiconductor Crystal Growth. n.p., n.d.

9. "Research Institute News," Science & Technology in Japan Jul.-Sept., 1984, pp. 36-38.

10. "Science and Technology Forum, '84," Science & Technology in Japan. Jul.-Sept., 1984, pp. 20-21.

11. "Solar Cell R&D in Japan," Science & Technology in Japan. Jul.-Sept., 1984, pp. 22-25.

12. "Technical and Research Reports," Japan Science & Technology News, Volume 3, Number 3, June, 1984. pp. 20-36.

13. Technocrat, Volume 16, number 11. November, 1983.

Articles related to opto & microelectronics are:

"Elastomers for Electronic Industry."
"New Ion Beam Implantation."
"Large Area SOI Formation Techniques."
"Three Dimensional IC Circuit Technology."
"Development Status of Vertical Magnetic Recording."

C. Semi-Open or Closely-Held

1. K. Asakawa, "Dry Etching Process for Optoelectronic Integrated Circuit," Paper L5-11. Collection of Research Papers, (January-December, 1983), Optoelectronics Joint Research Laboratory. Printed in English.

2. K. Asakawa and others, "Plasma Physics Considerations on Reactive Dry Etched GaAs," Paper L5-1. Collection of Research Papers, (January-December, 1983), Optoelectronics Joint Research Laboratory.

3. Kazunori Chida, "Improvement in VAD Optical Fiber Preform Fabrication System," Review of the Electrical Communication Laboratories. NTT. May, 1984 (Volume 32, Number 3) Special Issue: "Optical Fiber Fabrication Techniques." pp. 395-402.

4. T. Fukuda & T. Iizuka, "LEC GaAs Crystal Growth for Optoelectronic Integrated Circuit," Paper L1-12. Collection of Research Papers, (January-December, 1983), Optoelectronics Joint Research Laboratory. Printed in English.

5. Tsuguo Fukuda & others, "Growth of GaAs Single Crystal by Magnetic Field Applied Technique," Oyo Buturi. Volume 53, issue 1, January 1984. pp. 42-47.

6. T. Fukuzawa and others, "Disordering in Superlattice Due to Zn Diffusion," Paper L4-13. Collection of Research Papers, January-December, 1983), Optoelectronics Joint Research Laboratory.

7. T. Fukuzawa and others, "GaAlAs Buried Multi-Quantum-Well Laser Fabricated by Diffusion Induced Disordering," Paper L4-19. Collection of Research Papers, (January-December, 1983), Optoelectronics Joint Research Laboratory. Printed in English.

8. S. Furukawa, "Techniques for producing SOI Structures," Oyo Buturi. Volume 53, issue 1, January 1984. pp. 28-32.

9. Ryoichi Ito, "Power and Short-Wavelength Limits in Semiconducting Injection Lasers," Oyo Buturi. Feb. 1984 (Vol. 53, No. 2). pp. 124-127.

10. T. Judo & others, "Limits of VLSI," Oyo Buturi. Volume 53, Number 2. February, 1984. pp. 120-123.

11. Noburo Kawamura, "An Overview of Developing High Speed Transistor Technologies," Oyo Buturi. Feb. 1984 (Vol. 53, No. 2). pp. 124-127.

12. M. Kawashima, "Present Status of Future Electron Device Project--An Introduction," R&D Association for Future Electron Devices. 3RD Symposium on Future Electron Devices. Abstracts of the Third Symposium on Future Electron Devices, July 4-5, 1984

13. Matsuyama Kentaro, "Optical Fiber Research--Present State & Future Problems," Review of the Electrical Communication Laboratories. NTT. May, 1984 (Volume 32, Number 3) Special Issue: "Optical Fiber Fabrication Techniques." pp. 393-394.

14. Toshihiko Kitagawa & others, "Properties of Surface Passivation Films for Gallium Arsenide Integrated Circuits," Paper ED83-96. Shingaku Giho, Volume 83, Numbers 244-255, January 24-25, 1984.

15. Y. Matsuoka and others, "Uniformity Evaluation of MESFET's for GaAs LSI," Atsugi Electrical Communication Lab, NTT.

16. J. Nishizawa, "Level Up of the Operating Frequency of and the Out-Put Power of the SIT" Oyo Buturi. Special Issue: "Expected Limitations to the Future Advance of Semiconductor Devices," Volume 53, Number 2. February, 1984.

17. Kinji Hoshi & others, "Growth of Silicon Monocrystals in a Magnetic Field," Oyo Buturi. Volume 53, Issue 1, January 1984. pp. 38-41.

18. Akihiro Shibatani, "Fabrication Process Technologies of Gallium Arsenide Integrated Circuits," Oyo Buturi. Volume 53, issue 1, January 1984. pp. 33-37.

19. Sugada, "Formation of P Layer in GaAs by Be Ion Bombardment," Paper L5-4. Collection of Research Papers, (January-December, 1983), Optoelectronics Joint Research Laboratory.

20. S. Takagishi & H. Mori, "MOCVD GaAs Crystal Growth in the TMG System Under Reduced Pressures," Paper L2-15,L2-16. Collection of Research Papers, (January-December, 1983), Optoelectronics Joint Research Laboratory.

21. S. Takagishi & H. Mori, "Effect of Operating Pressure on the Properties of GaAs Grown by Low-Pressure MOCVD," Paper L2-19. Collection of Research Papers, (January-December, 1983), Optoelectronics Joint Research Laboratory. Printed in English.

22. S. Takahashi and others, "Self-Aligned Ohmic Contact Technology." Central Research Lab, Hitachi, Ltd. n.p., n.d.

23. N. Tanno & others, "Spectral Measurement of Cancer Cells by Microscopic Fluorescence Method with Laser Excitation," Oyo Buturi, Vol. 53, No. 3, (March, 1984) p. 241-248 .

24. N. Tsukada and others, "Periodic Structure on GaAs Surface Generated by Laser Annealing and Laser Photochemical Etching," Paper L5-2. Collection of Research Papers, (January-December, 1983), Optoelectronics Joint Research Laboratory.

25. N. Tsukada and others, "Enhancement of Dry Etching by Laser Irradiation," Paper L5-5. Collection of Research Papers, (January-December, 1983), Optoelectronics Joint Research Laboratory.

26. N. Tsukada and others, "Grating Formation on Gallium Arsenide by One-Step Laser Photochemical Etching," Paper L5-25. Collection of

Research Papers, (January-December, 1983), Optoelectronics Joint Research Laboratory. Printed in English

27. K. Tsutsui & S. Furukawa: "Ion Implantation Effects on Pt/GaAs System and Its Application," Paper ED83-95. Shingaku Giho, Volume 83, Numbers 244-255 (January 24-25, 1984)

28. Bulletin of the Electrotechnical Laboratory, Special Issue--Josephson Computer Technology, April, 1984. Abstracts. pp. 129-131.

29. Electrical Communication Laboratories Technical Journal, Special Issue: "Wide Area Dedicated Subscriber Fiber Optics Communication System." NTT. Volume 33, Number 3 (1984) Title pages and English abstracts of articles are appended.

30. Electrical Communication Laboratories Technical Journal. Special Issue : "GaAs LSI Technology." Volume 33, Number 4 (1984). Abstracts published in English.

31. GaAs/GaAlAs MQW High-Power Lasers, by Hitachi. n.p., n.d.

32. "GaAs MESFET Technology (SAINT) for Ultrahigh-Speed LSI's," Electrical Communication Laboratories Technical Journal. Special Issue : "GaAs LSI Technology." Volume 33, Number 4 (1984). pp. 643-654.

33. Japan Society of Applied Physics (JSAP), 1984 Spring Meeting, Abstracts.

34. Japanese National R&D Projects in Microelectronics: Supercomputers, Future Electron Devices, Optoelectronics. n.p., n.d.

35. List of technical papers published in 1982, Optical Measurement & Control System, Technical Reports (I), Optoelectronics Joint Research Laboratory. March, 1983.

36. "Mixed Crystal Phase Devices Survey Report I," JEIDA survey, March, 1984.

37. Ōyo Buturi Volume 52, Number 8 (August 1983) - Volume 53, Number 4 (April 1984). Translation of tables of contents.

38. "Present Status and Future Development of Optical IC & OETC," Special Seminar on "Frontiers of Materials & Process Technology in Optoelectronics," March 13-14 1984. n.p.,

39. R&D Progress Report of Optoelectronics Joint Research Association, June, 1984. Part A: R&D Results of Optoelectronics Joint Research Laboratory.

40. R&D Progress Report of Optoelectronics Joint Research Association, June, 1984. Part B: R&D Results of Component Technologies

41. Sample pages and articles of Japanese journals related to the microelectronics area:

Transaction of Electronic and Communication Engineers of Japan, Part C.
Journal of the Surface Science Society of Japan
Journal of the Vacuum Society of Japan
Electronic Materials

42. Shingaku Giho, Volume 83, Number 244-255. January 24-25, 1984. Translation of tables of contents.

43. Supercomputer Project, ETL.

Part A: R&D on Superspeed Computer System, Summary of Progress in 1982.
Part B: R&D Progress in 1983 & Plans for 1984 on HEMT and GaAs FET Devices. March, 1983.

44. Survey Reports on Materials for Opto Information Processing Materials, JEIDA (Japan Electronic Industry Development ASsociation). March, 1984.

45. Symposium on New Functional Devices (Future Electronics Devices), Abstracts. July, 1983. (Consists of three sessions on Superlattiçe, Hardened IC's, and Three-Dimensional IC's.)

46. Third Symposium on Future Electron Devices, July 4-5, 1984. Research and Development Association for Future Electron Devices, MITI. Compilation of titles and short English abstracts.

# Part III

# Mechatronics

The information in Part III is from *JTECH Panel Report on Mechatronics in Japan,* prepared by J. Nevins, J. Albus, T. Binford, M. Brady, N. Caplan, M. Kutcher, P.J. MacVicar-Whelan, G.L. Miller, L. Rossol, and K. Schultz. The study was prepared for Science Applications International Corporation under contract to the U.S. Department of Commerce, March 1985.

# Acknowledgements

The Japanese Technology Evaluation Program (JTECH) is indebted to the panel members for their efforts in completing the assessment in the short time allotted while continuing to meet other commitments and pursuing ongoing research interests.

We appreciate the contributions made by Mr. George Mu from the Department of Commerce, International Trade Administration, Office of Japan. His help and guidance as the DoC Contracting Officer was invaluable. We express our gratitude to the National Science Foundation for their supplemental financial support of this study, and particularly to Mr. Frank Huband of the Division of Policy Research and Analysis for his contributions to the program.

# About the Authors

**James L. Nevins (Chairman), Charles Stark Draper Laboratory**

Mr. Nevins received his B.S. in Electrical Engineering from Northeastern University, his M.S. from Massachusetts Institute of Technology and is presently the Robotics and Assembly Systems Division Leader at the Charles Stark Draper Laboratory, Inc. At present, he directs the laboratory's activities in a variety of automation projects sponsored by industry, the National Science Foundation (NSF), and the Office of Naval Research. The principal focus of his work is applied research on advanced robotics, intelligent systems, and programmable automation/assembly systems. His past responsibilities include co-principal investigator of the NSF-sponsored Product System Productivity Research study, the direction of an ARPA-DOD project on computer-controlled manipulators, the Space Nuclear Systems Office (SNSO) project in support of the development of the NERVA Engine Instrumentation and Control System, the SNSO project to develop a multimoded remote manipulator system, and the man-machine design and implementation for the Apollo and Lunar Module guidance, navigation, and control systems. He is Vice-Chairman of the International Federation of Automatic Control (IFAC) Technical Committee on Manufacturing Technology, Chairman of the American Automatic Control Council (AACC) committee of the same name, and the IEEE delegate on the AACC Social Effects of Automation Committee. His previous professional activities include being a member of the Automation Research Council, the past Chairman of the Technical Committee for Space for the Institute of Navigation.

**James S. Albus, National Bureau of Standards**

Dr. James S. Albus is presently Chief of the Industrial Systems Division and Manager of the Programmable Automation Group, Center for Manufacturing Engineering, National Bureau of Standards. He is responsible for robotics and automated manufacturing systems interface standards research at NBS, and designed the control system architecture for the Automated Manufacturing Research Facility. He has received the Department of Commerce Silver Medal for his work in control theory and manipulator design, and the Industrial Research IR-100 award for his work in brain modeling and computer design. Before coming to the Bureau of Standards, he worked 15 years for NASA Goddard Space Flight Center where he designed electro-optical systems for more than 15 NASA spacecraft. Dr. Albus is the author of numerous scientific papers which have been published in technical journals, conference proceedings, and official government studies. He has also written for popular publications such as Scientific American, Encyclopedia Americana, OMNI, BYTE, ENTERPRISE, Metal Working News, and The FUTURIST. He has written two books, Brains, Behavior and Robotics (BYTE/McGraw-Hill 1981), and Peoples' Capitalism: The Economics of the Robot Revolution (New World Books 1976).

**Thomas O. Binford, Stanford University**

Thomas O. Binford is Professor, Computer Science Department at Stanford University. He holds a Ph.D. in Physics from the University of Wisconsin, 1965, and a B.S. in physics from Pennsylvania State University, 1957. He is a Leader of the Computer Vision and Robotics group at the Stanford Artificial Intelligence Laboratory. From 1967-1970 he was Research Associate, Artificial Intelligence Laboratory at Massachusetts Institute of Technology. Research topics included artificial intelligence, computer vision, representation of shape, and LISP programming systems. Professor Binford has been a member of the NASA committees on Automation and Future Missions in Space and Machine Intelligence and Robotics and a member of IEEE Pattern Recognition Technical Committee. Professor Binford has chaired four conferences on Computer Vision and has served as chairman or co-chairman on other major conferences. He has contributed extensive written works to a variety of professional forums.

**J. Michael Brady, Massachusetts Institute of Technology**

J. Michael Brady is Senior Research Scientist in the Artificial Intelligence (AI) Laboratory of the Massachusetts Institute of Technology (MIT). He has held that position since joining MIT in March 1980. Prior to joining MIT, Michael Brady was on the faculty of the Department of computer Science of the University of Essex, England. Dr. Brady is the founding editor with Professor Richard Paul, Purdue University, of the International Journal of Robotics Research. Dr. Brady received a first class honors B.Sc. degree in Mathematics in 1966 and the degree of M.Sc. in 1968 from the University of Manchester, England. He completed his Ph.D. at the Mathematics Department of the Institute for Advanced Studies of the Australian National University, Canberra, Australia, in 1970. Michael Brady's research interests are in Artificial Intelligence and, more particularly, Image Understanding and Robotics. In Computer Vision, his current research concerns the representation of two and three dimensional shapes and he has developed a representation of two-dimensional shape called Smoothed Local Symmetries. In Manipulation, Dr. Brady's robotics research is the application of techniques for depth perception to the problem of picking objects out of a bin. Dr. Brady is the author of The Theory of Computer Science: a Programming Approach, and has edited several other books. He has had many articles published in professional journals, conferences and other forums.

**Michael Kutcher, IBM**

Michael Kutcher, C.Mfg.E., is with IBM at their Kingston, New York facility. He has 35 years of industrial experience in automation and manufacturing systems. He is founding chairman of IBM's Corporate Automation Council and is currently consultant to the IBM Corporate Manufacturing Staff. He is the author of numerous papers, articles and presentations particularly on "Data Driven Manufacturing and Integrated Manufacturing Systems". He has several basic patents on systems and robots. He frequently serves as advisor and consultant to U.S. federal agencies on productivity and technology assessments. He is a senior member of the Society of Manufacturing Engineers & past chairman of the Factory Automation Council of its Computer and Automated Systems Association (CASA). He is a member of the IEEE Components hybrids and manufacturing technology Society and Representatives to the IEEE Robotics and Automation Council and is a past member of the U.S. Air Forces Computer Assisted Manufacturing Advisory Group (CAMAG).

**P.J. MacVicar-Whelan, Boeing AI Center**

P.J. MacVicar-Whelan has a Ph.D. in Physics from the University of British Columbia. He is a principle scientist at in the Mechatronics system section at Boeing AI Center and he is responsible for the development of AI aided product systems. Activities include integration of technology applicable to engineering and manufacturing. Applications of this capability will initially be to manufacturing and the space station. Projects include data and knowledge bases, integration of CAD and CAM technology, identification of applicable of AI technology, its transfer to a test bed, and application to specific projects of interest to manufacturing research and development groups.

**G. Laurie Miller, AT&T Bell Laboratories**

After receiving an MSc. in Mathematics and a Ph.D. in Physics from London University, G. L. Miller worked at Brookhaven National Laboratory from 1957 to 1963. Since then he has been in the research area of AT&T Bell Laboratories at Murray Hill, NJ, working on a wide range of instrumentation and measurement problems in the physical sciences. Since 1982 he has been head of the Robotics Principles Research Department. Dr. Miller has published over 80 technical papers and holds 22 patents.

**Lothar Rossol, GMF Robotics**

Lothar Rossol, Vice President of R&D, joined GMF Robotics in 1982. He is responsible for the development of new products, including robots, vision systems, robot programming languages and software, networking and computer integrated manufacturing systems, and off-line robot programming products. A pioneer in vision and robotics, he was a founder of the vision and robotics research group at the GM Research Labs and was responsible for the development of a number of well-known pioneering robots and vision systems, including Consight and others. He holds several patents in robotics, and is co-editor of the book, Computer Vision and Sensor-Based Robots. While at General Motors, he also led research work in computer graphics, solid modeling, and a number of other computer related activities. Prior to joining GM in 1963, Rossol was a development engineer at IBM where he designed optical character recognition equipment. Mr. Rossol holds two electrical engineering degrees -- a B.S.E.E. from Wayne State University in 1962 and an M.S.E.E. from University of Michigan in 1964.

**Karl B. Schultz, Cincinnati-Milacron**

Karl B. Schultz holds a B.S. in managerial science from the University of Cincinnati and is presently Manager, Systems Control

Engineering, U.S. Plastics Machinery Division at Cincinnati-Milacron. Since joining Cincinnati-Milacron in 1964 he has had managerial responsibility for the system design and development of software systems for large flexible manufacturing systems as well as programming language and systems for control of individual machines. His primary research interest is the design, analysis and application of software systems for computer integrated manufacturing. Mr. Schultz is the author of several articles for professional journals and a frequent speaker at seminars on flexible manufacturing.

# 1. Summary

This study was performed to evaluate the status and direction of Japanese R&D in Mechatronics.

The term "Mechatronics" was originated by the Japanese to describe the union of mechanical and electronic engineering needed in producing the next generation of machines, robots, and smart mechanisms for applications such as manufacturing, large-scale construction, and work in hazardous environments. In this study, Mechatronics was divided into nine areas for analysis:

- o Flexible manufacturing systems (FMS)
- o Vision systems
- o Non-vision systems
- o Assembly/inspection systems
- o Intelligent mechanisms
- o Software
- o Standards
- o Manipulators
- o Precision mechanisms

Comparisons between the United States and Japan were made in each area according to: basic research, advanced development, and product implementation.

Technical Summary

Figure 1-1 shows the assessment summary for each technical area. The symbols indicate Japan's current status, and the arrows show the trend.

The assessment summary shows Japanese basic research to be equal to the United States in all areas except vision and software. Further, the Japanese research is staying even with the U.S. in spite of their lack of large DoD and NASA programs. In vision they are not much behind, and they will probably catch up in the near future.

| CATEGORY | BASIC RESEARCH | ADVANCED DEVELOPMENT | PRODUCT IMPLEMENTATION |
|---|---|---|---|
| FMS | 0 → | 0 ↗ | + ↗ |
| VISION | – → | + → | + ↗ |
| NON-VISION | 0 → | 0 → | 0 ↗ |
| ASSEMBLY | 0 → | > ↗ | > ↗ |
| INTELLIGENT MECHANISMS | 0 → | + → ↗ | 0 → |
| SOFTWARE | < ↘ | – → | – ↗ |
| STANDARDS | 0 → | 0 → | |
| MANIPULATORS | 0 → | + → | + ↗ |
| PRECISION MECHANISM | 0 → | + → | + ↗ |

CODING SYSTEM – JAPAN COMPARED TO U.S.:

| PRESENT STATUS | RATE OF CHANGE |
|---|---|
| < FAR BEHIND | ↑ PULLING AWAY |
| – BEHIND | ↗ GAINING GROUND |
| 0 EVEN | → HOLDING CONSTANT |
| + AHEAD | ↘ FALLING BEHIND |
| > FAR AHEAD | ↓ LOSING QUICKLY |

FIGURE 1–1. MECHATRONICS ASSESSMENT SUMMARY

In basic research, the Japanese are behind and falling further behind only in artificial intelligence (AI) software techniques. However, it should be noted that AI is not the only path to future intelligent systems, and Japan appears to be embarked on a broad approach. Thus, one can argue that an information-control versus AI approach to advanced robotics/process problems may be the method used by the Japanese to accelerate their basic research programs for intelligent machines. The Toshiba Software Factory (see JTECH Computer Science Panel Report) has already demonstrated the Japanese ability to create high-quality software at productivity levels some seven times greater than those produced by Americans.

In advanced development and product implementation, the Japanese are ahead or equal to the United States, and the rate of change is definitely in favor of Japan.

Conclusions

- In all categories of Mechatronics except software, Japan is holding constant or gaining ground over the United States, and there is evidence that the lag in software is closing.

- The Japanese integrated approach to manufacturing systems coupled with government planning and support, which has been very successful in the past, is now being applied to the area of Mechatronics.

- Japanese progress in Mechatronics is of major importance because it addresses the very means of production. Mechatronics itself has a regenerative effect on manufacturing industries. Further, it identifies the need for, and offers a wide range of, new products which are key to the advanced/automated manufacturing systems of the 1990s.

Observations

- The Japanese have decided that software is indeed a key issue in the development of intelligent systems.

- Vertical integration significantly aids major Japanese companies in Mechatronics development.

- Japan is able to focus national resources through the activities of the Ministry of International Trade and Industry (MITI).

- The major U.S. government funding of robotics and AI is currently through military agencies. This provides some commercial spinoff benefits to the U.S. economy, but is vastly less effective than the Japanese model of direct MITI-industry collaboration on Mechatronics.

Recommendations

- An appropriate national response to the Japanese challenge in the area of Mechatronics needs to be formulated.

- A formal process of collecting and disseminating technical information on Japanese Mechatronics activities is needed.

# 2. Introduction

## 2.1 Purpose of Study

The principle goal of this study was to evaluate Japanese R&D in Mechatronics and to estimate the impact on U.S. industry. A second purpose was to comment on the availability of Japanese literature related to Mechatronics.

## 2.2 Mechatronics Defined

Mechatronics is a term coined by the Japanese to describe the union of mechanical and electronic engineering. It is used to emphasize a multidiscipline, integrated approach to product and manufacturing system design. Mechatronics encompasses the next generation of machines, robots, and smart mechanisms necessary for carrying out work in a variety of environments. The environments are primarily manufacturing, but extend to hazardous regions such as space, underwater, and nuclear -- as well as for disasters like fire, chemical, explosion, and nuclear emergencies. Further, the Japanese are pursuing applications in the construction and service industries.

By both implication and application, Mechatronics represents a new level of integration and approach to manufacturing systems and processes. The intent is to force a multidiscipline approach to these systems as well as re-emphasize the role of process understanding and control. It can only speed up the already rapid Japanese process for transforming ideas into products.

Currently, Mechatronics describes the Japanese practice of using integrated teams of product designers, manufacturing, purchasing, and sales personnel acting in concert to design both the product and manufacturing system with minimal technical complexity.

For the future, Mechatronics offers a means for implementing advanced processes and production technology. The Japanese are already succeeding in specific areas. For example, the integration of VLSI technology and machine vision software algorithms is creating cheaper, more capable vision systems.

Mechatronics will have an impact on higher education. Engineers are needed who understand both processes and systems. There is a further need for business schools to be aware of new issues, options, and tradeoffs that Mechatroncis will create.

## 2.3 Brief History and Background

The current Japanese Mechatronics effort should rank high as an area of U.S. national concern for several reasons:

- The pragmatic Japanese approach, coupled with their particular infrastructure, allows them to create systems very rapidly with present technology. That is, using minimal on-line sensors they create systems by coupling product design to present technology and use people whenever the technology is missing or fails them.

  Since most large Japanese firms are supported by large production technology centers, they can produce the needed systems rapidly and independently of the supplier market place. In the U. S. on the other hand, users depend on suppliers, and suppliers do not generate new systems unless they are assured of a market place with a good return on investment. In a sense this is an institutional problem. Mechatronics addresses the issues related to the way in which product design is carried out, the coupling of designers to manufacturing, and the coupling of users to suppliers.

  In contrast, in the U. S. the "Factory-of-the-Future" (FOF) approach taken by a number of companies offers integration of product design and manufacturing systems, but does not address the user-supplier infrastructure issue. Several large U.S. companies, notably IBM and GE, are addressing the infrastructure issue by acting as their own supplier--but this does not help the mid-size company.

Until the user-supplier infrastructure question is addressed, the U.S. may continue to lag behind Japan in implementing these new systems.

- o Japan appears to be embarked on a broad spectrum approach to intelligent mechanisms. One can argue that an information-control versus an AI approach to advanced robotics/process problems may be the method used by the Japanese to accelerate their basic research programs for intelligent machines. Currently, they lag significantly behind the U. S. in intelligent systems and complex autonomous system research*. But, they have started such work, and the question is which strategy they will use.

- o Previous history of the Japanese for rapid implementation of products together with the potential of many new products and systems available from Mechatronics is expected to provide strong advantage to Japan in world-wide economics.

## 2.4 Report Organization

In the next section each of the Mechatronics subtopics selected for analysis is discussed. The depth to which each subtopic was explored is not uniform due to several factors -- such as Japanese activity in the area, panelist's experience, and availability of relevant Japanese data. In most of the areas, the Japanese status and rate-of-change compared to the U.S. is summarized using the coding system noted previously in Figure 1-1.

---

* See also the JTECH Computer Science Panel Report for more substantiation on these issues.

# 3. Technical Analysis

## 3.1 Flexible Manufacturing System (FMS) Development

In Japan, any multimachine system with some material handling ability and a degree of flexibility may be considered a flexible manufacturing system (FMS). The Japanese have been installing FMS of all types for over a decade. A wide range of educational, cultural, and economic pressures have supported and pushed this movement to FMS, along with central planning and goal setting.

Installed FMS is growing rapidly. In 1981, Yamazaki opened a new plant with two fully automated FMS systems. Since then, two larger FMS systems have been installed. Toyoda, SNK, Niigatta, Makino, and other Japanese machine tool companies have also installed FMS and are competing strongly in the world-wide FMS marketplace.

Japanese companies have been implementing FMS with available technology. They have not delayed implementation while more advanced systems were being developed. Consequently, the Japanese have become experienced users of FMS. The technology is not necessarily the most advanced, but the systems are being fully utilized. They have reached a point where they employ new FMS without major disruptions to their organization.

Dr. Kenjiiro Okamuar of Kyoto University points out four levels of production engineering which are visible in a Japanese plant:

1) Working Level - Suggestions are encouraged to increase productivity. Money can be spent with a foreman's approval to purchase reasonably priced changes that improve productivity.

2) Assignment of Entry Level Engineers to the Foreman - Engineers are trained in problem solving. They spend 3-5 years in an apprenticeship assignment and work closely with

the foreman and workers. They are authorized to make changes.

3) Production Engineers - Have had "Level 2" experience and deal with improvements in existing processes. They deal with vendors of equipment and design new processes.

4) Advanced Manufacturing R&D - This group may or may not have "Level 2" experience. Their job is to develop processes. Almost all of the larger Japanese industries have a "Level 4" group. We have seen examples of work on "wireless" carts, adjustable boring bars, vision, inspection for chips, automatic assembly, software for FMS, and more. The R&D at this level relates to pragmatic manufacturing problem solving.

This view of Japanese production engineering shows how management has embraced technology changes. R&D activity takes place in both the supplier and user sectors. R&D is heavy in the user sector, and it tends to drive the suppliers' activities. "Level 4" involves the implementation of existing technology. The goal for "Level 4" is to replace current equipment and methods on a seven (7) year cycle. The evolution of manufacturing in Japan is continuous. Japan has learned from its experience in the steel industry. They built the industry on the most modern technology of the time, and they took much of the world market. But did not continue to update their production technology. Today, Korea and Taiwan are underselling Japanese steel products. The Japanese do not intend to let this happen in other areas.

MITI has sponsored the "Flexible Manufacturing System Complex" at the Tsukuba test plant since 1977. This program is commonly known as the "FMS WITH LASER." This is a national effort concerned with advancing the manufacturing above "Level 4." The Japanese have two goals:

1) Develop new processes such as improving how a laser may be used in manufacturing.

2) The development of flexible machine tool components which can be mated to create different types of machine tools. For example, a grinder and a turning center might be the same basic machine with different heads.

The FMS with Laser project is expected to provide the Japanese with a wide range of new developments which can be commercialized.

The trade press is reporting a 40% to 70% increase in Japanese robots this year. General Motors, GCA, General Electric, and IBM are using and marketing Japanese robots. A simple manipulator can be called a robot in Japan; therefore, the number of units being reported as sold may be a little deceiving. Robot technology in Japan is comparable to that of the U.S. The U.S. may have a lead in robotic control and software, but that lead is closing.

Professor Sata of the University of Tokyo believes that the next major target for FMS is in Flexible Transfer Units, which is for producing a high volume family of parts. The line will allow flexibility for design changes and new parts, and is applicable, for example, to the appliance, automobile, and tractor industries. Nippondenso produces a range of automatic radiators using flexible transfer units and assembly stations. Professor Sata points out that flexible assembly systems and flexible transfer lines are very similar.

CAD/CAM integration in Japan is growing, but it is not as advanced as in the U.S. and Western Europe (see Figure 3-1). The Japanese tend to buy U.S. CAD and production control systems. The Japanese tend to push toward standardized designs with firm production schedules. They press their distributors to absorb inventory when business is slow to keep plant productivity flowing.

| | BASIC RESEARCH | ADVANCED DEVELOPMENT | PRODUCT IMPLEMENTATION |
|---|---|---|---|
| FMS | 0→ | 0↗ | +↗ |
| CAD/CAM | 0→ | 0→ | 0→ |

Figure 3-1. FMS-CAD/CAM Development.

## 3.2 Assembly/Inspection Systems

### 3.2.1 Summary

One area of Mechatronics in which the Japanese excell is assembly systems. They have achieved this by choosing a system design strategy that is deterministic (geometrically driven). By deterministic it is meant that a successful assembly can be accomplished by an open-loop control strategy (generally no sensing or feedback) where the parts are simply pushed together. Quality of product and high yield are direct result of part quality and tooling precision. Through careful product redesign, motions to assemble are restricted to one- or two-degrees of freedom, thus simplifying the system design. Further they have achieved a high level of integration of the product design, function, production technology, and vendor/supplier control.

Part of the product assembly is done by people and the rest by automation. Tasks like attaching and mating fine wires are done by people. If the wire laying is a structured task (like attaching leads to a chip) then the process is automated.

The interesting thing is the sheer number of new systems that have been implemented since 1980 and the variety of system architecture being explored. Each system architecture appears unique to each company. There appears to be little interest to determine which system design might be optimal across a variety of product lines. System types range from the modular, highly parallel operation, Sony FX units to the Seiko HIKS linear system with an uptime goal of 99.5%. Sony and Hitachi are using totally dissimilar architectures to assemble the same basic VCR mechanism. Hitachi uses a distributed system composed of people, robots, and fixed automation whereas Sony uses groups of FX modules and people.

The Japanese also benefit from the fact that their production technology centers can create the necessary system as well as being an integral part of the product-manufacturing system design. In the United States, only a few companies like IBM, G.E., or AT&T are mounting comparable efforts. Many companies are certainly capable, but they have chosen the external supplier method to implement systems. If the market dynamics are slow, then this is the most cost efficient method. Engineering costs for systems developed in production technology centers appear to be twice that of U.S. assembly equipment suppliers. Actually the new Japanese systems are doing things that are more complex than comparable U.S. systems. The lower engineering cost for U.S. sytems is because the incremental engineering done for each system is quite small. There is also much dependence on standard modules and standard tooling.

In a rapidly changing marketplace, there is a question as to the viability of the U.S. method.

### 3.2.2 Present Systems

In the late 1970s, a number of Japanese companies independently reached the conclusion that fixed automation assembly systems would not be adequate in the 1980s. This conclusion was based on two needs (ref. 1):

- The requirement to automate the assembly of products with fine precision parts.
- Satisfying a market place with a wide range of products and with large fluctuations in product volume.

The Hitachi system for assembling tape recorder mechanisms (ref. 2) is an example of older type automation. Whereas the Sony FX-I and II Phoenix-10 Assembly Center (ref. 3 and 4); and the Hitachi Assembly Line for VTR mechanisms (ref. 5) are examples of the new systems.

The arguments for the Japanese move from fixed automation to a more advanced, flexible automation system can be found in ref. 1, 3, and 6. Where reference 6 argues the societal-institutional pressures for increased quality and productivity, reference 3 is an excellent case study of how, why, and what one company did in response to the perceived needs; and reference 1 gives a general summary of a number of activities and systems.

The activities currently taking place can be categorized into three efforts, namely:

Major Trends

- Higher Precision Requirement
- Multikind products with wide ranges in annual volumes
- Many different system architectures
  - Modular - Sony FX (ref. 3)
  - Distributed - Hitachi Assembly Line (ref. 5)
  - Continuous Flow - Seiko HIKS (ref. 7)

Technology

- Minimal system complexity (hardware and software)
- Open-loop control strategy for most assembly

Strategies

- System design driven by market needs
- Highly integrated multidiscipline effort
- Products extensively redesigned
- Systems generally hybrids

The following figures from a Sony FX-I system presentation (ref. 3) illustrate these issues. Figure 3-2 shows the relation between the number of product models and production systems for Sony Tape Recorder mechanisms over

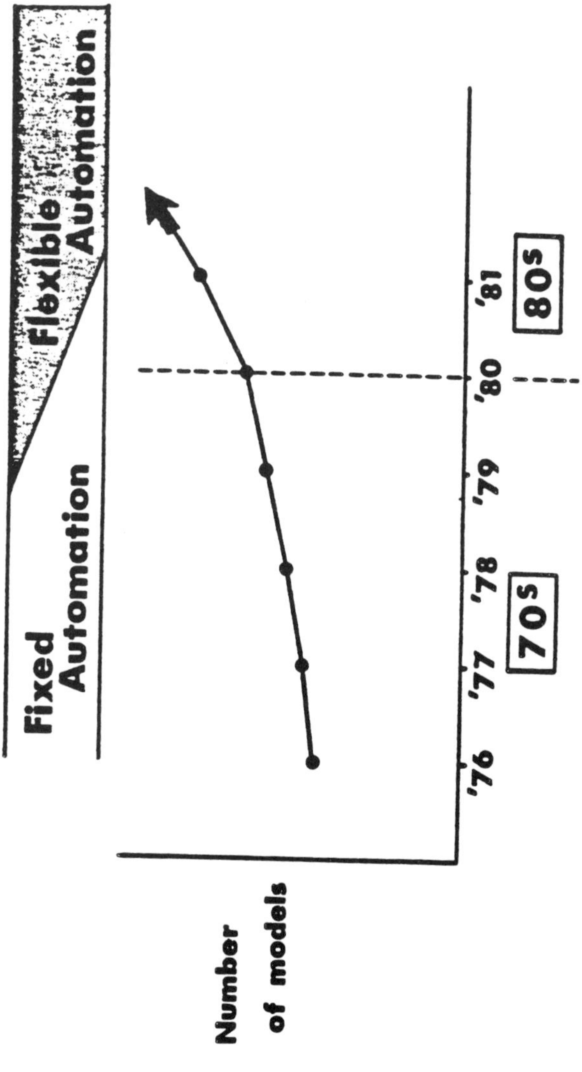


Figure 3-2. Trend in Numbers of Models and Related Production Systems in Sony Tape Recorders.

the time period of 1976-81. Figure 3-3 shows the assembly process elements and the relative time required for each element operation. The requirements of product design and production requirements are shown in Fig. 3-4. The results of this integrated effort was simplicity in product design (Fig. 3-5) and the unique FX-1 modular assembly system (Fig. 3-6). The FX-1 system consists of an x-y table, a frame from which tools are hung, pallets for parts, and a pallet transporter all controlled by an Intel 8080. One system configuration built with FX modules for assembling the Sony Walkman is shown by Figure 3-7.

The development of an automatic assembly line for VTR mechanisms is described in the Hitachi paper (ref. 5). This paper, as well as the paper by Taniguchi (ref. 1), describes the Hitachi Assemblability Evaluation Method (AEM) used to quantify and control product design.* The Taniguchi paper also describes a Versatility Indicator for evaluating the degree of flexibility of assembly systems prepared by Professor Makino who is also the designer of the world famous SCARA robot.

Although Sony and Hitachi products are similar, the companies chose different system architectures. The Hitachi choice is a highly distributed system composed of robots, fixed automation stations, and people. The Sony system is composed of modules and people. Both systems have highly integrated flexible parts supply and product handling systems.

The earlier fixed automation systems had positioning accuracies of the order of 0.1 mm. The newer systems are capable of positioning accuracies of 0.01 mm (ref. 2). Note: A more complete description of their precision mechanism work can be found in Section 3.3.7. Most assembly systems use an open-loop control strategy and depend on this high positioning accuracy and good piece parts to achieve high yields. High-speed closed-loop vision systems have been implemented for automatic wire bonding machines for I.C.'s (ref. 1).

---

* NOTE: This system is currently being marketed in the U.S. by G.E.

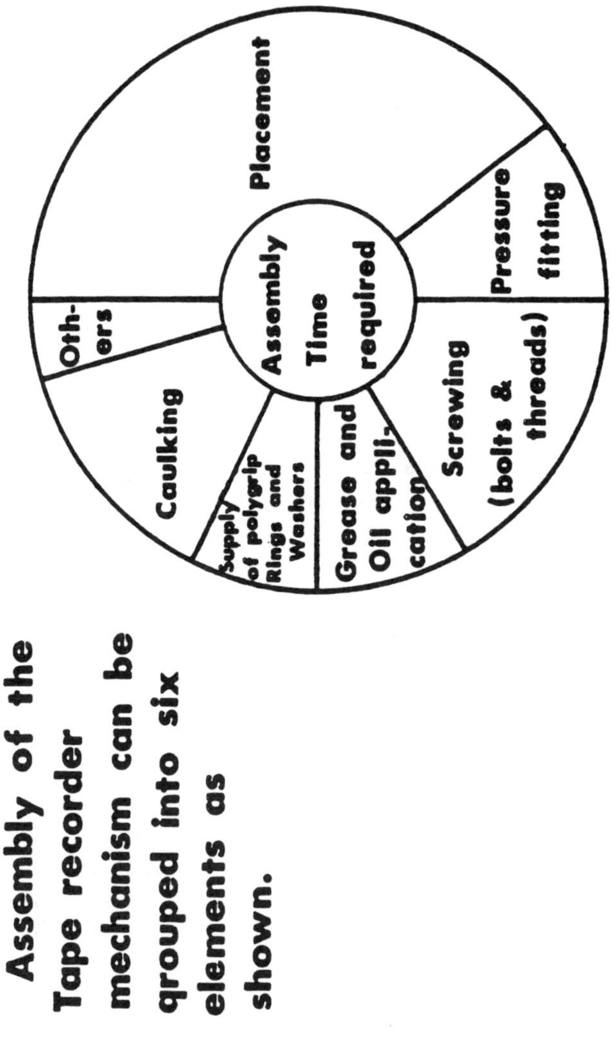


Figure 3-3. Assembly Process Elements.

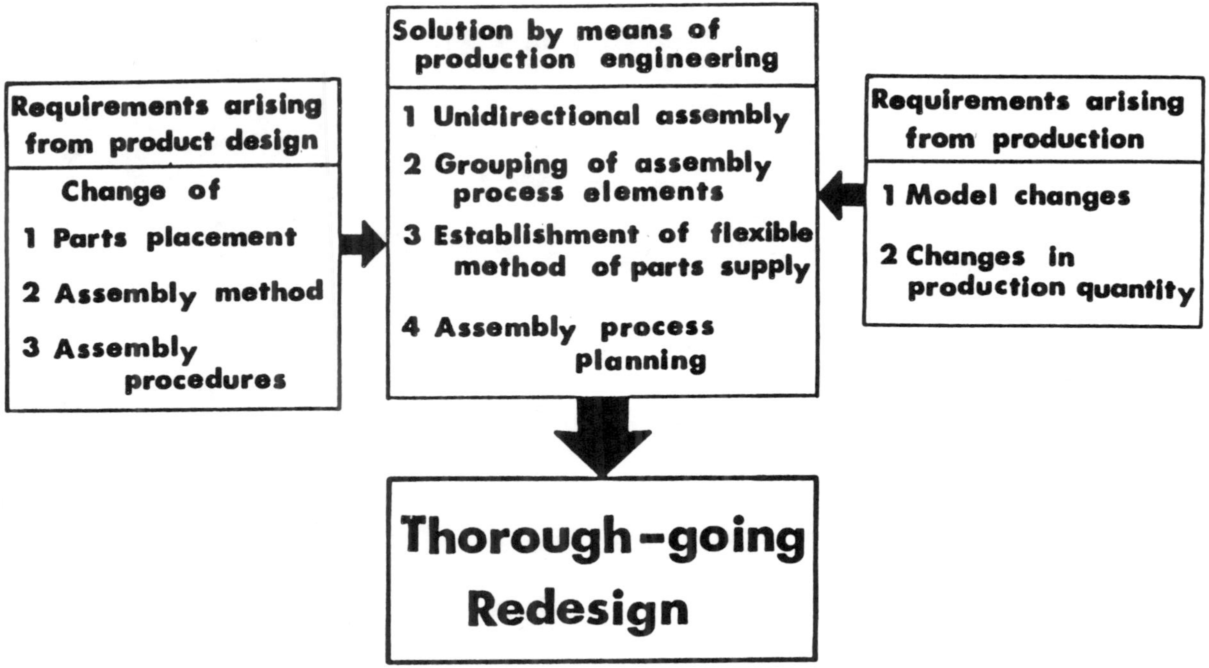
Requirements arising from product design
Change of
1 Parts placement
2 Assembly method
3 Assembly procedures
Solution by means of production engineering
1 Unidirectional assembly
2 Grouping of assembly process elements
3 Establishment of flexible method of parts supply
4 Assembly process planning
Requirements arising from production
1 Model changes
2 Changes in production quantity
Thorough-going Redesign

Figure 3-4. Requirements and Measures for Flexible Automation.

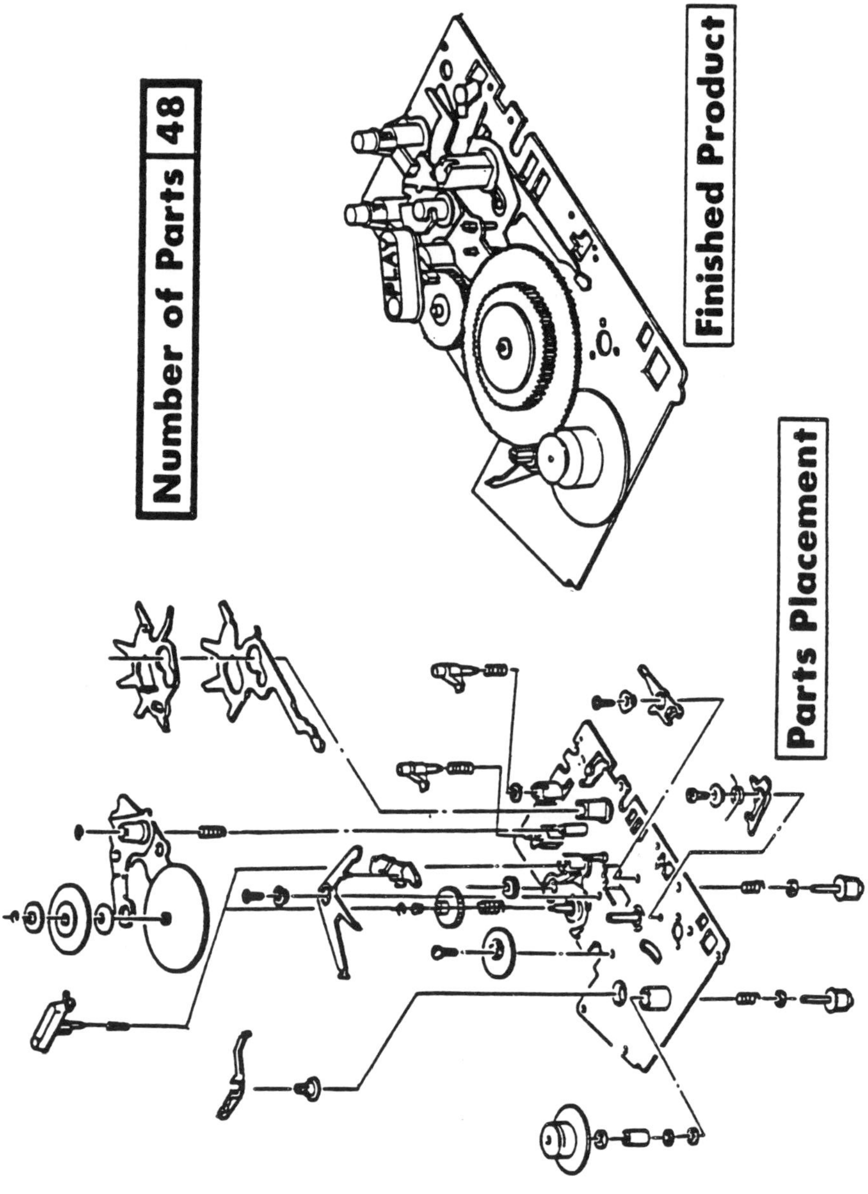


Figure 3-5. Unidirectional Assembly.

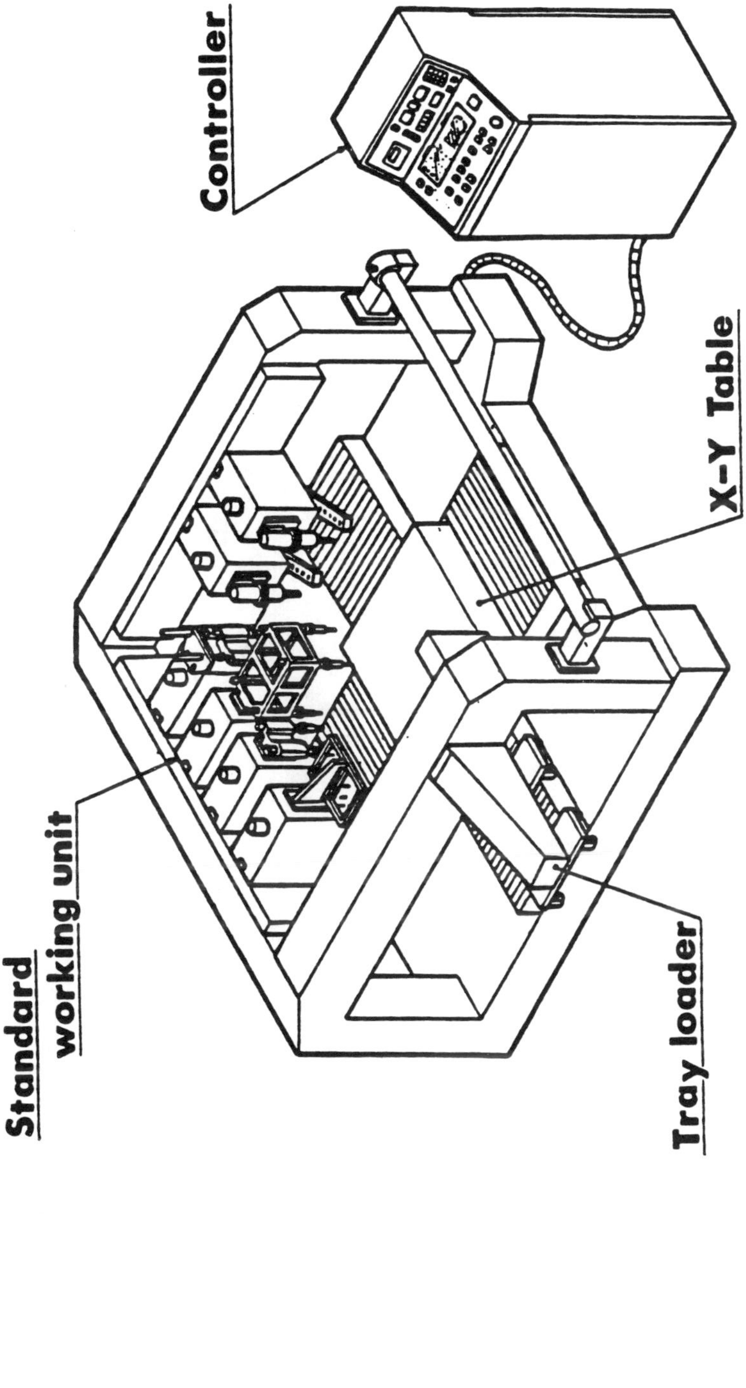


Figure 3-6. Assembly Center.

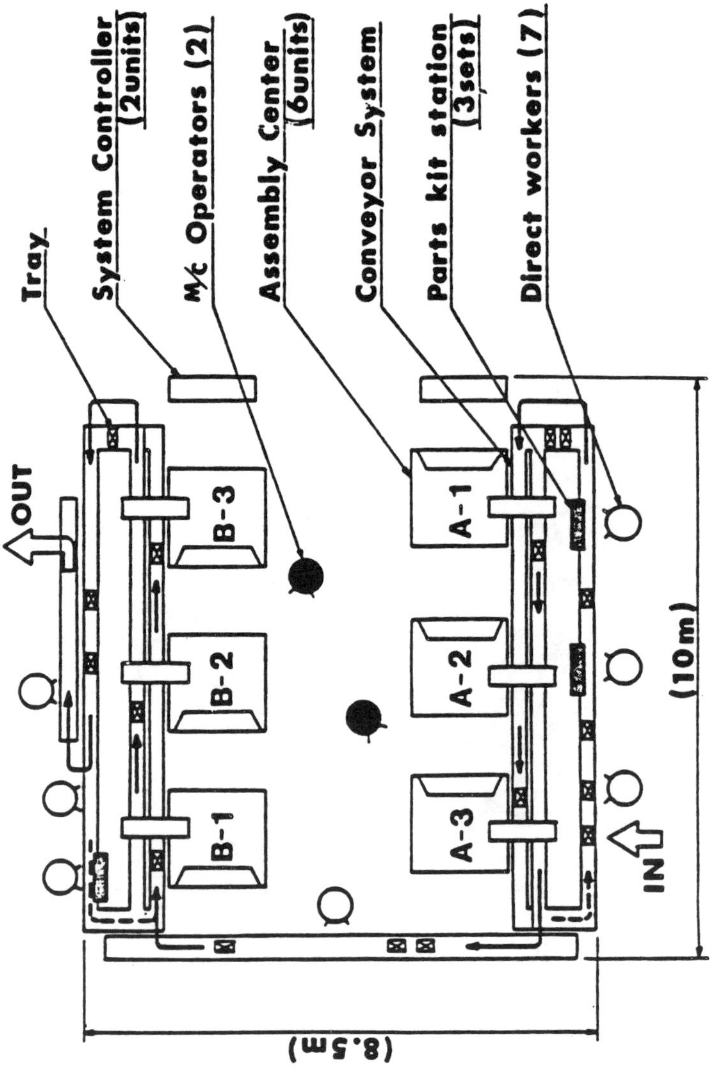

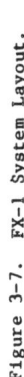

Figure 3-7. FX-1 System Layout.

The extension of these techniques to continuous flow systems is being done by Seiko with their HIKS (Hoop Integrating Assembly System) with cycle times of approximately 1.2 sec. (ref. 7). To achieve near continuous flow they have designed the assembly system to have an uptime of 99% for each 10 process steps. The associated reliability of the part feeding units is 99.9%. A line can be switched from one product to another in 10 minutes. To achieve this they have designed totally new part feeding mechanisms. Small parts are captured in plastic strips that are reeled. Larger parts are captured on 'short-braid' strips with about 24 pieces on a strip.

Each company pursues a different system strategy depending on their product and marketplace. Nippondenso has three basic strategies. For hi-volume lines they tend to use continuous flow lines (cycle times approximately 1.0 sec.) integrated to extensively redesigned products. The classic one is the instrument gauge for automobile dashboards (ref. 1 and 6). With product redesign 4 or 5 components can be assembled into 100 different product types on a line with a cycle time of 1 sec that can be switched in one complete machine cycle by a dummy base unit cycling through the system. The line is capable of 200 model changes per day. Similar work has been done for relays used in automobiles.

For lower volume lines, Nippondenso has developed a family of robots. These robots are used to assemble such things as automobile instrument clusters, air-conditioner modules, etc.

For small batch manufacturing like radiator cores, they use adjustable dedicated systems interconnected with flexible parts handling systems.

One of the unique aspects of these activities is that new production technology is developed. For example, Sony, after 5 years of developing and testing these new systems, offered them for sale late fall 1984. A similar pattern has been followed by Hitachi, Fujitsu-Fanuc, and others. It is part of the reason why so many Japanese robots are licensed in the United States instead of being developed by U.S. industry.

### 3.2.3 Future Systems

Much research is being done on intelligent systems in Japan. With their experience in integrating product design, manufacturing process, and system design, they are in a good position to capitalize on these advanced systems. Further, their proven ability for rapid development and implementation of advanced product technology will make them a strong competitor for advanced production technology and systems.

| | Basic Research | Advanced Development | Product Implementation |
|---|---|---|---|
| Deterministic or Geometrically Driven Processes | 0 | > | > |
| Intelligent Systems | - | - | - |
| System Architectures | 0 | > | > |
| Software Architectures | - | 0 | 0 |
| Smart End Effectors | 0 | 0 | 0 |

Figure 3-8. Summary of Relative Japanese/United States Status in Assembly System.

## 3.3 Sensors

### 3.3.1 Vision Systems - Inspection and Computer Vision

#### 3.3.1.1 Summary

Japanese firms lead in in-house vision applications and have serious applications efforts. They threaten to dominate vision applications although U.S. firms hold their own now in several major areas.

R&D in VLSI devices for vision is especially important. Japanese firms lead in VLSI for vision; they have stronger development efforts leading to low cost, high performance vision devices. VLSI devices will have a major impact on vision applications.

Company research centers, not government labs and universities, are developing applications and VLSI in Japan. U.S. labs lead in areas of fundamental vision research, models of biological vision, optical flow, and stereo. The U.S. has a slight lead in vision systems. In the U.S., research is conducted primarily in university laboratories.

### 3.3.1.2 Introduction

The Japanese began major development effort in vision for applications early on. Hitachi first reported on PC board inspection in 1973. They reported on transistor lead bonding in 1974.

Initial applications used simple binary vision. Initially, Japanese firms developed dedicated hardware for applications of binary image processing. Applications have become increasingly sophisticated. To support sophisticated vision with increasing computation power, Japanese firms have invested heavily in development of special purpose hardware for computer vision. Now they are developing VLSI architectures for gray scale vision.

The major efforts in applications and dedicated vision hardware have been in private companies. Hitachi Production Engineering Research Lab, Hitachi Central Research Lab, NEC, Toshiba, Fujitsu, Mitsubishi Research Lab, and Komatsu have been leaders.

Japanese have had a lead in commercial image sensors as a result of their dominant position in consumer electronics, especially TV. Sony, Hitachi, and Panasonic (Matsushita) have all marketed low-cost solid-state cameras.

Their work in applications was backed up by research in sophisticated vision. In 1971, Hitachi demonstrated an impressive system which duplicated an assembly from drawings.

Research in vision at the Electrotechnical Laboratory (a MITI laboratory) and at Kyoto and Osaka universities is excellent, but the research at these locations appears to be quite separate from industrial research and applications. At Electrotechnical Laboratory, there seems to be some internal questioning of their proper direction in order to justify a place between universities on the one hand, and industry on the other.

Considerable impetus was given to vision development in Japanese companies by the PIPS project. PIPS provided broad industry participation, which aided in bringing about image technology transfer into companies.

#### 3.3.1.3 Applications

A relevant contrast is that in vision applications: it is Japanese electronics giants with large in-house efforts compared with American startup companies. The applications targets have similarities in mask and reticle inspection in semiconductor manufacturing, IC lead bonding, pc board inspection, and automotive inspection.

Another contrast is between commercial firms with goals to develop products for high-volume, low-cost applications vs military, low-volume, high-cost applications. Commercial and government support of computer vision R&D in Japan has had commercial applications goals. Much computer vision research in the U.S. has had military support for future applications in mapping, photointerpretation and surveillance, and weapons. Funding for civilian research in the U.S. has been inadequate.

Japanese participants in vision applications are the electronics giants, Hitachi, Mitsubishi, NEC, NTT, Toshiba, and Fujitsu. In the United States, giant companies have played some role, but a lesser one. GM developed a handful of applications and has a credible research laboratory but has had trouble transitioning systems from research to production. GE has had a substantial research effort, a few applications, and some commercial efforts. IBM has had little research in vision and presumably had traditional applications in lead bonding and mask inspection. TI had some applications in computer vision. Westinghouse, Honeywell, CDC, and Hughes have participated in research and development of military projects like the Auto-Q system for film imagery.

One of the major applications for vision has been in lead bonding. Hitachi developed a system for automated lead bonding of transistor chips in 1974 and automated lead bonding for integrated circuits a few years after. GM, TI and IBM developed systems at about the same time. A small vendor, View Engineering developed a visual control for lead bonding which has been successful.

A second area of applications is inspection of masks and reticles for IC manufacturing. Masks are crucially important in IC production because faults in masks are replicated in the product. Binary vision is adequate because masks are inspected in transmission. Inspecting masks is low volume as opposed to inspecting products, hence computation requirements are not extreme. In the United States, KLA is dominant. In Japan, NJS and Seiko are competitors, while major electronics companies have in-house programs, i.e., Hitachi and NEC. Automated mask inspection is used in only about 50% of production, even though this is a standard and cost-effective process.

PC board inspection, both bare and populated boards, has received considerable attention in Japan and the U.S. In Japan Hitachi, Fujitsu and Toshiba, have conducted extensive development efforts in this area. The first system reported was in 1973 at Hitachi, based on a design rule method. It was said to have been used in production. By 1981, it was not used in production. Hitachi had developed another system which was used in production. It is difficult to get information about whether systems are used in production, how many, and what their performance is. Hitachi reports that they use inspection systems in production, although they are closed about saying anything further. In the United States, about five vendor companies have developed systems in the last year. This is an area of rapid development, especially in inspection systems for loaded boards. Because there are many forms of defects on pc boards, no system appears capable of doing the total inspection job for bare boards. Because inspection of pc boards is inspection of product, computation requirements are much higher than for mask inspection. Because defects on automatically loaded boards are infrequent, reliability requirements for inspection are high.

Inspection of solder joints has attracted some attention. An infrared method was developed by Vanzetti. Hitachi PERL has developed a production system for solder joint inspection. They sent samples to Vanzetti to test their IR method and found it inadequate. Based on a visit to PERL and a brief discussion of the system, it appears capable only of crude analysis of solder joints by a plane of light 3d method. It does not seem adequate to determine cold solder joints. A member of PERL stated that cold solder joints are not a problem in wave soldering if the system is operating right, that if there are any cold solder joints there are many. My conclusion is that the solder joint inspection system does only part of the job. The defect rate for solder joints is low and it is not clear how many defects can be distinguished visually. Solder joint inspection is difficult.

Wafer inspection systems have been developed at Hitachi Production Engineering Research Lab (PERL) and other firms. This is a simple visual problem.

Inspection of diode pellets was demonstrated long ago at Hitachi PERL. Inspection of pellets for drugs has been a real application. Mitsubishi reported commercial classification of fish and fruit using vision. There has been some effort on inspection of hybrid circuits and in-process inspection of products which has proven to be very difficult and not yet within reach.

Industrial OCR is another substantial inspection application. Cognex and GE specialize in reading markings on products, reading characters in instruments, LEDs, and LCDs. There are in-house efforts at major firms. Some applications now under investigation are handling packages and inspecting currency. Japanese firms are concerned with reading addresses, a special problem because of the Kanji characters. A variety of companies have worked on visual input of drawings, with some success. Different domains have different drawing conventions and require development of separate systems. Design automation for electronics has been a special project at Fujitsu. This work has been combined with expert systems for design automation.

Inspecting arc welds is a large market being contested for by many Japanese firms. these firms use mechanical tracing of parts, eddy current techniques, and vision. Komatsu uses sophisticated through-the-arc vision and looking at the arc puddle. Arc welding is one of the major vision applications.

Komatsu, a major manufacturer of automated vehicle systems for warehousing, is concerned with visually-guided vehicles which could adapt to changing warehouses configurations and navigating around obstacles. Their current vehicles follow wires in the floor.

Mitsubishi, considers visual control important for assembly, particularly in electronics. Vision is regarded as even more important for assembly of mechanical parts.

Project Jupiter is a Japanese effort to develop robots in dangerous work, and it includes research in vision.

Image sensors have been developed by Sony, Hitachi, and Panasonic (Matsushita). These firms have major consumer products efforts and have developed solid state cameras for the VCR market. They have reportedly introduced them at artificially low prices based on a projected increase in volume. Prices are about $1000 for 380x480 arrays. US firms include Fairchild and RCA which market products and TI which has built special systems. They have had some military sales but do not seem cost-competitive, although their camera specs are similar.

Much of the competition in the U.S. is for the general vision module consisting of frame grabber, frame buffer, ALU, display, and general microcomputer.

3.3.1.4 Research Computation

High performance computation is recognized as a major requirement for making inspection and computer vision fast and inexpensive. Even modest vision systems are computationally expensive, two orders of magnitude beyond the capabilities of high performance micros like the MC68000. Japanese firms are aggressively developing high performance parallel vision computers, for a range of devices. Examples are:

a. Current Products: Hitachi and Fujitsu incorporate special logic for inspection systems based on simple binary vision. Several inspection systems in the U.S. incorporate similar special logic. Toshiba offers high-speed image processing systems for research in the TOSPICS and TOSPICS II systems. High-end image processing systems in the U.S. have processing power roughly equivalent to TOSPICS II. Systems such as IIS, Comtal, Vicom, Westinghouse AUTO-Q, and IRI IP-256 are representative.

b. New Products: Hitachi has announced the availability of two new chips. It has made one of these chips available to leading Japanese research laboratories. The chip contains four processors each with 8x8 multiple. A result is output each 167 nsec, giving 186 MOPS per second. A 4x4 convolution over a 256x256 image requires 11 msec. NEC announced a vision dataflow machine designed for parallel operations. There are no equivalent products available in the U.S. They are significant for potentially low cost. Hughes RADIUS chip may be comparable, but Hughes is not a commercial IC manufacturer. CMU is developing the WARP processor, a systolic parallel processor with 10 elements, each at 10 MFLOPS. It is relatively expensive, however.

c. Near-Future Devices. NTT has made an 8x8 subarray of bit-serial processors on a chip containing 110,000 transistors. Martin-Marietta has announced the GAPP chip which is a 6x12 array with 128 bits per processor. NCR is marketing the GAPP chip for nonmilitary markets. Goodyear has an effort comparable to that which leads to the MPP, a one-of-a-kind processor in the multimillion dollar range.

Hitachi, NEC, and other firms have above critical-mass efforts in architectures for computer vision, especially chips which appear suitable for low cost products. It seems clear that they will push hard on development of systems based on these chips and develop new concepts for VLSI. The Japanese have a moderate edge in the first two levels of systems, especially in having in-house systems in operation and with VLSI systems in production and available as products. Because of their involvement in VLSI architectures, it is likely that their efforts will have more widespread impact than those of U.S. firms.

3.3.1.5 Systems

The leading research systems in Japan have been developed by Nagao and Matsuyama and that of Ohta, all from the University of Kyoto. They share much with current work in expert systems. These systems are different from the ACRONYM system of Stanford. Here there may be a slight edge in U.S. research.

3.3.1.6 Image Understanding

Japanese research is strongest at Electrotechnical Laboratory, the University of Kyoto, and the University of Osaka. These programs have been strong since the late 1960s. Their research programs are somewhat parallel to those at U.S. research centers. The Electrotechnical Laboratory has considerable exchange with MIT and investigates similar problems such as photometric stereo and extended gaussian images. The University of Kyoto has investigated vision systems, and the University of Osaka has investigated motion and is now involved with mobile robots.

One area followed in the U.S. is analysis of biological perception based on an AI perspective. This is particularly strong at MIT. Many of the U.S. leaders in perception have an interest in this area.

The U.S. has a much stronger program in optical flow and motion. The U.S. also leads in spatial organization and image interpretation.

### 3.3.2 Nonvision Sensor Systems

The Japanese sensor industry is active with over 300 firms supplying specialized sensors of nearly every type. Apart from this activity most of the larger robot manufacturers have their own sensor development groups producing devices specifically tailored to robotic application. The Japanese practice is to divide such sensors into internal and external devices. The former include all angular and linear encoders that relate to the robot configuration, as well as internal temperature sensors to correct for arm thermal expansion effects. All sensors that allow the robot to interact with its task are regarded as external sensors.

Only a small subset of sensor topics, primarily those that relate directly to robotics, will be considered in the following discussion.

#### 3.3.2.1 Position and Angle

As is the case with the majority of the robots made today, most Japanese units have incremental optical encoders associated with each drive motor. The motors drive the various degrees of freedom through suitable reduction gearing, often of the harmonic type. Knowledge of the motor revolution, the gearing details, and the arm geometry allows calculation of the end effector position.

A variant of this scheme has been the recent development of small high-precision cartesian assembly robots by Panasonic and others. These also employ incremental optical encoders, but they are used in conjunction with high-precision, preloaded helical ball screws to provide accurate cartesian motion at amazingly low cost, e.g., ~\$10K for an actuator with 25μ accuracy.

This represents a price/performance breakthrough. Other well known and widely used position measurement technologies include resistive strip potentiometric systems (both linear and angular), synchro resolvers, absolute optical encoders, and LVDTS (Linear Variable Differential Transformers) of various designs. Some of the highest precision machines, such as VLSI lithographic equipment, employ laser-based interferometers. All of these systems are comparable to U.S. practice.

There are also areas in which Japanese practice is unique and apparently has no direct U.S. equivalent. One is the SONY Magnescale series of high resolution digital magnetic position measuring devices. Another is the use of photolithographically patterned InSb Hall effect elements as contactless potentiometers. Others are the gas jet "Gyrocator" of Honda and the "beam rate sensor." Both are intended for low cost inertial guidance systems.

Position sensing techniques, as distinct from measuring techniques, employ well known optical, inductive, capacitive and Hall effect methods. These are widely used for limit sensing and safety applications. Another widespread application of magnetic sensors is for determing the position of fiducials in rotating elements such as those in VCR tape transports and automobile ignition systems. Japanese and U.S. practice appears similar in these areas.

### 3.3.2.2 Ultrasonic Sensors

Significant work using ultrasonic sensors has been reported by MEL at Tsukuba in connection with navigation aids for mobile robots. This has also been extended to their MELDOG work on a mobile prosthetic guiding robot for the blind. This work is similar to that of the Polaroid Corporation in sensors and CMU in robot guidance in the U.S. (Similar work has also been reported with regard to the French HILARE robot and the "Sonic Cane" for the blind in England). A significant early goal for a simple, truly autonomous robot may well be a "robot sentry" for building security.

The biggest area for ultrasonics at the present time is in medical imaging. A conceptually similar area is that of ultrasonic imaging of defects in nuclear reactor pressure vessels and monitoring weld integrity. Mobile robots provide a convenient way to make such measurements on large structures.

It may turn out that ultrasonics will play a significant role in local ranging and imaging in assembly and other robotic applications. Mitsubishi, among others, has significant activity in the ultrasonic sensor and imaging areas.

### 3.3.2.3 Force,Torque,and Pressure Sensors

Force and torque sensing, usually implemented by means of silicon strain guages at the robot wrist, have been demonstrated to be useful information gathering techniques. Most of such work has been reported in research endeavors although some U.S. manufacturers, notably IBM, have already incorporated such a capability into production robots. This technique is beginning to be used commercially in Japan. For example, both Kobe Steel and Kawasaki have recently announced deburring robots where force feedback information is employed. Fujitsu has developed a precise active-compliance device for precision assembly.

Modest and readily implementable extensions of such schemes can lead to useful capabilities. One example is the simultaneous measurement of both the torque and the normal force exerted by a drilling robot which provides information on the sharpness of the drill bit. It can be expected that similar techniques will be undertaken by Japanese in the near future.

A related area is the problem of pressure sensing. Here a highly developed technique is that of using diffused silicon strain gauges integrally fabricated on a thin silicon pressure-deformable diaphragm. Techniques of this kind are well advanced in Japan and fit naturally into programs for "smart sensors" in which both the sensing and information processing is implemented on the same semiconductor chip. Similar work is carried out by U.S. corporations such as Honeywell.

### 3.3.2.4 Tactile Imagers

Much has already been written regarding the potential of touch sensors (or better "tactile imagers") in robot assembly. Opinion is divided as to how such sensors should best be made. European and most U.S. work has so far favored various resistive sheet schemes in which the local electrical resistance of a suitable sheet of material is modified by the applied pressure pattern. Some believe that a multiplexed capacitor approach is more favorable. Significant work of this latter type has been reported in both the U.S. and Germany, although the basic idea was apparently first disclosed in a 1979 Japanese patent application.

A number of other schemes have been suggested, though very few have actually been demonstrated. These include piezoelectric effect arrays (which have the disadvantage of lacking a DC response) and arrays of small magnets flexibly mounted above corresponding arrays of multiple permalloy magnetoresistive readout elements. The latter scheme holds the potential of providing both shear and torque information, along with the normal force, but is complex and suffers from the fact that the magnetoresistive effect is very small (only 2% total resistance change for a 90 rotation of the applied magnetic field). Various optical fiber schemes have also had an airing both in the United States and Japan.

A topic that resurfaces regularly is that of microscopic tactile imagers. Many tactile imaging schemes can in principle be married to some form of silicon VLSI to form a tiny tactile imaging device. Several attempts have already been made, (it is neither easy or inexpensive), but none has so far succeeded. The problem is in part due to economic scale. Silicon microcircuit elements are only inexpensive if fabricated in very large numbers. The need for significant numbers of robots with tiny tactile imagers is not yet apparent. Until such a need is demonstrated, it is doubtful that there will be much Japanese work in this area.

The only tactile imager commercially available in the U.S. is manufacturered by the Lord Corporation. This imager uses an array of small, spring-loaded mechanical vanes which interrupt the light beam between an LED and a photodiode at each location of interest. This scheme has the advantage of completely separating the mechanical and electrical features of the system, but it does so at the expense of a complicated sensing arrangement with a large number of separate components.

Most of the preceding approaches have been proposed or developed in the United States. The Japanese have concentrated on simple single degree of freedom torque or force information sensors, and they have actually implemented them in production lines. Honda, for example, uses robots with simple one dimensional tactile sensors in automobile windshield insertion.

The Japanese will probably hold off from significant high resolution tactile imager work until there is a demonstration of the utility of such devices.

#### 3.3.2.5 Speech Sensors

Machine recognition of speech is an important area that is being actively pursued around the world and particularly in Japan. The development of practical speech recogniton devices would have a significant impact on robotics and mechatronics. However, speech recognition can be viewed as an add-on technology and will not be discussed here. As NTT and other Japanese leaders develop this technology, it will become increasingly important in mechatronic applications.

#### 3.3.2.6 Sensors for Navigation

Autonomously guided vehicles or "smart carts" are becoming important for material transport in factory and warehouse environments.Sometime in the future they may also become widely used in semiconductor VLSI clean room facilities. At present, all of the commercially available carts are simple cable (or alternatively paint-stripe) following devices. They have no autonomous navigational capability. A significant area of endeavor is to make these carts free ranging and that requires a navigational sensing capability.

Among a variety of possibilities, one of the most attractive is the use of inertial guidance techniques in conjunction with odometry (integrating off the wheel rotations). The key element here is a low-cost, low-drift gyroscope. Superb ring laser gyros have been developed in the U.S. for military programs. Unfortunately, these gyros are far too expensive for factory applications.

Honda has produced a low cost gas jet inertial guidance system for automoblie navigation called a "Gyrocator." This type of technology is tailor-made for factory smart cart applications. A number of Japanese organizations are now working on the factory cart navigation problem using both laser beam guidance and vibrating beam PZT inertial guidance systems among other approaches.

### 3.3.2.7 Other Sensors (The JUPITER project)

Virtually any type of sensor can be used in conjunction with a robotic system. The MITI-organized program on robots for hazardous environments is a good example. Here one can expect that robots for radiation environments will employ scintillation counters, ion chambers, p-n junction radiation detectors and so forth, all adapted to the specific robotic use. Fire fighting robots will need smoke detectors, thermometers, optical pyrometers, and chemical vapor sensors. Robots for hazardous environments will also make extensive use of remote video and other links to human guidance.

The last point raises the question of autonomy versus teleoperation for such mobile robots. While teleoperation is clearly advantageous, there are a number of situations where autonomous operation is mandatory. The work by the Jet Propulsion Laboratory on the Pioneer spacecraft is an example. Other areas where autonomy may also be crucial could involve situations such as traversing a hazardous path where an umbilical cord was impossible while metallic shielding precluded radio communications.

We can expect rapid Japanese progress in this area. Successful Japanese developments of this type will have an impact far beyond the obvious technical or financial contributions. Consider, for example, the national pride and international impact when the first human beings are rescued by a Japanese robot.

### 3.3.2.8 Sensor Information Processing

The question of how best to handle and use the information from robot sensors is a central issue. Languages such as VAL II and IBM's AML have sensor input capability. Advanced Japanese software systems, such as that provided for the Mitsubishi MELFA series of robots, are also being considered.

The whole area of intelligent industrial robots centers on closed-loop feedback control of the mechanical actuators. How the robot observes its world and how it makes decisions based on these observations are basic questions. This issue is sharply defined in the area of robot control through machine vision, and makes contact with many areas of artificial intelligence.

Basic U.S. research in these areas is substantially ahead of the Japanese and likely to remain so for some time, see Figure 3-9.

## 3.4 Intelligent Modules/Autonomous Machines

"Intelligence" as applied to mechanisms, means different things to different people. A simple interpretation of "intelligence" refers to the replacement of a fixed controller by a program that incorporates logical branching. As microprocessors become smaller, cheaper, and use less power, and as they increase in speed and complexity to the level of Motorola's 68000 range of 32 bit 10 MHz machines, so more and more functionality is assumed by a controlling computer. Disk drives and other computer peripheral equipment are now dubbed "intelligent" according to this interpretation. The incorporation of powerful controlling microprocessors into Mechatronics has been a logical step. Progress is rapid both in the U.S. and in Japan with no discernible leader.

| | Basic Research | Advanced Development | Product Implementation |
|---|---|---|---|
| Position/Angle Encoders | 0→ | 0→ | 0→ |
| Force/Torque/ Pressure Sensors | 0→ | 0→ | +↗ |
| Tactile Imagers | -→ | -→ | -→ |
| Proximity/Range Detectors | 0→ | 0→ | 0→ |
| Smoke/Chemical/ Radiation | 0→ | 0→ | 0→ |
| Navigation Sensors (low cost) | 0→ | +↗ | +↗ |
| Sensor Information Processing | <→ | -→ | 0↗ |

Figure 3-9. Summary of Relative Japanese/United States Positions in Sensors

Microprocessor control allows better modeling of external events and "disturbances," built-in test and diagnostic capabilities, information pooling (as in the popular factory-of-the-future (FOF) proposal), and sensory processing. Microprocessor control represents a step along the continuum that stretches from adaptive control, through micro-processor control, to artificial intelligence.

Mechatronics spans the union of mechanical and electrical engineering and emphasizes a multidisciplinary integrated approach to product and manufacturing system design." There is considerable activity in Japan aimed at the development of increasingly intelligent machines, but only a fraction of that activity is specifically aimed at Mechatronics. Best known among the other projects are:

- o The Fifth Generation project that hopes to propel Japan into world leadership in computing.
- o The Jupiter project that aims at flexible, sensor based, robot control with emphasis on hazardous environments.
- o The Language Translation project that aims at useable machine translation between Japanese and English with automatic relational database entry.

The Fifth Generation project marks a major commitment to software by the Japanese, who have traditionally been regarded as strong in hardware development but weak in software.

Significantly, the Fifth Generation project represents a major leap forward in software practice in Japan. Traditionally, software has been written in assembler or in Fortran. The languages chosen for the Fifth Generation project are LISP and PROLOG. This commitment to advanced software systems represents two important commitments by the Japanese:

- o AI is viable. The Japanese have reached this consensus and have begun development.

- AI is the basis of future software systems. An early example of this trend can be seen in "intelligent" aids for VLSI design.

The Fifth Generation project can already take credit for the following:

- It has been a learning exercise for industry. There is now a much keener awareness of what AI can achieve.

- It has caused large numbers of AI programmers from industry to be trained. U.S. companies have tried to promote AI efforts by hiring programmers straight from college. These programmers have little understanding of the disciplines and requirements of industrial programming. U.S. companies have also hired programmers from other companies without adding to the total pool of talent. The Pattern Information Processing System (PIPS) project of the 1970s was considered to be a scientific failure in the United States; but it achieved the goal of placing thousands of well trained image processing personnel in industry. Many were trained in U.S. universities.

The applications of "intelligent mechanisms" in Mechatronics include:

- Hazardous environments. Through the Jupiter project and other efforts, the Japanese are pulling away from the United States.

- Warehousing. There is no strong effort in this area in Japan. The U.S. and Japan appear approximately equal.

- Intelligent design aids. The Japanese have started from a position behind the U.S. However, the growing concentration on advanced software and the strong base of expertise provided by the PIPS project is enabling the

Japanese to start to pull away. Examples include: VLSI design systems, systems that can produce finished machine drawings of circuits and other linear data from hand-drawn sketches, and interactive CAD systems for building and plant layout.

- Natural language interfaces. The Japanese are starting from behind the U.S. There is no equivalent of ATN-based products such as LIFER. However, through national direction that combines the power of universities such as Nagao's group in Kyoto with companies such as Toshiba, the Japanese appear to be pulling away. They are concentrating initially on translating abstracts of technical papers.

- Robot programming. The Japanese have only recently produced programming languages such as IBM's AML, Automatix's RAIL, and Unimation's VAL n. They have never developed an analogue of Stanford University's AL. Recently, AL has been adopted as a standard in Japan to run on VAX-like computers. A key to higher level languages building on AL is spatial reasoning and compliance. The Japanese have organized several workshops on these issues. Nevertheless, it is safe to rate Japan as being behind and staying behind.

- Speech input. The Japanese have assumed a position of leadership in speech input. They seem to be pulling away. Chips and recognition systems can now be purchased at modest prices from several Japanese corporations.

- Dynamic scheduling. The situation in Japan appears to be on a par with that in the U.S. and staying that way.

Figure 3-10 summarizes these assessments.

| APPLICATION | STATUS/CHANGE |
|---|---|
| Hazardous Environment | 0↗ |
| Warehousing | 0→ |
| Intelligent Design Aids | -↗ |
| Natural Language Interface | <↗ |
| Robot Programming | -→ |
| Speech Input | +↗ |
| Dynamic Scheduling | 0→ |

Figure 3-10. Applications of Intelligent Machines in Mechatronics in Japan Relative to the United States.

There are two major areas of product development:

- Speech input systems. The Japanese are starting from a position ahead of the U.S. and are pulling away.

- LISP machines. Given the relative disadvantage of the Japanese in higher level software such as LISP, there has been less expertise in building efficient machines. There has been almost no expertise in building special purpose hardware to support symbolic processing. Thus the Japanese start from a position behind the U.S. However, Fujitsu and others are now marketing LISP machines that perform competitively against most U.S. products such as Xerox

> Dolphins. The University of Kyoto has completed a COMMON LISP compiler that is now available on SUN microsystems in the United States. The Fifth Generation project work on PROLOG machines will add to the skill base for developing special purpose architectures for pointer chasing, suggesting that the Japanese will narrow the gap in LISP machine development.

Much of the work that falls within this section of the report is also covered in other sections. There is considerable overlap with the vision and software sections. Sensory processing is difficult, computationally demanding, intimately involved in intelligent behavior, and requires the deployment of domain-specific models to be practicable. Hence much of the development of sensing is involved with intelligent machines.

There is enormous activity in vision ranging from chip through innovative algorithm development for three-dimensional vision. Overall, the Japanese are slightly ahead in product development, but they lag in research. Device development includes fast convolution hardware, such as Hitachi's Sobel chip, and entire image processing stations such as Toshiba's TOSPIX 2. These devices are used to reduce to real time well-understood processes that previously took minutes in microcode. Experimental systems based on these devices include motion (U. Osaka), stereo (U. Osaka, Toshiba), and photometric stereo (ETL). ETL have emphasized research on matching computed three-dimensional representation (surface patch models and Extended Gaussian Images) to stored object models ("CAD models"). In a similar vein, domain specific knowledge has been used by Hitachi, Fujitsu, and U. Kyoto to guide segmentation of multispectral (typically landsat) images.

Parallel to the signal processing aspects of the research is a substantial government sponsored project aimed at natural language understanding. One application is the translation of abstracts of engineering reports from English to Japanese and Japanese to English. The key component is a language-independent representation of the meaning of the abstract. The representation supports a semantic information retrieval system whose aim is to keep engineers fully informed of U.S. and European developments. Similar systems exist in research prototypes in the U.S., but there has been no coordinated government sponsored and directed effort since the abortive machine translation project of the 1950s. This author had one of his own articles translated by machine at University of Kyoto. The translation to Japanese and back to English compared well with the original.

There is also work on more complex mechanisms such as multifingered hands, walking machines, and wheeled vehicles. Overall, the work is about on a par with similar efforts in the U.S. There is work on control in Japan, and the MIT-UTAH hand compares favorably with hand designs in Japan. Recent work in Japan has demonstrated coordinated hand control in a world frame, essentially equivalent to the system developed by Salisbury at JPL, Stanford, and MIT. Multijointed arms have been developed at Toshiba, for inspecting nuclear piles. The Toshiba System is snake-like with 17 degrees of freedom and has a camera mounted at the end. Processing is carried out centrally, not at the remote hand. Other work on multijointed arms has led to the development of soft grippers at Tokyo Institute of Technology. Current research explores shape memory alloys as actuation sources for tiny, intricate devices in which joule heat is a sufficient power source. A succession of wheeled and wire-guided devices have been built throughout Japan. They are on a par with U.S. efforts. For example, MEL has developed a series of "guide dog" robots to steer a blind person along uncluttered sidewalks using sonar.

One consequence of the current fascination with the expert system program methodology is its application to mechatronics problems such as process planning. Japan appears to lag behind the U.S. and Europe in these areas. It can be predicted that attention will turn to intelligent reasoning systems for mechatronics as appropriate computer systems become widely available. More likely is the use of LISP machines. LISP machines are already available from Japanese manufacturers such as Fujitsu, and they are a cornerstone of the natural language work. Japanese companies are likely to be attracted to the newly adopted industry standard Common Lisp, which replaces several incompatible dialects. Finally, Japanese research on the Fifth Generation projects already suggest, the weaknesses of PROLOG have been rediscovered.

## 3.5 Software for Mechatronics

Japanese efforts in Mechatronics have been concentrated almost entirely on hardware. Software received low priority.

For example, the highly publicized project "Flexible Manufacturing System Complex Provided with Laser" made virtually no effort to coordinate or manage the development of software. In fact, it has been reported that the Ministry of Finance was upset over the amount of money spent on software development in previous projects; and, as a result, this FMS project was not even allowed to include software in its budget. What happened was that each company on the project was responsible for developing its own software and simply "padded" its hardware budget to cover some software costs. Thus, the FMS with Laser project developed as little software as possible and that was mostly assembly language code. The system had no geometric models for producing NC tapes. NC programs were written manually and at a very low level. Little programming was done in APT. The strategy was to build flexible hardware first and worry about the software later. This is perhaps an extreme case, but, for it to happen at all on a project as large and prestigous as FMS With Laser, indicates the attitude toward software.

There are many software development systems currently in use for automation systems throughout Japanese industry. But, for the most part, these are not considered advanced systems by American standards. For example, Toyoda Machine Works, Japan's second largest machine tool company, uses geometric modeling for NC program development for turned parts only. They plan to extend this capability to milled parts in the future. Currently they use an automated scheduling system similar to the IBM CAPOSS system.

A more advanced system is the FMS system at the Yamazaki Machinery Work which uses the MAZATROL tool software system for turned parts. The software development environment here allows an operator to input the geometry of the part, put up a graphics display of the part, and prompts him with various menus for revelent information. The display then simulates the cutting sequence, shows the cutter pathways, and outputs the program to a milling tool to make the part. There is also a new MAZATROL system for milling which was in use earlier than the comparable General Electric GE2000 system. Yamazaki also uses the CADCAM system on an IBM 4341 computer.

The most impressive Japanese software development project that has come to our attention is the Software WorkBench (SWB) factory of Toshiba Fuchu (Ref. 8). The SWB factory consists of about 2000 programmers sitting at individual workstations in a large one-room building. It specializes in process-control software for systems used in steel mills, nuclear power plants, and flight guidance. The development of software in this factory environment, using rigid software engineering principals for a narrow problem domain, has produced some remarkable results. The factory delivers 4,000,000 lines of assembly language equivalent code per month, which works out to about 2000 lines per programmer per month. This is about four times the output for similar code in the United States. In addition, the SWB factory achieve a reuse rate of about 65%, which means that for every 1000 lines of code delivered, 650 lines are lifted from previously written programs. This is unheard of in America. Finally, the code is of extremely high quality. Error rates of 0.3 bugs per 1000 lines of code are reported. Typical U.S. error rates are ten times greater. The Toshiba code is so good that it comes with a ten-year warranty. Any bug discovered within ten years will be fixed at no charge to the customer.

### 3.5.1 Universities

Some of the most advanced work is being done in the universities. For example, the GEOMAP CAD modeling system developed at Tokyo University has the ability to automatically generate NC data from models for hole cutting, face milling, and contour milling. It also generates robot programs in VAL. GEOMAP contains about 200,000 lines of FORTRAN code. Its geometric modeling part is similar to the PADL2 system developed at the University of Rochester. GEOMAP has been supported by about 30 companies, but, so far, it remains a university research project. There is no indication that GEOMAP has been used in industrial production.

There are 15 public universities with computer science departments in Japan each of which have approximately 50 students. The private universities admit several hundred computer science students per year. This is a small number by American standards.

### 3.5.2 Fifth Generation Project

The Japanese efforts in software development may be considerably advanced by the recently initiated Intelligent Robotics project as well as by the highly touted Fifth Generation Computer project. On the Fifth Generation project, however, it appears that software is not the primary emphasis. The effort so far has been in developing the hardware for an artificial intelligence workstation that can perform logical inferences at 100 to 1000 times as fast as presently possible. Much emphasis is being placed on a natural language interface.

PROLOG has been chosen as the language for this project. Many American computer scientists feel that a more logical choice would have been LISP. It has been suggested that one of the primary reasons for using PROLOG was that it would distinguish the Fifth Generation Project from American artificial intelligence research which is virtually all done in LISP. It is generally thought that this was an unfortunate choice because of the relative scarcity of software written in PROLOG, the fact that few computer scientists are trained in PROLOG, and the relative scarcity of documentation and textbooks on PROLOG.

In any case, there probably will be little or no impact of the Fifth Generation project on mechatronics because there is no indication that robotics or automated manufacturing software is high on the list of candidates for Fifth Generation Project application software.

There is some indication that the Japanese are planning to use ADA extensively for applications in mechatronics. This is because they believe that most computers in the future will have an implementation of ADA. However, this remains to be seen since production quality ADA compilers are not yet ready. In any case, such plans do not indicate any lead on the part of the Japanese over the Americans.

### 3.5.3 Next Generation Application Software Project

This project is focused on development of production software for the whole process of manufacturing including assembly, CAD/CAM, and control. It has already begun with a two-year survey which was scheduled to be completed by the end of March 1985 when a proposal for a national software project will be developed. There is no indication of what the magnitude of the new project will be. The study is being chaired by Professor Yoshikawa of Tokyo University. It is under the direction of the Agency for Industrial Science and Technology, which is one of the branches of the Ministry of International Trade and Industry (MITI).

### 3.5.4 Robot Planner and Off-Line Programming

The Japanese are active in the Computer Aided Manufacturing - International (CAM-I) Robot Planner project. CAM-I's proposed work statement for a contract to develop the Robot Planner was written by Dr. Norio Okino of Hokkaido University. This Robot Planner will be an off-line programming system that will define robot motions for multiple robots and environments. It will include the possibility of moving objects. The Planner will have a simulator and interactive graphics system to facilitate the man-machine interface and provide for program verification. It will be independent of any specific robot, producing output similar to CL data which can be post-processed into specific robot instructions. It also will be computer independent, written in FORTRAN or C under a Unix operating system. This system will attempt to have a convenient interface to CAD data so that robot programs will be able to be generated automatically (for certain specific tasks such as arc welding or spot welding) from the dimensional data in the CAD database. For example, the path of an arc welding robot can be determined from the intersection of two planar parts with the angle of the welding rod defined from the normal vectors of the two surfaces. The system will also provide for intelligent responses to sensor inputs and will have convenient interfacing to FMS systems.

The CAM-I effort is relatively small, less than $1 million, which translates into something around 10 man/years of programming effort. This is probably too little to produce a very large advance in the state-of-the-art, but it is quite possible that the companies bidding on this project will put their own money into the effort. The development of such a system could have great commercial value, and the company which does the work will have a distinct competitive advantage despite the fact that the results of this work will be shared by all of the participants in the Robot Planner project. There are many American firms represented in CAM-I, so this effort is far from being an exclusive Japanese project. However, the contractor selected to produce the software will probably be a Japanese firm.

### 3.5.5 Japanese Research Compared to the United States

The Japanese are three to five years behind the leading edge of software for robotics compared with the most advanced work in America. The robot programming goals of CAM-I are somewhat advanced over U.S. systems which are already in operation in the Air Force sponsored MCL system or its derivitives at McDonnel-Douglas, Grumman, and GCA. But MCL systems have been in existance for over two years, and the CAM-I project has not yet begun. The Japanese have nothing on the market comparable to the Westinghouse VAL-II system, the IBM AML language, the Grumman MCL System or the McDonnel-Douglas Computer-Vision, or General Electric CALMA robot programming systems. Their published research does not reveal work equivalent to the experimental robot programming environments being developed at the National Bureau of Standards, Purdue, MIT, Stanford University, Carnegie-Mellon, and elsewhere. The Japanese are aware of these more advanced systems, they understand the basic principals involved, and they intend to pursue these issues vigorously. Most of the technical leaders in Japanese software research laboratories have been educated at the best institutions such as Edinburough, MIT, Stanford, and Carnegie-Mellon.

The Japanese are less behind in CAD and machine tool programming systems. The General Electric CALMA system, Computer-Vision, Evans and Southerland, Autotrol, CADDAM, Catia, and a host of other systems developed and marketed in the United States are more advanced than competing Japanese products. There are, however, several areas where Japanese systems seem to have some competitive advantage. THE MAZATROL machine tool controllers have interactive operator interfaces with graphics displays that compare favorably with products available in the United States. Also the Tokyo University GEOMAP work is equivalent to, or better than, comparable efforts in the U.S.

Although the U.S. appears to have a significant lead in most areas of software development for Mechatronics, particularly in high level languages and systems software development, the Japanese are achieving some remarkable results in the implementation of software engineering for specific applications. In the case of the Toshiba software workbench factory, they have not only closed the gap, but they are far ahead. The SWB factory is a uniquely Japanese phenomenon. The SWB factory environment would probably not be acceptable to the American worker. Yet the productivity of this installation is far superior to comparable American software houses.

The implication is that our lead in software is substantial but, by no means decisive. The software factory concept may enable the Japanese to overcome their relatively poor position in software research and system development.

The relatively small number of computer science graduates in Japan is another factor that will slow them down, but their use of software factories to produce workable code may enable them to overcome their shortage of formally trained computer scientists. Certainly, the current American lead in software research and development is no reason for complacency. The Japanese have finally recognized the importance of software and are now making great efforts to excell in producing good manufacturing software.

There is every indication from the research literature that the Japanese know how robot and machine tool software should be written. There appears little likelihood that the development of their Mechatronics effort will be retarded for long by a temporary lag in software research.

## 3.6 Manipulator/Actuator

### 3.6.1 Introduction

Basic research in manipulator technology is being conducted in universities, research institutes, and industrial laboratories. Research in public institutions is concentrated on theroetical problems--sensory preception, positional accuracy, modularization, and simplification. Industrial laboratories have concentrated on problems related to applications--increasing speed, reducing weight, miniturization, and computer control.

Two thirds of the manipulators being produced are still of the fixed sequence class (sequence not easily changed). Intelligent manipulators, employing active sensing in the control loop, constitute only 0.5% of the manipulators being produced.

A manipulator is a system (open kinematic chain) of links, joints, actuators, sensors, and its controller. One link of the chain, the base, is fixed so that the last link, the end effector or hand, can be described in terms of a coordinate system fixed with respect to the base. The manipulator's work space is defined by the points reached from the base by the end effector. Actuators move one link relative to another at their common joint. By sensing the state of the links and joints, the manipulator configuration (position and orientations of all links) can be determined. The actual manipulator configuration, planned configuration sequence, and their rates of change provide input data for the manipulator control algorithm. The manipulator system function is to change an object's configuration (position and orientation).

We first discuss standard manipulators or lumped parameter systems. The-state-of-the art in manipulators is 1 micron in precision and 5 microns in accuracy which is exhibited by the Fujitsu MicroArm. The SCARA robot, an entirely Japanese project, is fast becoming the major assembly manipulator in both the U.S. and Japan. Direct-drive manipulators have low-speed, high-torque motors, and they do not have transmissions (speed-reducers/ torque-multipliers). Multijointed manipulators with greater than 6 degrees of freedom are being developed for inspection in entangled spaces such as the human body, nuclear reactors, and jet engines. Teleoperated manipulators are being developed for the nuclear industry but also have applications in other hazardous environments. Manipulator systems containing flexible links require the most sophisticated control techniques. Manipulators using active sensing in the control loop, known as the intelligent robot, are necessary for small-lot production to reduce the long programming times of industrial robots.

A discussion of particular manipulators is followed by a discussion of activity in the development of the basic system components: links, joints, actuators, and actuator control. This is followed by a section on manipulator control and a section on space applications. Finally we summarize the findings and include comments on the literature searched and recommend future activity in the assessment of Japanese technology.

### 3.6.2 Manipulators

Japanese manipulator research and development activities at Tokyo University include basic design, new material development, control improvements, autodiagnosis, and system integration. The top priority is increased speed. Other high priority objectives include greater compactness, greater load to weight ratios, and modularity. By considering the manipulator as an effective mass attached to a rod of effective length which is moved through some angle at constant angular acceleration, we can show that the heat energy dissipated in motor windings will be proportional to the second power of length times mass and inversely proportional to the cube of the transit time or the cube of the speed. This simple modeling illustrates the importance of the design objectives. The system integration objective (Factory Automation Integration) has been met by the Mitsubishi RH Series manipulator.

Precise position accuracy has been demonstrated with the Fujitsu 6-DOF (degrees of freedom) MicroArm which was developed to assemble optical fibers. As compared to the typical accuracy of 50 microns for commercial models, their report of 7 micron precision (very recently improved to 1 micron) represents a significant advance.

The SCARA (Selective Compliance Assembly Robot Arm) 4-DOF manipulator has been a Japanese design and development project at Yamanashi University since 1978. This robot is available from a number of suppliers and has been steadily gaining in popularity for assembly tasks in both Japan and the U.S. Its limitation to vertical assembly has recently been eliminated in the form of Pentel's SF-70.

Another major new type of manipulator is the direct-drive manipulator being developed in both the U.S. (CMU & MIT) and Japan (ETL, Mitsubishi, Shinmeiwa, Yokogawa-Hokushin). Its actuators are low-speed, high-torque electric motors that are practical due to the availability of the new high-magnetic energy materials. This type of motor eliminates the need for power transformers and hence avoids their problems. It has significant increases in speed and acceleration over the current generation of manipulators. Direct drive manipulators are in the second generation of development at ETL. The multijointed manipulator project at Chuo University is directed at emulation of lumped parameter manipulators as special cases.

Multijointed manipulators of seven or more DOF are being developed for inspection tasks in entangled spaces such as the human body, nuclear reactors, and jet engines. A variable length manipulator constructed from rigid plates connected by lengths of spring is being developed at the Tokyo Institute of Technology which can enter a small opening and surmount an obstacle to arrive at a goal position. A fixed length multijointed manipulator mounted on a track has been developed by Toshiba. The manipulator's configuration changes constantly with motion along a supporting track enabling the device to thread itself through a maze of pipes. The Agency of Industrial Science and Technology has developed a 12 joint manipulator for tasks where force control of a lightweight manipulator is desirable.

Teleoperated manipulation is an active area despite its being cumbersome and tiring to use. Furthermore, it is a logical step from human manipulation to the complete automation of manipulation tasks. Considerable effort is being made to improve teleoperated arm pairs for applications in space, hazardous environments, and as an aid for the handicapped. The Power Reactor and Nuclear Fuel Development Corporation (PNC) has developed BILARM83A which has a low weight to load ratio. Further improvements in speed and load capacity are being sought.

Manipulator systems containing flexible links (beams), the distributed parameter system regime requiring sophisticated control algorithms, are being studied at Stanford and at the Science University of Tokyo. Both groups have demonstrated that a moving link can be brought to rest in roughly twice the natural period of the beam. This implies that more real-time computing is required during a control interval. This concern leads to investigation of various mathematical approaches, improved designs yielding decoupled motion, a simpler dynamic model, and more powerful computer control.

The intelligent manipulator/robot has been an object of Japanese research for 20 years. The word "intelligent" refers to real time use of information from sensors in the dynamic control algorithm of the manipulator. Intelligent use of sensor information requires knowledge of the robot, its environment, the tasks and goals, and strategies for handling unexpected states of the robot's universe. The Japanese focus is on assembly related projects (grasping and positioning parts, fastening, and inspection). Intelligent manipulation is considered essential to economically manufacturing lots of one unit automatically as they move away from hard automation.

### 3.6.3 Links, Joints, Actuators

As a step in developing strong, lightweight, manipulators whose efficiency approaches that of human muscle, active work is underway on links, joints, and actuators (primary conversion of energy into mechanical motion). This goal and the need for more energy-efficient devices has led to a reexamination of fluid, electrical, and thermal motors.

A major link technology development is the use of CRFP (Carbon Reinforced Filament Plastic) material. Hitachi, Mitsubishi, and Shin Meiwa are reported to be developing this link material. Such material is lighter for a given strength and will yield more energy efficient operation of the manipulator since heat losses scale roughly proportional to the square of the mass being moved. The Hitachi links are reported to be 20% lighter than their aluminum equivalents.

The use of two and three dimensional joints is being reported. The main impetus for using such joints is greater manipulator flexibility (degrees of freedom) which is required when manipulating in crowded spaces. Toshiba uses a two degree of freedom joint in its multijointed manipulator and Tokyo Institute of Technology multijoint manipulator has three degree of freedom joints.

Hitachi has developed interleaved stacks of pizoelectric elements as a motor with wide output range. The Tokyo Institute of Technology developed a Polyvinylidene Flouride (PVDF) piezoelectric actuator in 1980 and applied it to micromanipulation (0.5mm reach). Sumitomo's development of NEOMAX (6 times as efficient as Samarium-Cobalt) places it as the leader of this field. A direct-drive motor has been developed by Hokushin Electrics Co. which equals the performance of a DC motor with speed reduction. AC servo motors are fast becoming available in commercial manipulators, such as Fanuc and Toshiba, due to recent advances in digital control technology.

Improvements in fluid motors (pneumatic and hydraulic) have been reported by Hitachi and Chuo University. Improvements in digital control technology have been applied to this area to yield better real-time valve control.

Thermal motors made from Titanium-Nickel (Ti-Ni) wires, whose thermo-mechanical operating time is of the order of a second, are in the beginning stages of development at the Tokyo Institute of Technology, the Tokyo Electric College, and the Hitachi Mechanical Engineering Laboratory. Hitachi achieved faster heating with thinner wires but has not yet achieved the faster cooling goal.

Transformation of motor output to speeds appropriate for manipulators has been developed at a number of laboratories. Studies by Fujitsu on gear trains in a repeated operation showed a continual improvement in positional accuracy, due to improvement in meshing, until 500 hours of continuous operation had elapsed. Accuracy deteriorated after this time due to gear wear. Harmonic drives, which are small and light, suffer from hysteresis effects. Fujitsu achieved cancellation of hysteresis effects by mounting one harmonic drive on top of another. Application of this technique yielded the high precision of the MicroArm 150. The Electrotechnical Laboratory has used a spur gear and a sprocket and chain drive as well as direct drive in their ETA-2 manipulator. Toshiba uses a ball screw device in their parallel link and SCARA manipulators. In the course of developing their multijoint manipulator, the Tokyo Institute of Technology group developed an actuator to control the tension in tendon wires. They used a compensating spring and pulley wheel mounted eccentrically, to reduce the working motor torque by a factor of 8.

### 3.6.4 Manipulator Control

Increasing manipulator operating speed implies that more detailed modeling will be required since inertia terms will be more dominant. Increasing task speed implies that a manipulator spends more of its duty cycle in the transient mode with rapid starts and stops which excite vibration in the system. Effort is being directed at achieving rapid acceleration without vibration. The control problem is severest in the case of very flexible members where wave propagation phenomena in a distributed parameter system dominates. In practical applications, the manipulator itself is a subsystem so that its controller needs to interface with a higher level controller. Mitsubishi has designed the RH series SCARA robot to be easily integrated into higher level systems.

A number of groups are actively involved in minimization of vibration through more sophisticated control algorithms. A Yaskawa team has developed a real-time vibration damping system for the Motoman L-10. We note the similarity of this approach with that taken for distributed parameter systems and the contrast with the Virtual Cam Curve SCARA control system. Inclusion of the flexibility of links in the dynamic model is part of the approach to control taken at the University of Tokyo. Once the rigid link constraint is relaxed, manipulators with lighter links and the accompanying savings in energy consumption may be controlled effectively. A more detailed dynamic control model implies that more real-time computing is required during a control interval. Alternatively, a manipulator design which eliminates coupling among joints results in a simpler dynamic model and hence real-time control problems.

Absolute position accuracy and repeatability itself may become less important as sensor based control becomes more practical. Almost 10 years ago Hitachi reported the insertion of parts with clearances in the 10-20 micron range from initial position errors as large as 2 mm with their HI-T-HAND system. This performance has since been improved to the 3-10 micron range from initial errors as large as 3.5 mm. These clearances correspond to a clearance ratio of 50 microns. This manipulation requires several seconds so that other approaches, such as the remote compliance center device with characteristic mating times of 0.1 seconds, are preferable when they can be used.

Considering tracking errors to be dominated by the limited bandwidth of the control system, and this limited by the lowest resonant period of the manipulator, then the errors will be proportional to the square of the lowest period. Reducing this period is a goal of the new link material development efforts. For given material, these considerations predict a tracking error proportional to the fourth power of the manipulator's characteristic length. The optimal manipulator control system is considered to be the combination of feed-forward and feed-back gain adjustment based on sensed data.

The tracking problem has been studied in Japan for some time. Studies in the late fifties were directed at automobile control. Extensions to manipulator control date from the early sixties. The Tokyo Institute of Technology has developed an empirical nonlinear feedback law to reduce tracking errors by dividing the control space experimentally into two regions having different model parameters. Hitachi reported last year on a PI control algorithm whose parameters were empirically adjusted to yield improved performance by a factor of 3.

The Yaskawa team dynamic control algorithm produced reduced tracking efforts as well as vibration reduction which were demonstrated for trajectory speeds up to one-half meter per second along rectangles and arrow shaped figures.

The development of active force control includes the ETL project. Force control is included in both the manipulator and fingers of the end-effector. Work in this area has continued at MEL since the mid 1970s.

### 3.6.5 Manipulator Applications in Space

Applications of manipulator technology in manufacturing are well known but those in space are just beginning. Other areas where applications are planned and work is in progress include the life sciences (agriculture, forestry, livestock, fishery, medical care, and social welfare), mining in the earth and sea, transportation, and maintenance (buildings, vehicles, routes, pipelines, transmission lines). A number of these applications include situations where humans cannot work (e.g., at ocean depths greater than 300m) or where conditions are very hazardous. Many applications which are successful in one area may be applied in other areas. For example, the dual bilaterally controlled manipulators being developed in the nuclear industry will also be applicable to space tasks. As time progresses and advances in flexible automation are made, the role of humans and specialized machines will be diminished.

Mechatronics activities in space will be occuring in orbits at altitudes of about 500 km. Robot development is encouraged because of the obvious environmental hazards and the cost of life support systems. Initially manipulation tasks will be relatively simple, but a space factory is being planned and partnerships with U.S. aerospace firms are being formed. Construction projects include experimental stations and solar power stations. Related tasks are radioactive analysis, geological sample acquisition, and surveying.

In terms of task importance, judgment is considered to be the most important function (22%). Physical displacement of objects account for an additional 24% (transportation 9%, manipulation 15%). Remote control is rated at 15% and sensing at 11%.

### 3.6.6 Summary

Japanese manipulator technology is summarized as follows:

The current thrust of Japanese Mechatronics is in the direction that extends the capability from hard automation to flexible automation for lots of one. This implies the increasing need for intelligent manipulation and knowledge based computing support. Japanese activity ranges from basic research and development to applications of existing technology. Their manipulators are widely used in the United States and will continue to gain strength as newer developments in control and intelligent manipulation become available.

## 3.7 Precision Mechanisms

### 3.7.1 Status of Japanese R&D

Precision mechatronics is divided into three areas: high precision robots, semiconductor device fabrication, and computer peripherals.

The word "precision" is used here in keeping with current practice but is not particularly well defined. Better terms are "accuracy" (for absolute error), "resolution" (for fractional error) and "repeatability" (for absolute error in relocation). Precision is sometimes used interchangeably for any of the preceeding terms, but most frequently refers to accuracy.

The development of precise intelligent mechanical-electronic systems lies at the core of many emerging technologies. Accurate mechanical motion is at the ~ 1μ level or better, as evidenced by optical storage discs and VLSI camera autoregistration systems, while fundamental physical research is at the ~ 1Å level. The trends toward higher electromechanical precision, together with its electronic control, will be a continuing and increasingly important theme.

### 3.7.2 High Precision Robots

High precision robots can be used in demanding assembly tasks. Examples of such tasks are the assembly of Winchester disc drives, optical disc drives, video tape transports, optical communication elements, hybrid semiconductor circuits, and a wide variety of small electromechanical systems. At present, all of the high-precision robots are made in Japan. One example is by the remarkable ~ 5μ accuracy and ~ 1μ repeatability achieved by a preproduction Fujitsu arm. A repeatability of ~ 5μ has also been claimed by Seiko for some of their smaller assembly robots. By way of comparison the most precise U.S. robots currently exhibit an accuracy of ~ 25μ.

The SCARA (Selective Compliance Assembly Robot Arm) geometry, a purely Japanese development, is finding increasing use in precise robotic assembly. The invention of this arm was motivated by the practical consideration that in many tasks it is advantageous if the manipulator is very stiff in one direction, the direction of part insertion during assembly, while maintaining acceptable compliance in the two perpendicular directions. This led to the highly cost-effective SCARA design which is now being actively exploited by virtually all Japanese assembly robot vendors. By contrast few U.S. vendors currently manufacture SCARA arms, though several sell imported Japanese units.

Early work in precision assembly tasks was exemplified by the development of the RCC (Remote Center Compliance) device at Draper Lab. However, while passive RCC devices are available in the U.S., sophisticated industrial work has recently been reported in Japan. For example, Fujitsu has demonstrated an active two dimensional compliance-controlled actuator for use in the assembly of Winchester disc drives. (This is the same actuator that has been reported in press releases as writing three Japanese characters on a grain of rice.)

Comparable two dimensional actuators were developed at Draper Lab as early as 1972 but did not find commercial applications. In 1984 IBM reported a 2D "planar fine positioning" device, conceptually quite similar to Fujitsu's, and capable of related applications. The development of small very precise actuators of this general type can be expected to be a growing activity, particularly in connection with the future production of very small assemblies.

Related work on a much larger scale has been reported by the Carnegie Mellon Robotics Institute. This involved the construction of a computer controlled RCC device that exhibits software controllable compliance but without the feature of high speed 2D feedback. Instrumented, but still completely passive, RCC devices have also been produced by Draper Lab.

The excellent performance of Japanese high precision robots is not primarily attributable to superior angular (or linear) position encoder performance, but rather to painstaking and original mechanical design and construction. The robots are usually constructed by a group which is closely connected with the needs of the precision application which has a direct and beneficial bearing on the design process.

The Japanese have been among the leaders in improving electric motor performance through the development of brushless techniques, AC servo motors, coreless motors and related developments. The improvements realized have been incremental and cumulatively significant, particularly in the areas of improved reliability and reduced maintenance costs. (Essentially all precision robots are driven by electric motors. Even for large, high-power robots, electric motor drives are making steady inroads on hydraulic systems.)

Permanent magnet motors are particularly attractive in small sizes of the type employed in high precision mechanisms. Here the most significant advance is the recent development of the Neodymium-Iron-Boron permanent magnet materials. The Japanese have done ground-breaking work in this area, and Sumitomo is the acknowledged world leader with their NEOMAX material. Not only does this provide twice the energy of Cobalt-Samarium, but it does so at a much lower, (perhaps one third), cost per unit weight. This translates into a factor of six in the critical "flux per buck" figure of merit.

A further point to be made is that the development of better magnets leads to lighter electric motors of the same torque. Improvement in motor performance therefore carries with it a large premium in terms of overall actuator performance.

This is an example of basic and original Japanese materials research leading to a significant technological improvement. The closest U.S. work on magnetics is probably that of the GM Research Lab, although the majority opinion is that Sumitomo holds a significant lead.

There are a number of basic considerations and contraints concerning high precision robots. One is that for any given robot design the tracking error scales like $L^5$, where L is the linear dimension of the system. This indicates that small robots should be used for small, precise, applications, and several Japanese vendors are supplying small high precision robots. (The only miniature robots manufactured in the U.S. are very low precision units, primarily intended for hobby or instructional purposes.)

A similar result relates to speed. It can be shown that the energy required for a given actuator motion scales like $T^{-3}$, where T is the total time of execution. This indicates that significant speed improvements, for a given actuator design, are costly in terms of energy. This of itself is not important, but the associated temperature rise can be a significant problem. (This is particularly true for NEOMAX which has a Curie temperature of only 130°.) This again ultimately argues, through thermal and scale-effect considerations, for small high precision robots for small high precision tasks.

The whole area of actuator design for small precise motions is an open one that can be expected to become increasingly important, particularly if smaller and smaller elements and systems are to be fabricated. Electromagnetic elements may not be the only basic actuators of interest for such applications. One example is the Japanese work reported on the use of nickel-titanium "Nitinol" shape-memory alloys for robotic actuators. Another is the Hitachi development of very large multiple layer interleaved stacks of piezoelectric elements to provide an entirely electrostatic wide range mechanical drive capability. Such approaches could prove significant for numbers of future precision applications.

Remarkable work on miniature 2D "walking" piezoelectric actuators has also been reported by IBM Zurich. Although this work was actually carried out in connection with fundamental physics research on vacuum tunnelling microscopy, it may find applications in other areas concerned with micro manipulation. (This class of device is related to the well known PZT inchworm actuators, though carried out in two dimensions.)

Structural stiffness and rigidity are important factors in robots and in high precision devices. Here the development of high strength lightweight composites, such as monofilament carbon reinforced plastics, is an emerging area of great potential. Numbers of robot manufacturers both in the U.S. and Japan are experimenting with such materials. Mitsubishi, Shinmeiwa and Hitachi have all reported the construction of complete CFRP arms. This also is expected to be an area of increasing activity. The use of stiffer, lighter materials raises the lowest mechanical resonant frequency of the arm, while the tracking error scales like the square of the reciprocal of that frequency.

Similar remarks can be made with respect to large direct drive motors for "gearless" robots. These have the potential to produce high-speed, highly accurate robots without the backlash and cyclic nonlinearities exhibited by reduction gears. One U.S. manufacturer, Adept, is already manufacturing such a robot and significant U.S. research effort is underway at both CMU and MIT among other places. In Japan two such arms have already been reported by ETL Tsukuba, while Mitsubishi, Shinmeiwa and Yokogawa-hokushin have all developed preproduction direct drive arms.

Sony FX-I, FX-II, and Phoenix-10 systems could be described as general purpose robots. However, they are exceedingly efficient devices and have yielded impressive accuracy (~ 15μ) in continuous high-speed production use. The underlying strategy of providing separately controlled precision motions of both the workpiece and the insertion device, is applicable to a wide range of small assembly tasks.

### 3.7.3 Precision Mechanisms for Semiconductor Device Manufacture

For this report, semiconductor devices will be divided into silicon

VLSI and all others (III-V, II-VI, modulation band-gap, optoelectronic, etc.). The silicon devices are the dominant items both in terms of numbers and commercial value. Gallium arsenide devices, particularly in modulated band-gap realizations, are exhibiting the highest speed performance in both logic and RF applications, while a variety of compound and heterojunction semiconductors find widespread use in optoelectronic areas. The basic fabrication steps are similar for all these devices and will be discussed with reference to silicon VLSI since this is the dominant technology.

Semiconductor device fabrication, with one exception to be discussed subsequently, is not an area that naturally lends itself to the introduction of general purpose machines. This is because the process depends on the repeated application of small number of highly specialized basic steps, e.g., oxidation, photoresist spin on, lithographic exposure, etching, cleaning, ion implantation, diffusion, etc. In a typical case, device fabrication involves ~ 120 steps achieved by suitably and repetitively cycling through the ~ 10 basic processes.

Semiconductor device fabrication is the most demanding and complex mass produced item made today. The price of entry for a new state-of-the-art silicon VLSI plant is currently in excess of $100 million and rising rapidly as device dimensions continue to shrink. This is an area in which Japan is making a tremendous investment, and they are bidding to become the dominant VLSI producers.

Among all of the difficult issues involved in this technology, four problem areas stand out: lithography, pattern transfer, inspection, and cleanliness. Mechatronics has a significant bearing on three of these four topics.

Lithography involves mask production and feature printing. The most widely used method of making masks is with E-beam machines. These consist essentially of computer driven electron microscope columns mounted on laser-interferometer controlled precision x-y tables. The masks are typically chromium on glass or quartz. Both the electron beam spot size and the basic address structure of modern machines is ~ 1/8μ. Machines of this type are produced by a small number of U.S. manufacturers (EBT, Perkin-Elmer, Varian) in a $2 to 3 million price range, and are generally considered to be equal or superior to comparable Japanese products. The existence of such mask making machines is critical, and their production is a substantial engineering feat, but the numbers needed are small compared with the required number of feature printing machines. Optical laser reticle generating machines are also under development. (This possibility arises since one need only print 5X or 10X features on a reticle for reduction step-and-repeat lithography.)

Nearly all semiconductor printing is carried out by optical lithography. Some activity exists in competing technologies such as direct E-beam writing and x-ray lithography, but these efforts are modest by comparison. Older technology employed optical contact printing while the present trend is optical projection printing with an increasing emphasis on projection step-and-repeat lithography. Minimum feature sizes, defined by corresponding "design rules," have shrunk monotonically with time and are now below 2μ in the most advanced production facilities. Low volume preproduction capability exists at the 1.0 to 1.25μ level. (Much finer elements can be fabricated, e.g., 80A metal lines have been reported, but this has nothing to do with a complete VLSI production capability.) Advanced optical lithographic research centers on the use of all quartz optics and step and repeat cameras with deep UV excimer laser or other exotic sources. These systems should yield submicron factory production capability over the course of the next five to ten years. U.S. companies such as GCA, Ultratech, and Perkin Elmer, holds a strong position in the entire VLSI optical lithography area. The closest Japanese contender is probably Canon. Canon provides much of the Japanese industry with its lithographic capability, but has not penetrated the U.S. market significantly. The Japanese VLSI industry has been successful in spite of the fact that they do not hold a significant technological edge in any of the basic fabrication processes. This success lies in the Japanese superior inspection, cleanliness, and production controls that has resulted in consistently higher yields. (These production controls apparently have less to do with machine technology than with the fact that Japanese VLSI plants have dedicated engineers assigned to the factory floor and sometimes to individual machines.)

All VLSI photolithographic hardware involves an extensive mechatronic component. Modern projection step-and-repeat printers, for example, use complex autoregistration and autofocus techniques that provide a fractional-micron mask re-registration capability in each ~ $1cm^2$ field in a time of ~ 1 second. These techniques will be stressed to their limit as feature sizes continue to shrink. The Japanese, with their competence in high precision engineering (they probably already make the best high resolution electron microscopes in the world) can be expected to be strong contenders in the photolithographic area. In view of the very high cost of these machines, price/performance will be a decisive factor.

Prediction in this area is risky but a plausible case can be made for the following scenario. Optical lithography will be pursued relentlessly over the next five to ten years to the point where ~ 0.5μ critical feature dimensions will be achievable in the factory and ~ 1.0μ will be routine. The technology of mass VLSI production will then level off for two reasons. First, the complexity and cost of more superior mass fabrication capabilities will be staggering, and, second, the resulting devices will be within striking distance of their physically limiting capabilities (~ 1000 to 2000Å channel lengths).

At the same time nonoptical techniques (direct E-beam writing and x-ray and synchotron radiation lithography) will be pursued to their limits to provide the ultimate performance for cost insensitive markets such as the military. The resulting use of semiconductors will then be bimodal, with the vast bulk being room temperature silicon devices with ~ 0.5 to ~ 1.0μ features, while an additional tiny fraction will be ~ 2000Å feature silicon or compound semiconductor devices for highly critical cost insensitive applications. Since the U.S. spends some $300 billion annually on the military while Japan spends only $12 billion, it is easy to project which nation will be dominant in the cost-insensitive high technology area.* The economically

* The "spinoff" argument that military R&D helps civilian R&D is often advanced to justify such expenditures; however, genuine examples are hard to find and counter examples abound.

important area, however, will be the room temperature - 0.5μ to - 1.0μ silicon one since these VLSI devices will be used throughout the economy and have the greater impact.

A concomitant issue to that of VLSI lithography is that of VLSI inspection. This involves the issues of inspecting the original masks and inspecting device fabrication at different masking levels. These problems are difficult at present and are becoming more difficult as device dimensions shrink. Future automated inspection systems will probably be increasingly reliant on scanning E-beam electron microscope techniques. Inspection is now performed by people in clean-room factory areas. Since people are significant contamination producers in such environments. it may be that only by solving the inspection problem will it be possible to remove people from the clean rooms and thereby achieve the better than class 10 capabilities needed for the upcoming submicron devices. This whole area of cleanliness is a critical one, and it raises the issue of wafer transport.

It was mentioned at the outset that only one area in semiconductor device fabrication seemed to lend itself to the use of general purpose machines. This area is that of the physical transport of wafers between the various processing stations. At present such transport is primarily by handcarried boxed cassettes. One idea is to automate transport. This was done some years ago by Hitachi at their Musashino VLSI plant using small "smart carts." (Veeco in the U.S. has recently announced a similar smart cart product, while OKI has completed a VLSI plant using an unconventional linear motor wafer transport system.) However, Hitachi is reported to have discontinued the use of their carts and returned to hand carrying. This may be for the previously mentioned reason that little is to be gained in plant cleanliness by totally automated wafer transport until the inspection problem is solved and people are removed from the fabrication areas.

A halfway measure exists in which cassette-to-cassette wafer processing stations are linked by individual robot transport systems. This poses little in the way of technical problems since any electric powered robot can be configured to produce less contamination than a person, but it does not solve the general problem. One final area is the issue of automatic wire bonding. The surge in acquisition of offshore semiconductor facilities in the 70's was primarily driven by the need for low-cost labor. This trend has reversed in the 80's with the development of high-speed, completely automated wire bonders. Machines of this type use computer vision and pattern recognition for accurate parts and contact-pad recognition and location, coupled with high speed wire bonding heads. The state-of-the-art is an incredible 7 bonds per second, and it is generally considered that the best machines are those produced by Hitachi. These machines are not for sale.

This technology has important potential in connection with that of miniature hybrid microcircuit assembly with particular significance for the rapidly growing field of lightwave communication. More than likely, optoelectronics will remain a hybrid semiconductor technology for many years because the light sources and optical detectors have different semiconductor material requirements. It follows that lightwave systems will probably be hybrid, and the automated fabrication of these hybrid systems will become an important activity. A concomitant feature of these systems, particularly in "pigtail" configurations, is that they also require highly accurate automatic mechanical alignment of the optical fibers and the semiconductor optoelectronic elements. This whole area is a natural one for advanced miniature precision mechatronic assembly. Toshiba-Seiki has recently targeted this area of the production of precise miniature robots specifically designed for miniature semiconductor assembly.

### 3.7.4 Mechatronics in Computer Peripherals and Related Areas

The cost per function of semiconductor elements has dropped by more than a factor of $10^4$ in the last 20 years, while precision electromechanical devices declined in price much more slowly. Even in mechanical areas the pace has been quickened in the last few years due to the introduction of a high degree of automation in production.

A remarkable area of development has been that of computer disc drives. The OEM purchase price for 5 1/4" minifloppy drives, for example, is now less than $50. This is an area in which U.S. manufacturers of all types of drives, from large Winchesters to microfloppies, have remained fully competitive with Japanese products. This has involved a high degree of U.S. plant automation and even the use of U.S. produced sensor-based robots. The pace of these developments can be expected to be maintained as the industry turns to the still higher density vertical recording technology.

A similar situation exists in the high precision area of computer printers. U.S. manufacturers are maintaining a position of parity in spite of aggressive Japanese competition. The situation in precision mechatronics for entertainment is well known. Only Japanese tape transports are used in U.S. VCRs, and these transports come from just two Japanese manufacturers.

A similar situation is now developing in both the video and digital-audio entertainment optical disc area. Domination of low cost production in this high precision (~ 2μ tracking) technology will lead to a corresponding control of the developing interactive computer-video disc and large data-base area.

The status of the Japanese work in precision mechatronics is summaried in Figure 3-11.

| | BASIC RESEARCH | ADVANCED DEVELOPMENT | PRODUCT IMPLEMENTATION |
|---|---|---|---|
| High Precision Robots | 0 ↗ | + ↗ | > ↗ |
| Mechatronics for Semiconductor Fabrication | - → | - ↗ | 0 ↗ |
| Computer Peripherals | 0 → | 0 ↗ | 0 ↗ |
| Optical Discs | 0 → | + ↗ | + ↗ |
| Improved Motors | + → | + → | + → |

Figure 3-11. Relative Japanese/U.S. Positions in Precision Mechatronics

### 3.8 Standards

The Japanese Industrial Standards Committee was established in 1949 as a part of MITI. It employs 94 people and has 154 additional people working for it but paid by other organizations. At the end of 1981, the committee had published about 7700 standards in the fields of industrial and mineral products. A search of this list shows 2046 standards listed under mechanical engineering, many of which deal with machine tools. The following standards deal specifically with industrial robots:

| | |
|---|---|
| B0134 | Glossary of terms |
| B0138 | Symbols |
| B8431 | Standard form for indicating characteristics and functions |
| B8432 | Measuring methods for characteristics and functions |
| B8433 | General code for safety |
| B8434 | Identification symbols and colours for operator controls |

There is a fundamental difference between the U.S. and Japan with respect to standards. Under the direction of the government, Japanese companies cooperate to form product standards. They are not threatened with antitrust actions and, in fact, are encouraged to cooperate to capture the market in specific areas. Families of products with interchangable parts are one way of achieving this goal.

The Japanese have been active in the International Standards Organization Technical Committee. They have submitted many papers on safety and classification on the subcommittees on Numerical Control of Machines, Robots, and Requirements for Standards to Enable System Integration.

The Japanese Industrial Robot Association (JIRA) has established a committee on the standardization of robot language with Toshio Sada of Tokyo University as director.

A recent paper by Tamio Ari of Tokyo University outlines a proposed standard robot language called STROL. This work is a preliminary academic project, but it does reveal a thorough knowledge of work on robot programming languages in the United States and Europe.

The CAM-I robot programming language project claims to be attempting to develop a standard robot language. However, the size of this effort is probably too small to achieve a good off-line programming system. There seem to be little likelihood that a standard robot language will emerge from this project.

There are many U.S. efforts to develop standards for robots. In fact, there is so much activity in this area that there is an obvious need for coordination. A committee called the Industrial Automation Planning Panel was established in an attempt to provide this coordination role, but it has been unsuccessful. The Robot Industries Association and the National Bureau of Standards have contracted with a private company to generate a comprehensive report on what committees are currently working toward robot standards and what coordination is required. This effort is complex because standardization in robotics is intertwined with other areas. The most important example is the GM MAP protocols. The MAP protocols are being extended to include the application layer for numerical control machines and robots. There are three such efforts including IE 1393A, MIFAS, CCITT.4 and CCITT.6.

# 4. References


1. Taniguchi, N.J., "Present State of the Arts of System Design on Automated Assembly in Japan," (paper presented at the 4th International Conference on Assembly Automation, October 1983).

2. Hashizume, S., Matsunaga, M., Sugimoto, N., Miyakawa, S., Kishi, M., "Development of an Automatic Assembly System for Tape-Recorder Mechanisms," Research and Development in Japan, 1980. Awarded the Okochi Memorial Prize.

3. Akiyama, J., "Flexible Assembly Center System--FX-1 System," (presentation made at the Charles Stark Draper Laboratory, 2nd Annual 3-day Seminar, Robotics and Advanced Assembly Systems, Cambridge, MA, November 1981).

4. "Sony Compact Assembly Center (AC) Phoenix-10," (material provided to the Charles Stark Draper Laboratory, 4th Annual 3-day Seminar, Robotics and Advanced Assembly Systems, Cambridge, MA, November 1983).

5. Ohachi, T., Miyakawa, S. Arai, Y., Inoshita, S., Yamada, A., "The Development of Automatic Assembly Line for VTR Mechanisms," (paper presented at the 15th CIRP International Seminar on Manufacturing Systems, Amherst, MA, June 1983).

6. Aoki, K., "High Speed and Flexible Automated Assembly Line--Why has Automation Successfully Advanced in Japan?," (paper presented at the 4th International Conference on Production Engineering, Tokyo, August 1980).

7. Kaneko, K., Tatsuji, S., "A Newly Developed Unit Feeder Used in the Analog Quartz Watch Assembly System," (paper presented at CIC 1984).

8. D. Brandin, et al., JTECH Panel Report on Computer Science in Japan, JTECH-TAR-8401, December 1984.

# Appendix: Literature Support to Mechatronics Panel *

An objective of the JTECH Program is the identification of relevant and timely Japanese Science and Technology (S&T) literature. Panel members were requested to evaluate the usefulness of the various types of source materials provided. Panel members were also encouraged to recommend Japanese source material which should be acquired and dissiminated to the U.S. technical community on a continuing basis.

In this appendix, we summarize:

- the JTECH approach in providing literature/translation support to the panels, and a description of the general types of material provided,
- open-source Japanese literature available in the Mechatronics area,
- literature provided to the Mechatronics panel by the JTECH staff, and
- summary and recommendations

A.1 Approach

The following approach has been followed in providing panel members with source material:

1. At the panel "kickoff" meeting, the JTECH staff provides general source material for panel members' review. This material is in English and provides an overview of the subject.

2. Each panel member then ascertains the particular material needed for his topic.

* This appendix written by the JTECH Program Staff; the staff member responsible for literature and translation support is Dr. Y. Kim.

3. The JTECH staff then collects and translates the requested material.

(Steps 2 and 3 are iterative, with several interactions taking place during the panel assessments together with trips to Japan by Y. Kim to collect information.)

4. Selectivity. In providing the panelists with translated material, the following approach is used: (1) The titles of papers from Japanese documents are translated and distributed to panelists; (2) panelists request certain abstracts from the list of titles be translated (this step is usually done by telephone); and, (3) based on the abstracts, the full translation of selected papers is provided. The Mechatronics Panel required only about 10% of the total Japanese source material collected for full translation.

A.2 Categories of Source Material

Japanese S&T source material is divided into three categories. These categories are defined below, and the material provided to the Mechatronics panel in each category is listed in the next section.

A. General Material (mostly in English)

This material assists panel members in identifying additional data needed for their particular area of inquiry. This material lacks technical depth but provides a general background.

B. Open Material

This category consists of more specific articles from Japanese technical journals. A list of the technical journals pertinent to Mechatronics is given in the next section.

C. Semi-Open or Closely-Held Material

Japanese academic societies hold biannual or annual meetings, and participants are provided with the abstracts of the meeting. These abstracts are frequently hand-written; they are not reproduced in regular journal publications. They contain current and useful information, but they are obtained through privately-held sources. In this sense, the meeting abstracts are regarded as semi-open.

In the Japanese high tech community, seminars and conferences are organized to pool and disseminate research information. Meetings are usually attended by the invitees only, and the conference proceedings are printed in Japanese with most manuscripts being handwritten to allow entry of the latest information. Some of the more formal technical information contained in these proceedings eventually works its way through the appropriate professional journals published in English, but with a typical delay of six to eighteen months.

Japanese industries participate in the national R&D projects under a "Research Association" type of arrangement. The reports from these projects are disseminated first in the technical committees composed of member companies. The edited reports reach the public domain at a much later time.

A.3 Japanese Literature in Mechatronics

1. Society Publications

a. JIRA (Japan Industrial Robot Association Publication: Robot Monthly Robots in the Japanese Economy

b. Robotics Society of Japan, (1983+) Publication: Journal of the Robotics Society of Japan

c. Society for Instrument & Control Engineering (SICE) Publication: Transactions of...

d. Japan Electrical Society Publication: Journal of...

e. Japan Information Processing Society Publication: Journal of...

f. Japan Machinery Society Publication: Journal of ...

2. Technical Magazines

a. Measurements & Control

b. Automation Technology

c. Electronics Technology

d. Industrial Science & Technology, AIST

e. Promoting Machine Industry in Japan

3. Research Reports

a. National Projects
(1) Flexible Manufacturing System (FMS) Provided with Laser
(2) Hazardous Environment Robots

b. Laboratory Research Projects
(1) ETL (Electrotechnical Laboratory) Reports
(2) ETL Research Reports
(3) MEL (Mechanical Engineering Laboratory) Technical Reports
(4) Electrocommunication Application Laboratory Reports
(5) NTT Electrocommunication Laboratory Reports

c. In-house Industry Laboratory Reports

Most Japanese industrial laboratories publish periodical research reports. Some examples are: Hitachi Review, NEC Technical Reports, Toshiba Review, Sony Central Research Laboratory Review.

4. Popular Magazines & Promotional Publications

a. Nikkei Mechanical

b. Nikkei Electronics

c. Computer White Paper, JIPDEC

d. INS (Information Network System)

e. Frontier Technologies

A.4 Literature Provided to JTECH Mechatronics Panel

A. General


1. Albus (NBS), Trip Report to Japan, Oct. 1981.

2. Amber and Brady (NBS), Trip Report to Japan, Mar./Apr. 1980.

3. T.O. Binford, et al. (B.K.P. Horn, M. Raibert, P. Winston), Trip Report: A Visit to Japan.

4. Bloom, Haynes, Bettwy, and Hahn (NBS), Trip Report to Japan. Aug./Sept. 1982.

5. N.A. Bond, Jr. "Japanese Progress in Robotics: Tokyo Conference and Exhibition." ONR Scientific Bulletin, Oct.-Dec. 1983.

6. Brodsky, S.L. "Robot Control in Japan." ONR Scientific Bulletin. Volume 7, 1982.

7. Takashi Fukukita, "A Feature on Robots--A Comprehensive Survey on Assembling Robots; A Rush of High Performance and Low-Priced New Types of Robots; Limit Designs Based on Theoretical Backings," Tokyo NIKKEI MECHANICAL in Japanese, 20 June 1983, pp. 97-119.

8. Gevarter (NBS), Trip Report to Japan, Oct. 1981.

9. Hankinson and Branstad (NBS), Trip Report to Japan, Sept. 1982.

10. Kirsch and Gleissner, Trip Report to Japan. Oct. 1980.

11. Lyons and Sinnott (NEL), Trip Report to Japan. Sept. 1982.

12. Masubuchi, K. "Welding Robots in Japan," ONR Scientific Bulletin. Volume 7, 1982.

13. E. Nakano, "Potentialities of Japanese Robot Industry." Journal of Japanese Trade & Industry, November 1, 1982.

14. M. Tsuda, "Impact of Industrial Robots and Office Automation on Employment." 4 pages. Dentsu Japan Marketing/Advertising. Spr.-Summ., 1983.

15. Wakamatsu, Akahori, Shirai, & Kakikura. "Robotics Research at Electrotechnical Laboratory--R&D Program for Advanced Robot Technology."

16. S. Watanabe, "Future Developments in Industrial Robots," 3 pages. Dentsu Japan Marketing/Advertising. Spr.- Summ, 1983.

17. J. Woronoff, "Robots--East and West; The Tide May be Turning." Dentsu Japan Marketing/Advertising. Spr.- Summ, 1983.

18. K. Yonemoto, "The Economic and Social Roles Robots are Expected to Play," Dentsu Japan Marketing/Advertising. Spr.- Summ, 1983.

19. "Amusement Robots," Science and Technology in Japan, July-September, 1984. pp. 26-28.

20. Comparative Trends of U.S. Japan Industrial Robots Production. 7 page excerpts. Summary of Paul Aron Report (#226), September 7, 1983. Daiwa Securities America, Inc.

21. Competitive Status of the U.S. Machine Tool Industry (excerpts), National Academy Press, 1983. 9 page excerpts.

22. Competitive Status of the U.S. Auto Industry (excerpts), National Academy Pres,, 1983. 6 page excerpts.

23. Congressional Bills related to Mechatronics

    1. HR 4046, Sept. 30, 1983.
    2. HR 4047, Sept. 30, 1983.
    3. HR 4048, Sept. 30, 1983.

24. "Hazardous Environment Robot," Tokyo NIHON KEIZAI SHIMBUN in Japanese, 13 May 1983, p. 1.

25. High Technology Industries: Profiles & Outlooks. The Robotics Industry. U.S. Department of Commerce, International Trade Administration, April 1983.

26. Long-Term Demand Forecasting of Industrial Robots. (March, 1980) JIRA (Japanese Industrial Robot Association)

27. Mechatronics News, published by TECHNOVA, Tokyo, Japan

| Year | Issue | Title |
|---|---|---|
| 1982 | 11 | Mechatronics in Small Manufacturers |
| 1983 | 1 | Application of Mechatronics to FMS |
| 1983 | 2 | Application of Mechatronics in Nuclear Power Plants |
| 1983 | 3 | Mechatronics and the Labor Problem |
| 1983 | 4 | Applications of Mechatronics in the Auto Industry |
| 1983 | 6 | Mechatronics and Welfare Services |
| 1983 | 7 | Mechatronics in Construction Industry |
| 1983 | 8 | Applications of Mechatronics to Medical Equipment (executive summaries only) |
| 1983 | 9 | Mechatronics in Food Processing |

28. "R&D Robot Center," Tokyo NIHON KEIZAI SHIMBUM in Japanese, 21 April 1983, page 8.

29. Sample pages of Robot Monthly (in Japanese), JIRA (Japanese Industrial Robot Association) .

30. Sample pages of Robots in the Japanese Economy, JIRA (Japanese Industrial Robot Association)

31. Science & Technology Forum, '84, Science and Technology in Japan, July-September, 1984. pp. 20-21.

32. Stevenson-Wydler Innovation Act, Public Law 96-480.

33. "Two-Armed Bilateral Servomanipulator for Nuclear Fuel Cycle Plants," Science and Technology in Japan, July-September, 1984. p. 30.

34. U.S. Patent #4,369,563, January 25, 1983.

B. Open

1. T. Ara, "Standardization of Robot Languages," by Tamio Ara. Journal of the Robotics Society of Japan, Volume 2, Number 2.

2. Satoshi Hashino, "Data Structure of the Robot Language," by Satoshi Hashino. JRSJ, Volume 2, Number 2.

3. A. Higashimoto, "Research on the Element Technology on Product Inspection System for FMC," Journal of the Japan Society of Precision Engineering, Volume 49, Number 8, August 1983. pp. 1043-1049.

This special issue is on "Flexible Manufacturing System Complex Provided with Laser." pp. 991-1064.

The FMSC with Laser is a MITI-sponsored national project, and in this issue is a good account of the technical progress made on the project.

4. H. Inoue, "Towards Advanced Robot Programming," JRSJ, Volume 2, Number 2.

5. H. Suda, "Problem Oriented Robot Language--One Key-One-Function, by JRSJ, Volume 2, Number 2.

6. Y. Yamadori, "Office Automation in Japan," Science and Technology in Japan, Volume 3, Number 10, April/June, 1984. pp. 24-26.

7. T. Yoshikawa, "Multivariable Control of Manipulators," JRSJ, Volume 1, Number 2, pp. 10-15.

8. "Application of Modern Control Theory of Manipulator Control," JRSJ: Special Issue on Manipulator Mechanisms and Controls. Volume 1, Number 2, July 1983; pp. 16-21.

9. "Arc Welding Robot for Large Steel Construction MELFA-RW212," Mitsubishi Elec. Co., Robot. Volume 40, pp. 68-75.

10. "Control of Industrial Robot," JRSJ: Special Issue on Manipulator Mechanisms and Controls. Volume 1, number 2, July 1983. pp. 36-41.

11. "Dynamic Scene Analysis." JRSJ, Volume 1, number 4. pp. 23-26.

12. "FANUC ROBOT S Series for Special Job," Fanuc, LTD. Robot. Volume 39, pp. 88-94.

13. Industrial Science & Technology: Special Issue. Trends and Facilities in the Robot Industry. Volume 24, no. 11, 1983.

Table of contents translated into English.

14. "Intelligent Robot." Journal of Robotics Society of Japan. Volume 1, number 1. pp. 25-30

15. International Symposium on Design & Synthesis, Announcement of the conference to be held in Tokyo, July 1984.

16(a). Journal of Robotics Society of Japan.
   1) Title pages for Volume 1, no. 1 thru Volume 1, no. 4 & Volume 2, no. 1.
   2) Original Japanese Articles

16(b). Contents Pages of 6 issues of Journal of the Robotics Society of Japan. Volume 1, Numbers 1 through 4; Volume 2, Numbers 1 and 2.

17. "Multivariable Control of Robot Manipulators," JRSJ: Special Issue on Manipulator Mechanisms and Controls. Volume 1, number 2, July 1983. pp. 10-15.

18. "New Developments on an Information-Oriented Society,"--Outline of White Paper on Science and Technology 1983-- Science and Technology in Japan, Volume 3, Number 10, April/June, 1984. pp. 27-30.

19. "Package Auto-Assembly Robot," JRSJ: Special Issue on Manipulator Mechanisms and Controls. Volume 1, number 2, July 1983. pp. 47-50.

20. "Present State and Future of Manipulation Technology." JRSJ: Special Issue on Manipulator Mechanisms and Controls. Volume 1, number 2, July 1983. pp. 4-9.

21. Promoting Machine Industry in Japan. Volume 16, Number 9, 1983. Japan Society for the Promotion of the Machine Industry.

    Table of contents translated into English, 3 pages.

22. "Speech Synthesis & Recognition" JRSJ, February, 1983, Volume 2, Number 1.

23. "Speech Research in University and Laboratory," JRSJ. February, 1983, Volume 2, Number 1.

24. "Structural Synthesis of Robot Mechanism and Some Kinematic Problems to be Considered," JRSJ, Special Issue on Manipulator Mechanisms and Controls. Volume 1, number 2, July 1983. pp. 23-29.

25. "Vision System for Assembling." Journal of Robotics Society of Japan. Volume 1, number 4. pp.36-41.

26. Index pages of Robot Monthly, (Volume 37 through 41)

C. Semi-Open or Closely-Held

1. Hashimoto, and others, "SM Actuator and its Application to Bipedal Robots," December, 1983. pp. 207-208.

2. Hirose, and others, "Fourth Report on a Servo Actuator Using SMA: Modeling of Material Characteristics and its Experiment," First Symposium of the Robotics Society of Japan. Abstracts. pp. 205-206.

3. S. Ikeda (ETL), "Laser Processing Mechanims of FMSC," Symposium of Flexible Manufacturing System Complex, June 29, 1984, sponsored by AIST, MITI. pp. 37-42.

4. C. Soda (Director of Machinery Dept., MEL.), "Processing Mechanism of Raw Materials in FMSC," Symposium of Flexible Manufacturing System Complex, June 29, 1984, sponsored by AIST, MITI. pp. 13-36.

5. M. Kawai (Nippon Denso Co., Ltd.) , "An Example of FA for the Assembly of Electronic Equipment," Machine Society Seminar on "Recent Trends of Factory Automation in Assembly Operations."

6. H. Kimura (R&D Director, Toshiba Machinery), "Tsukuba FMSC Test Plant," Symposium of Flexible Manufacturing System Complex, June 29, 1984, sponsored by AIST, MITI. pp. 1-12.

7. Mizoguchi & others, "Structural Analysis of Parallel Manipulator," Paper 1409. First Symposium of the Robotics Society of Japan. Abstracts. December, 1983. pp. 107-108.

8. F. Ozawa & others, "Study on the Movement of Hydraulic Multiarticulated Robots," First Symposium of the Robotics Society of Japan. Abstracts. pp. 157-158.

9. Sakagami and Sugimoto, "Coordinant Transformation in Robot Speed and Its Application to Track Control," Paper 1404. First Symposium of the Robotics Society of Japan. Abstracts. December, 1983. pp. 97-98.

10. A. Sugimoto, and others, "Research of a Manual Manipulator by a Pneumatic Servo Motor: Cooperative Control Using a Small-Sized Pneumatic Servo System," First Symposium of the Robotics Society of Japan. Abstracts. December, 1983. pp. 159-162.

11. K. Tani, "Present State of Manipulator Technology and Its Future Outlook," Materials for the 10th Conference of Lectures & Research of AIST, Advanced Technology for Robots. Japan Industrial Technology Promotion Association.

12. Bell Labs Trip Reports, 7 pp.

13. Business Survey Report on Industrial Robots, JIRA. December, 1983.

14. Cincinnati Milacron Database. Information search of Japanese Work in Mechatronics.

15. Cincinatti-Milacron data base: Selections of Approx. 19 articles on Japanese work on Flexible Manufacturing Systems.

16. First Symposium of the Robotics Society of Japan (December, 1983). Abstracts. 226 pages. English translation of titles of 96 technical papers.

Abstract #1207 (written in English) is reproduced in full. 10 pages.

17.
a) INSPEC Database search of Japanese Robot Sensor activities,
b) INSPEC Database search keyed to Japanese vendors
c) INSPEC search keyed to "Mechatronics,"

19. 1. Joint U.S.-Japan Automotive Study
a) Policy Board Recommendations
b) Policy implications based on study suggestions
c) Excerpts from "Joint U.S.-Japan Automotive Study Final Report," Research Chairmen: Keichi Oshima and Paul W. McCracken. excerpts from 300 page report

20. Materials for the 10th Conference of Lectures & Research of AIST, Advanced Technology for Robots. Japan Industrial Technology Promotion Association. November 1983. 73 pages.

Table of contents translated into English.

21. "An Overview of Artificial Intelligence and Robotics, Volume 1--Artificial Intelligence, Part B-Applications," NASA Tech. Memo 85838, October 1983.

22. "An Overview of Artificial Intelligence and Robotics, Volume I, Part C-Basic AI Topics," NASA Tech. Memo 85839, December 1983.

23. "An Overview of Artificial Intelligence and Robotics, Volume II-Robotics," NBS Report NBSIR 82-2479, March 1982.

24. "Present and Future of Driving Mechanism: Control Technology in Particular," The 10th Research Meeting on "Advanced Robotics Technology" Held at MEL. pp. 33-44.

25. Progress Report Meeting of "Research Association of Robotics in Hazardous Environment," June 22, 1982. pp. 3-19.

26. Seminar on Robotics & Advanced Assembly Systems, November 8-10. 1983, Charles Stark Draper Lab., Inc.

27. Research Association of Robots in Hazardous Environment. (MITI Working Paper)

28(a) "Survey Reports on the Utilization of Computer-Aided Manufacturing," by JMA Consultants. September, 1983. 215 page report. Translation of cover, abstract and table of contents.

28(b) "Survey Summary," Survey Reports on the Utilization of Computer-Aided Manufacturing, by JMA Consultants. September, 1983. pp. 1-23.

29. The 10th Research Meeting on "Advanced Robotics Technology" Held at MEL. 73 pages. Translation of Contents page.

30. U.S. information sources on Japanese high-tech activities, 1 p.

TABLE A.1 Statistical Summary of Literature Support to JTECH Mechatronics Panel

| | | Amount of Material (pages) | | |
|---|---|---|---|---|
| | Number of Source Material Items | Source Material in English | Source Material in Japanese | Translated Material |
| General | 19 | 858 | 1173 | --- |
| Open-Source | 15 | 7 | 1173 | 160 |
| Semi-Open | 25 | 510 | 653 | 228 |
| Closely-Held | --- | --- | ---- | ---- |
| TOTAL | 59 | 1375 | 1826 | 388 |

# Part IV

# Biotechnology

The information in Part IV is from *JTECH Panel Report on Biotechnology in Japan,* prepared by D. Oxender, C. Cooney, D. Jackson, G. Sato, R. Wickner, and J. Wilson. The study was prepared for Science Applications International Corporation under contract to the U.S. Department of Commerce, June 1985.

# Acknowledgements

The Japanese Technology Evaluation Program (JTECH) is indebted to the panel members for their efforts in completing the assessment in the short time allotted while continuing to meet other commitments and pursuing ongoing research interests.

We appreciate the contributions made by Mr. George Mu from the Department of Commerce, International Trade Administration, Office of Japan. His help and guidance as the DoC Contracting Officer was invaluable. We express our gratitude to the National Science Foundation for their supplemental financial support of this study, and particularly to Mr. Frank Huband of the Division of Policy Research and Analysis for his contributions to the program.

# About the Authors

### Dale L. Oxender

Dale L. Oxender received his B.A. degree in Chemistry from Manchester College in 1954 and his M.S. and Ph.D. degrees in Biochemistry from Purdue University in 1956 and 1959, respectively. He has been at the University of Michigan in Ann Arbor since 1958 and a Professor of Biological Chemistry in the Medical School since 1975. His primary research interest is the mechanism of amino acid transport in both microorganisms and mammalian tissues. He has several National Institute of Health grants for using recombinant DNA techniques to examine gene expression and its regulation. He is also doing research on protein engineering technology. He recently organized an International UCLA Symposium on protein design which attracted 500 scientists, many of whom were from industry. He is now editing a textbook on protein engineering. He has been an American Heart Investigator, a Macy Foundation Fellow and an American Cancer Society Scholar during sabbatical leaves at Stanford University.

Since 1982 he has been Acting Director of a campus-wide Center for Molecular Genetics at the University of Michigan.

### Charles L. Cooney

Charles L. Cooney received his B.S. in Chemical Engineering from the University of Pennsylvania in 1966 and his M.S. and Ph.D. in Biochemical Engineering in 1967 and 1970, respectively, from MIT. He has been on the faculty of MIT since 1970 and is Professor Chemical and Biochemical Engineering in both the Department of Chemical Engineering and the Department of Applied Biological Sciences. Dr. Cooney's research interests are bioreactor operation and control, recovery of biochemical products, fermentation and enzyme technology. He is on the editoral board of several journals, has published extensively in the biochemical engineering literature and served as a consultant to industry and government organizations.

### David A. Jackson

David A. Jackson received his B.A. cum laude in Biology from Harvard College in 1964 and his Ph.D. in Molecular Biology from Stanford University in 1969. He has been engaged in research and development in the area of recombinant DNA methodology for nearly fifteen years. As a postdoctoral fellow with Dr. Paul Berg at Stanford University, he developed the first generally applicable biochemical methodology for joining separate DNA molecules to one another, and was the senior author of the first published report describing recombinant DNA methology in 1972. In 1977, Dr. Jackson became associated with Genex Corporation as Chairman of its Scientific Advisory Board. In July of 1980, he joined Genex full time as its Vice President and Scientific Director. He is now Senior Vice President at Genex, with responsibilities for technology planning and corporate development. He has also been an active participant in the controversies surrounding applications of genetic engineering in our society. He has testified about these issues at a number of Congressional hearings and is co-editor of a book, The Recombinant DNA Debate, which explores these issues from many perspectives.

### Gordon H. Sato

Gordon H. Sato attended Central College, Pella, Iowa, received his B.A. degree in Biochemistry from the University of Southern California, and his Ph.D. degree from the California Institute of Technology in Biophysics, 1955. Dr. Sato is currently the Director of the W. Alton Jones Cell Science Center. He is a member of the National Academy of Sciences; President of the Tissue Culture Association; Member of the American Academy of Arts and Sciences; received the Rosenstiel Award from Brandeis University in 1982; Edwin J. Cohn Lecturer at Harvard; Honorary Professor of Biology, Tsinghua University, People's Republic of China. Previous appointments include Professor of Biology, University of California, San Diego and Professor, Graduate Department of Biochemistry, Brandeis University. Dr. Sato has authored over 150 publications.

### Reed B. Wickner

Reed B. Wickner received his B.A. from Cornell University and his M.D. from Georgetown University. He is a Medical Director and Chief of the section on the Genetics of Simple Eukaryots at the National Institute of Health. His main research interest has been the mechanisms of viral replications with emphasis on the role of the host, a subject he has studied extensively using Saccharomyces (yeast). He is a member of the American Society of Clinical Investigation and the American Society of Biological Chemists, and is the U.S. editor of Yeast. He has studied Japanese for the last 9 years.

### John R. Wilson

John R. Wilson received his B.S. degree in Chemical Engineering from the University of London, UK, and his M.S. and Ph.D. degrees in metallurgy and materials science from the University of Sheffield. After teaching materials science at the University of Birmingham, and at Queen's University, Kingston, Ontario, Canada for a total of 13 years, he joined Shell Development Co. in Houston, Texas in 1971 as a staff research engineer. From there, he moved to Bendix Corp. (Southfield, Mich.) in 1979, where he was responsible for new business development, with particular emphasis on biotechnology, and was involved in establishing the Bendix-sponsored protein engineering program and Genex Corp. Following the Allied-Bendix merger, he became Vice President, Research and Development for Allied Canada, Inc. Recently he moved to a new position as Vice-President, Corporation Research and Development for Lord Corporation in the Research Triangle area. Dr. Wilson's research interests have ranged from liquid metal structure to composite materials to biosensors for industrial applications.

# Executive Summary

The last 15 years or so have given rise to an explosion in knowledge in the area of life sciences. That explosion was fueled by the development of a new battery of technologies referred to as recombinant DNA technologies and hybridoma technologies. The application of these technologies to the development of commercially-important products is referred to as the new biotechnology. Global interest in the new biotechnology has been expanding rapidly in the last several years. Thus far, only a few products have come out of the application of new biotechnology; however, virtually every large company of the industrialized countries is now committing significant resources to biotechnology research and development. Estimates of the market range from $15 to $100 billion dollars worth of products coming from biotechnology worldwide by the year 2000. Specialty products of relatively high value are likely to appear first, with the production of chemicals, feedstocks and biomass conversion taking a longer time to develop.

## Early Development in the United States

Much of the early technology was derived from government-sponsored basic research carried out in the major research universities of the United States. In the mid to late seventies, over 200 new entrepreneurial companies were formed with readily-available venture capital to take advantage of the promise of biotechnology. In addition, many large manufacturing firms have recently developed biotechnology divisions. Even though the United States has enjoyed the lead in most areas of biotechnology, it faces a serious challenge on the world market from Japan, West Germany, Britain, Switzerland, Sweden, and France. The international competitiveness has been heightened in recent times by the declaration by several foreign governments of biotechnology as an important area for emphasis. Several countries have made substantial resource committments to research and development of biotechnology. Both West Germany and Japan now spend more than 2.8% of their gross national product on R&D.

## Challenge from Japan

The United States is expected to face its most serious challenge from Japan. The Japanese government has declared the commercialization of biotechnology a national priority. The Council for Science and Technology, which is the highest governmental agency concerned with R&D, released a report in late 1984 mapping out its policies for promoting science and technology over the next decade. One of the major areas listed is "life sciences," and it includes most of the areas traditionally referred to as biotechnology.

## Coordination of Government/Industry/University Programs

Japan has well-organized associations which coordinate the information, support research and development projects, and sponsor national and international symposia on biotechnology. For example, a government agency, the Ministry of International Trade and Industry, has catalyzed formation of the Bioindustry Development Center (BIDEC). This is a cooperative research association of over 130 major chemical and manufacturing companies that have joined together to enhance the commercialization of biotechnology. Such associations avoid duplication and share R&D related to biotechnology. The Ministry of Education is also supporting the development of several biotechnology centers at major universities in Japan.

## Other Factors Improving Japan's Competitiveness

Japanese firms are making licensing agreements with, and investing in, companies from the United States and Western Europe. These agreements usually provide badly-needed funds to a small R&D company and, at the same time, provide the Japanese company with production and marketing rights.

Finally, although there have always been many Japanese scientists trained in the west, there has been an increase in the number of Japanese industrial scientists sent abroad for retraining. As a result, there is an increase in the number of qualified Japanese scientists in technical areas, such as recombinant DNA technology.

Recent Reports

In 1983, a report entitled "Commercial Biotechnology, an International Analysis," prepared by the United States Congressional Office of Technology Assessment, concluded that Japan is likely to be the leading competitor of the United States. First, Japan has extensive bioprocess experience and technology; and, second, the government has targeted biotechnology as an important technology for commercial development. The report called for interdisciplinary cooperative research and the development of applied research capabilities in the United States National Laboratories. In June 1984, a report entitled "Biotechnology in Japan," prepared by Dr. Herman Lewis of the National Science Foundation, was released. This report contains a comprehensive description of who-is-doing-what in the commercialization of biotechnology in Japan. The report concludes that the quality of Japan's research activity is comparable to that in Europe and the United States.

A United States Department of Commerce Report, prepared by the International Trade Administration, was released in July 1984. This report contained a study of the competitive position of the United States in biotechnology and also concluded that the United States faces the most serious challenge from Japan.

This Report

This report is an extension of the above mentioned reports. It differs somewhat from earlier reports in that it is intended to be an assessment of Japanese research and development in several of the main areas of biotechnology using information gathered primarily from Japan. The five areas of biotechnology evaluated in this report are:

a) Biochemical Process Technology
b) Biosensors
c) Cell Culture Technology
d) Protein Engineering
e) Recombinant DNA Technology

The field of biotechnology is much broader than the subjects represented here. Additional fields, such as agriculture and waste treatment, are also the subject of significant applications of new technology.

## Evaluations of Individual Areas

### Biochemical Process Technology:

This area represents the application of new technology to the development of new products from fermentation or bioconversion. It was evaluated by Professor Charles Cooney of MIT. Fermentation has historically been a strong and important field in Japan. Since biochemical process and separation technology are so very important to the commercialization of biotechnology, it is this area where the Japanese are thought to have the greatest advantage. Japan has extensive experience in bioprocessing represented in the large companies, although they have not always penetrated the world market in the past. Japan has recently, however, made significant increases in the world market with amino acids, nucleotides, antibiotics, and chemotherapeutic agents. Significant attention is now focussed on specialty chemicals and processes by the petrochemical industries in order to decrease their dependence on petroleum and develop products with higher profit margins.

Some of Professor Cooney's conclusions from the evaluation of Japan's biochemical process industry (BPI) are the following: BPI is an important growth industry in Japan; their manufacturing technology is state-of-the-art, but not unique (i.e., there is room for the United States BPI to develop competitively); the BPI seems to be more technology-driven than market-driven (this observation may be more related to a longer time frame used for strategic planning in Japan, as will be discussed below); BPI in Japan needs to develop low-cost manufacturing to offset expensive feedstock, or to distribute the manufacturing closer to sources of low-cost feedstock and be closer to the market; several large Japanese companies appear to be moving in this direction by making agreements with other countries that have supplies of low-cost feedstocks.

Biosensors:

The four most common configurations of biosensors include potentiometric or amperometric electrodes, chemically-sensitive electronic devices, enzyme thermistors, and opto-electronic devices. This area was evaluated by Dr. John Wilson of Allied of Canada (presently at the Thomas Lord Research Center). The focus of biosensor development in Japan has been on techniques which are compatible with current semiconductor fabrication methods. This may not be "ideal" for medical sensors, although the ability to use proven technologies will be of major benefit in getting them on the market. As a result, the Japanese developments have a high chance of commercial success. Areas requiring biochemical technology represent Japan's greatest weakness in the development of biosensors. Immobilization of enzymes require only limited biochemical expertise, whereas immunosensors call for greater experience. The prediction is that the next generation of biosensors in Japan will be initially micro amperometric enzyme electrodes, and immunosensors will lag behind. Dry reagent chemistry strips offer significant competition to multifunctional microsensors, although the latter would be cheaper to make and would offer enough advantages to justify their development. Japan has high caliber academics working with industries in medical sensor research. They will offer stiff competition in medical sensors of the future.

Large-Scale Tissue Culture and Hybridoma Technology:

The use of animal cells to produce important products is a newly-emerging technology. Targeted cell culture products for the future include lymphokines, interferons, tissue plasminogen activators, and monoclonal antibodies for diagnostic use. Dr. Gordon Sato, of the W. Alton Jones Cell Science Center, evaluated this area. Many of the large Japanese companies are convinced that large-scale tissue cultures will be useful for the development of important products, although microorganisms offer significantly lower-cost technology at the present time. Several technical breakthroughs are required before large-scale cell culture technology can compete with microorganisms in the field of biotechnology. Efficient bioreactors and improved tissue culture strains that can use low-cost media must be developed. Many research groups are using technologies such as bead cultures, encapsulated cultures, and hollow fiber culture vessels. It is difficult to assess the competitiveness of Japan in this area since it is just developing; although, judging by the current interest and research activity in large-scale cell culture, they are likely to become significant competitors in the future. As an example, most of the large petrochemical companies in Japan are developing cell culture facilities and acquiring expertise in the techniques of tissue culture.

Protein Engineering:

This area concerns the application of new genetic techniques to produce new or improved proteins and enzymes. For example, increases in stability, activity, or turnover of an enzyme could prove to be commercially important. This area was evaluated by Dr. David Jackson of Genex Corporation. At the present time, there is very little existing protein engineering in Japan. A significant limitation in Japan is their modest facility for protein structure determination and small pool of trained personnel in this area. They are very interested in the application of protein engineering to develop industrial enzymes, pharmaceuticals, diagnostics, sensors, and microelectronics. Japan seems to be willing to acquire protein engineering technology from outside by forging agreements with other companies. They appear to be taking a relatively long view with respect to commercializing this technology.

Molecular Biology and Genetics:

This area is primarily related to the application of recombinant DNA technology to produce commercially-important products. Some of these applications include the production of hormones (such as growth hormone and insulin), industrial enzymes, vaccines, and interferons. This area was evaluated by Dr. Reed Wickner from the National Institute of Health. Even though many of the early developments in molecular biology were outside of Japan, they have a large number of young scientists being trained in laboratories of the United States and Europe, which provides Japan with considerable expertise in recombinant DNA technology. There is no real technology gap, since the technology can be acquired rapidly. Funding for basic research has lagged in Japan, but the current national focus on recombinant DNA technology should improve their competitive position in the future. It was also noted that Japan's competitive position is affected by their low rates of crime and drug use, a very high rate of literacy and a formidable work ethic.

## Overall Conclusions and Assertions

1. One of the significant general conclusions coming from this evaluation is that Japan's strategic planning in biotechnology is based on a 10- to 15-year time scale, whereas most similar planning and resource committment in the United States is based on a 3- to 5-year time scale. This difference in time scale is partially responsible for the view that Japan's resource committment in several of the areas evaluated seemed to be more technology-driven than market-driven.

2. Japan will rapidly become more competitive with the United States and Europe because much of the commercialization of biotechnology in Japan is being carried out by large established companies. These companies have extensive experience in necessary process technology and the financial backing so necessary for bringing products to market.

3. Japan has a well-established coordination of the development of biotechnology by industry, universities, and government groups.

   a. The identification of biotechnology by the Japanese government as an important technology and the development of a national strategy for its development for the next 10 years will likely significantly increase their competitive position.

   b. In addition, government agencies such as MITI (Ministry of International Trade and Industry) and the Ministry of Education, Science & Culture, are promoting biotechnology development by resource committment and serving as a catalysis in the formation of research associations involving large groups of private companies.

   c. The BIDEC (Bioindustry Development Center) research association promotes cooperative research, provides an information center, and convenes symposia for its 300 member companies.

From this evaluation of the development of Japan's biotechnology, there is a clear need to continue to monitor the progress in the commercialization of all of these areas of biotechnology, and possibly others, by a more formal arrangement.

Some of the things that may aid in the development of biotechnology in the United States are:

1) Cooperation and exchange of information with Japan on biotechnology would help minimize competitiveness.

2) The commercialization of biotechnology in the United States would very likely benefit from the encouragement and development

of mechanisms of achieving longer-range planning than the present 3- to 5-year planning.

3) Perhaps a government agency is needed to channel resources toward the development of biotechnology and to catalyze the formation of cooperative research associations among the industrial corporations that are committed to the use of biotechnology for product development.

4) United States scientists, familiar with the Japanese language, should be encouraged to attend scientific meetings in Japan, such as the Industrial Fermentation Society meetings, which are largely conducted in Japanese.

5) Government support for basic research should continue to be a high priority, since it was largely the government supported university research effort that spawned the new technology.

6) The encouragement of partnerships between large established companies and new biotechnology companies, where their special resources are complemented, may enhance the commercialization of biotechnology in the United States.

7) Continued strong support of the Small Business Innovative Research (SBIR) grant program will also aid new product development.

# 1. Introduction

Dale Oxender, University of Michigan

## 1.1 Long-Term Strategy for Biotechnology in Japan

In December 1984, the Japanese government released a major report on the promotion of science and technology over the next ten years. The report, entitled "Tackling Changing Conditions; Comprehensive and Fundamental Policies for Long-Term Promotion of Science and Technology," is the work of the Council for Science and Technology. The Council is the highest policy-making body in Japan whose chairman is the Prime Minister. The report, which was drawn up by 150 experts from universities, government research laboratories, and private industry, calls for increased creativity and for Japan to become one of the top nations in basic research. The Council tries to identify key research areas to enable ministries concerned with science to implement policies in accord with government's overall aim. The basic research themes that make up Japan's strategic objectives over the next 10 years include materials science, information science, life science, and earth science. In the area of life science, the report emphasizes the following areas of basic research: gene regulation, analysis and synthesis of DNA, design and modification of proteins, chromosome engineering, cell and organelle modification, and alteration of components of cells and tissues. These areas include most of the fundamental technologies that are generally associated with biotechnology. The report reveals that Japan's R&D expenditure will reach 2.8% of gross national product in fiscal year 1984.

## 1.2 Background

Biotechnology refers to commercially-important new technologies, such as recombinant DNA, cell fusion, and hybridoma technologies that use living plant or animal organisms or their products to make or modify the cells or products and improve their economic value.

### 1.2.1 Development of Biotechnology in the United States

The past 10 years or so have seen a rapid development in the commercialization of biotechnology in the United States. Since 1977, over 200 new firms in the United States have responded to opportunities to develop new vaccines and diagnostic tests for a variety of plant and animal diseases, new hormones that may aid milk and meat production, crops that tolerate increased levels of salt, and microorganisms that eat toxic waste or leach metals from low-grade ores. It has been predicted that industrial activities using genetic engineering techniques will account for $40 billion in sales of new products around the world by the year 2000.

#### What Are the Events that Led to the Development of the New Techniques?

1944 Identification of DNA as source of genetic material, by Avery.

1953 Discovery of DNA structure, by Watson and Crick.

1961 Determining the genetic code, by Nirenberg.

1970 Discovery of restriction endonucleases, by Nathan and Smith.

1973 First gene cloned.

1975 United States guidelines for recombinant DNA outlined at Asilomar Conference.

1976 The start up of many new genetic engineering companies in the United States.

1980 United States Supreme Court permitted Chakrabarty to patent microorganism, Cohen/Boyer patent issued for recombinant DNA (rDNA) technology.

1981 Japan targets biotechnology.

1982 First commercial products of rDNA include insulin and animal vaccines.

1983 Private sector investments to commercialize biotechnology exceed $1 billion in United States.

As early as 1971, Japan's Science and Technology Council advised the government of Japan that the area of life science was an area worthy of special emphasis. Even though this area was identified early, systematic support and official targeting of biotechnology by the Japanese government did not occur until 1981. Two important developments in the United States were instrumental in the committment of Japan to target biotechnology. One of these events was the granting in 1981 of the Cohen/Boyer patent for the development of rDNA vectors and techniques for expressing products in microorganisms. This patent was seen by Japan as a potential barrier to their entry into the biotechnology market. This event, plus the successful manufacture of interferon and insulin by means of rDNA technology, convinced them to develop biotechnology as a strategic national industry. About this time, five leading Japanese chemical companies set up the Biotechnology Forum. Biotechnology research and the committment of resources for R&D in Japan has more than doubled in this area in the last four years.

### 1.2.2 Japan Takes a Different Pattern of Development than the United States

In contrast to the United States pattern of formation of new specialized entrepreneurial companies, it has been established companies in the chemical, pharmaceutical, and textile industries that are leading the push into biotechnology in Japan. As mentioned earlier, the recombinant DNA technology gave rise to over 200 new entrepreneurial companies in the United States since 1977. Many of these companies are relatively small and have not yet developed products.

The U.S. pattern of development is driven by the availability of venture capital resources. The gamble for venture capital lies in the seemingly limitless horizons of the application of genetic engineering to provide new industrially-important products. By contrast, biotechnology in Japan is being developed by more than 200 of the well-established industrial firms. In addition Japanese companies are investing in American manufacturing companies. At the end of 1984 a survey by the Japan Economic Institute in Washington showed that Japanese companies own all or part of 370 American companies.

The small entrepreneurial firms founded in the United States complement established companies, which has been acclaimed as a strength of the United States pattern of development biotechnology. It is not clear whether this arrangement will serve to provide sustained growth and development in a competitive fashion. A concern expressed in the recent OTA report was that Japan might repeat its performance in microelectronics, where it followed the U.S. initially and then overtook the U.S. during the commercialization phase of the development of the technology. Large integrated Japanese companies may have several advantages, such as continuing access to funds and a long history of experience in bioprocessing technology.

The Japanese fermentation industries already provide commercial sources of enzymes, synthetic amino acids, and antibiotics. It is clear that Japan has targeted biotechnology as an important future area of commercial development. In 1981, the Ministry of International Trade and Industry (MITI), which is one of the most prominent government organization in Japan involved in biotechnology, announced the "Research and Development Project of Basic Technology for Future Industries." The project is, in essence, a national program for biotechnology. The aim of the project is to develop basic technologies which are expected to play an indispensible role in the 1990's. Three technologies were selected for research during the next decade: (1) development of bioreactors, (2) development of large scale cell cultivation, and (3) development of recombinant DNA technologies. In 1982, MITI established the Bioindustry Office to coordinate the research and development of the three targeted areas. MITI is also carrying out research and development at its own national facilities such as its Fermentation Research Institute as well as cooperating with industrial and academic efforts. The developments in the field of fermentation will be detailed in the first chapter.

The central organization for research on biotechnology in the private sectors is The Research Association for Biotechnology, which was established by The Agency for Industrial Science and Technology of MITI. MITI's financial resources committed to biotechnology development over the 10-year period beginning in 1981 is around 130 million dollars.

The Ministry of Education, Science, and Culture supports university research and basic research in specialized national institutes in Japan via grants for specific projects. Most of this agency's support for research, estimated to be $1.3 billion in FY 83, goes to the national universities and research institutes such as The National Institute of Genetics. The Ministry of Agriculture, Forestry, and Fisheries is also a large agency, but as yet does not have much of an impact on biotechnology.

### 1.2.3 Development of Research Associations in Japan

The major mechanism by which MITI interacts with industry and academia is through the formation of research associations. These associations are formed by private companies, but MITI has initiated their organization by invitation. They operate under the Research Association Law, which allows cooperative research without violating Japan's antitrust laws.

### 1.2.4 Research Association for Biotechnology

In September of 1981, 14 member companies of the chemical industries were encouraged by MITI to form a research association to promote biotechnology. The purpose of the organization is to carry out research in the three areas targeted by MITI through collaborative efforts of government, industry, and academic scientists. These targeted areas are, bioreactors, large scale cell cultivation and Recombinant DNA. See the organizational chart shown in Figure 1-1. Most of the funds go to the laboratories of the participating

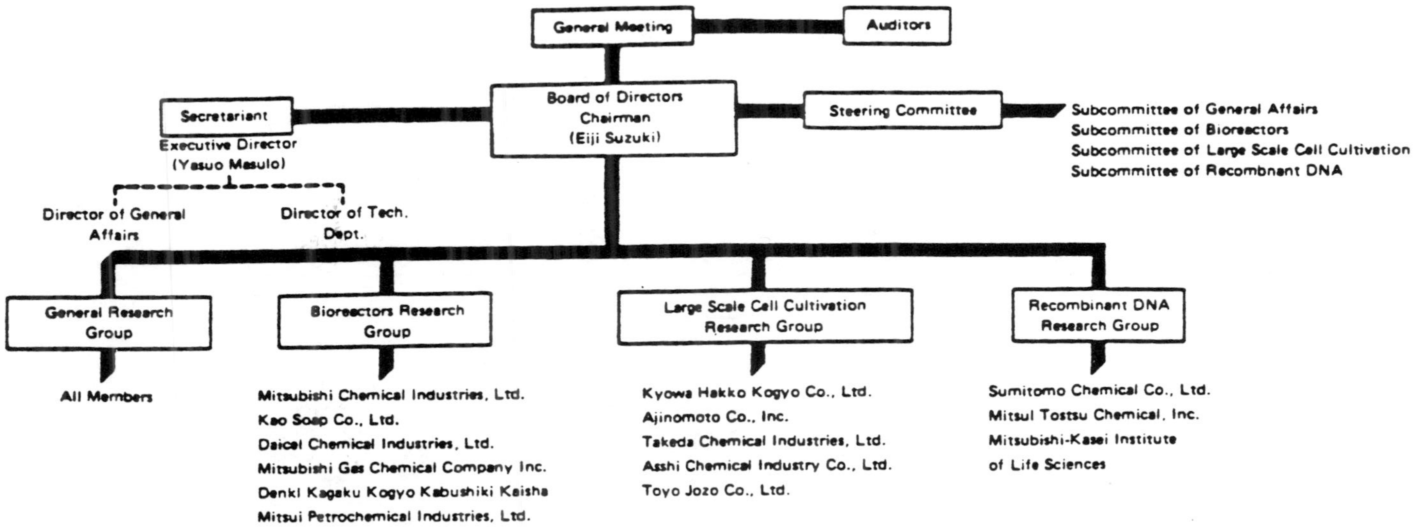


Figure 1-1. Organization chart of the Research Association for Biotechnology. The General Research Group is responsible for technology assessment in Japan compared to foreign countries, and for assessing other issues, such as safety. The other Research Groups responsible for the targeted biotechnology areas have developed plans for the ten-year period, 1980 to 1990. These plans are listed on the following page.

| Research Groups | STAGE I | STAGE II | STAGE III |
| --- | --- | --- | --- |
| Bioreactors | Screening of microorganisms and/or enzymes | Studies on optimum conditions of each bioreactor | Evaluation of new bioreactors in bench scale |
| Large Scale Cell Cultivation | Selection of cell lines. Research of media | Studies on optimum conditions of cell culture | Evaluation of cell culture process in bench scale |
| Recombinant DNA | Development of new host-vector system. Cloning of useful genes | Studies on efficient expression of new genetic materials | Evaluation of new genetic materials in bench scale |

Figure 1-1. (continued)

companies. This cooperation allows a member company to gain expertise in new areas.

### 1.2.5 Bioindustry Development Center (BIDEC)

The Japanese Association of Industrial Fermentation (JAIF) has been in existence for more than 40 years. It has over 300 sustaining company members and over 10,000 individual members. In April of 1983, with the support and encouragement of MITI, JAIF formed the Bioindustry Development Center (BIDEC). There are more than 120 companies affiliated with the Center. BIDEC's founding is a formal recognition of the contribution of biotechnology to the future of Japanese industrial development and the need for cooperation so that biotechnology develops rapidly in an orderly and healthy manner.

BIDEC operates through eight subcommittees listed below:

BIDEC Subcommittees

1) Human Resources Department
2) International Cooperation
3) General Information Search
4) Symposia and Congresses
5) Technology Development
6) Communication Promotion
7) Protection of Proprietary Research Results
8) Assigned or Entrusted Special Research Projects

BIDEC serves as a clearinghouse of information for the whole industry. In addition, it operates through the above eight subcommittees to promote international cooperation, collects up-to-date information, sponsors symposia and congresses, and publishs newsletters. Early in 1984, the Japan Bioindustry Letters began to appear. This newsletter is published in English by the BIDEC Subcommittee on International Cooperation. The newsletter contains general information on fields of biotechnology research from its member companies, specific news items extracted from industrial journals, as well as announcements of special meetings and events.

### 1.2.6 Other Reports Concerning Biotechnology

The United States Congressional Office of Technology Assessment released a report entitled "Commercial Biotechnology, an International Analysis" in 1983. The report documented the current leading position of the United States in the commercialization of biotechnology but indicated that this position may be threatened in the future by Japan. It concludes that Japan is likely to be the leading competitor of the U.S. for two reasons. First, Japanese companies have extensive experience in bioprocess technology and more established bioprocessing plants and engineers. Second, the Japanese Government has targeted biotechnology as a key technology of the future for commercial development. The government is acting to coordinate interaction among industry, universities, and government. The report concludes that the United States may compete favorably with Japan if it can direct more attention to research problems associated with scaling-up of bioprocesses for production. The OTA report cites five options to improve the competitive position of the United States. The options concern the encouragement of interdisciplinary cooperative research and the development of applied research capabilities in the National Laboratories.

Biotechnology in Japan is a report prepared by Dr. Herman Lewis of the National Science Foundation released in June 1984. This report is a comprehensive description of the commercialization of biotechnology in Japan. It describes the various government organizations and the interaction of industry through scientific associations promoted by the government. The report concludes that, although the commercialization of biotechnology in Japan is still at an early stage, the quality of Japan's research activity is comparable to that in the West and the infrastructure for R&D is strong.

The present report is intended as an extension of certain aspects of the previously-mentioned reports. Several important specific areas of biotechnology which are being developed in Japan were selected for an in-depth analysis. It is intended as an evaluation of the present status of a particular commercially-important technology in Japan, as well as the trends and future direction of Japanese R&D in that subject area. The uniqueness of the analysis is that the material for the evaluation comes mainly from within Japan. Each of the areas have been evaluated by recognized experts, and, in most cases, assessments of the future direction of the technology are offered. The areas selected include:

a) Biochemical Process Technology
b) Biosensors
c) Cell Culture Technology
d) Protein Engineering
e) Recombinant DNA Technology

Additional areas such as plant biotechnology were recognized by the panel; however, while important, this area was considered to require a longer time for commercial application than some of the areas selected.

Special historical and cultural factors that impact on the development of biotechnology in Japan have been treated in the final chapter of the report.

For each of the areas considered, an attempt was made to identify the principal laboratories, scientists, and sources of information that relate to that subject. This information will be relevant to monitoring the development biotechnology in Japan over the next several years.

# 2. Genetic Information Transfer

Dale Oxender, University of Michigan

DNA (deoxyribonucleic acid) is the molecule that contains all the genetic information present in a living cell. Therefore, it contains the codes for products which carry out all cellular functions, including its own replication. DNA is a linear polymer of deoxynucleotide repeating units. As diagramed in Figure 2-1, two complementary strands of DNA form a double helix. One strand of the helix contains regions of code called genes which direct the synthesis of cellular products such as hormones and enzymes. The first step in the information transfer within a cell is the transcription of the gene into a messenger RNA. The information in the messenger is then translated into the product of the gene. Some of the main products of genes are proteins which function as enzymes. The protein product is a linear polymer of amino acid repeating units, and the sequence of these amino acids is coded for by the particular sequence of nucleotides in the DNA. The functional enzyme is formed by specific folding of the linear form into a unique three-dimensional structure. The way the protein folds is determined by the specific sequence of amino acids. It is possible to take advantage of this information transfer pathway to develop new or altered enzymes by changing the information in the gene for a given enzyme. This process is referred to as gene splicing, recombinant DNA technology, or more often as genetic engineering. The commercial application of genetic engineering to produce an industrially important product is referred to as biotechnology.

The following section contains a review of the research discoveries that served as breakthroughs for the development of the field of genetic engineering in general and protein engineering specifically.

<u>What are the recent advances in molecular biology that give rise to the field of genetic engineering?</u>

One of the recent advances was the discovery in the early 1970s of a new class of enzymes that cleave DNA called restriction endonucleases. The diagram in Figure 2-2 illustrates how the restriction enzymes can be used to splice one type of DNA to another. These enzymes bind to the double-stranded

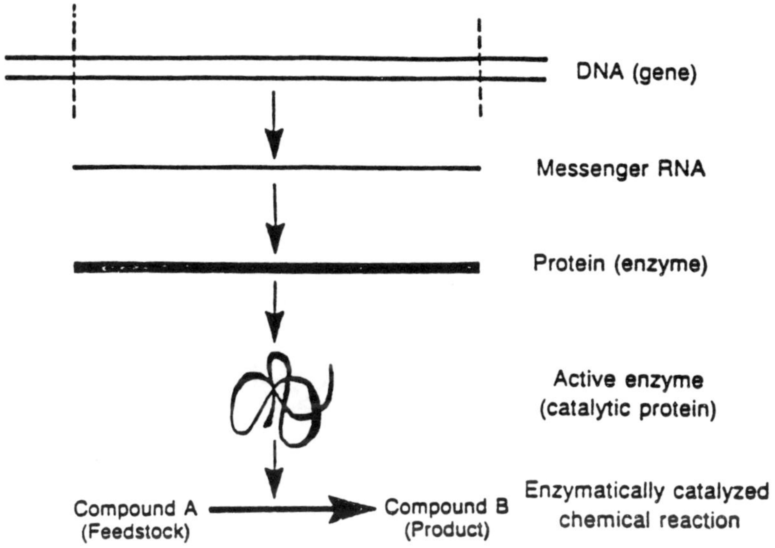


Figure 2-1. Information flow in biological systems.

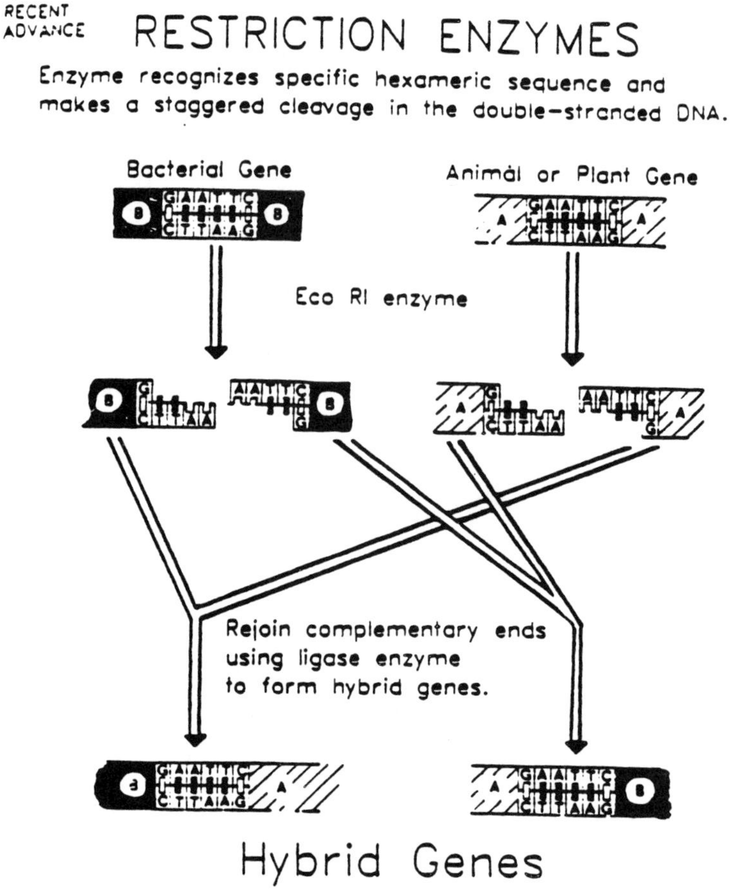


Figure 2-2. Restriction enzymes.

DNA molecule and search for a specific nucleotide sequence in the DNA, usually 4 or 6 nucleotides. When the enzyme finds the unique sequence, it makes a staggered cleavage of the strands of double-stranded DNA. Since the strands contain complementary overlapping ends, they can be reunited and rejoined by another enzyme called a ligase. As shown in the diagram, DNA from different sources or species can be spliced or combined with each other by breaking the DNAs with a restriction enzyme and subsequently rejoining them with the ligase enzyme.

A second advance has been the development of transferable plasmids which can be used as cloning vectors. As shown in Figure 2-3, a plasmid is a small circular double-stranded DNA, and it resides inside a bacterial cell. These plasmids are not a part of the bacterial chromosome and can replicate independently. Specialized plasmids for cloning have been constructed by adding genes which code for proteins that confer antibiotic resistance on the bacterial cell host and by adding cloning sites which are recognized and cleaved by restriction enzymes to permit the addition of DNA fragments or genes.

## CLONING OF A GENE

The use of gene splicing techniques, shown in Figure 2-2, to introduce a DNA fragment containing a gene into a plasmid or cloning vector is referred to as "cloning a gene." This procedure is illustrated in Figure 2-4 of this series. The key to the procedure is the use of the same restriction enzyme for both the plasmid and the preparation of the DNA fragment containing the gene to be cloned. Using the same restriction enzyme ensures that the ends of the DNA fragment will be complementary to the ends of the cleaved plasmid DNA. The DNA strands can be joined to produce a hybrid plasmid. The hybrid plasmid carrying the cloned gene can now be transferred into a bacterial cell where it will direct the production of large quantities of the desired protein or enzyme.

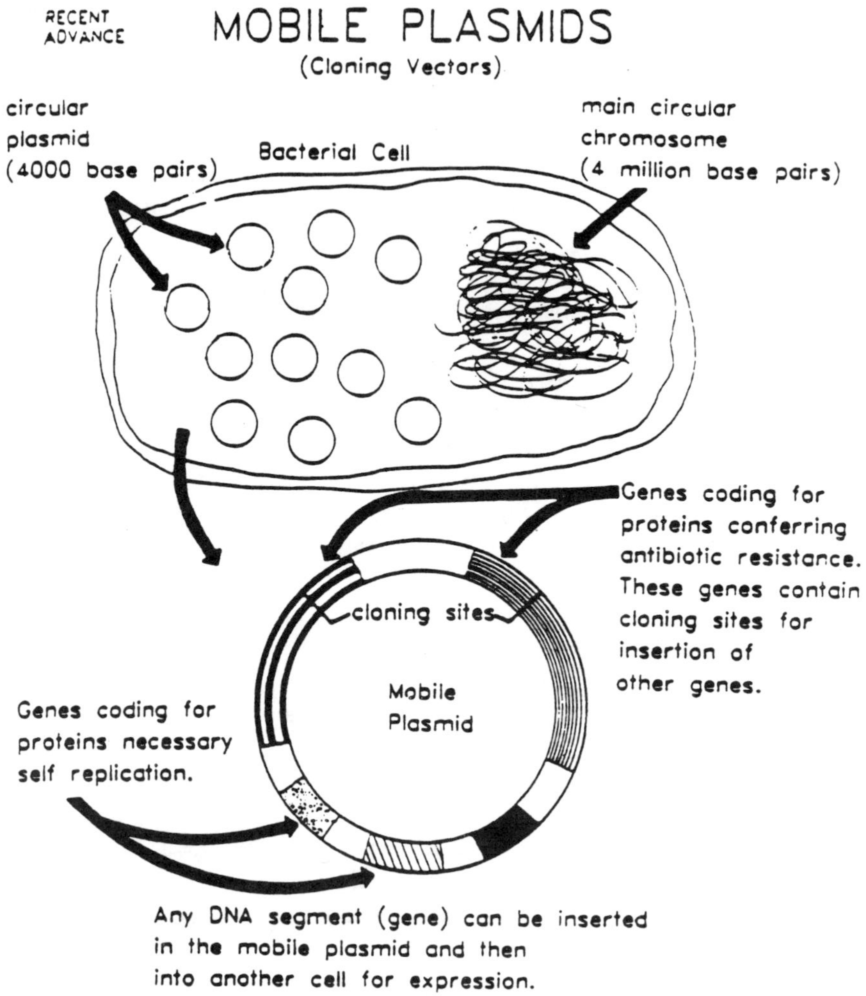
RECENT ADVANCE
MOBILE PLASMIDS
(Cloning Vectors)
circular plasmid (4000 base pairs)
Bacterial Cell
main circular chromosome (4 million base pairs)
Genes coding for proteins conferring antibiotic resistance. These genes contain cloning sites for insertion of other genes.
cloning sites
Mobile Plasmid
Genes coding for proteins necessary self replication.
Any DNA segment (gene) can be inserted in the mobile plasmid and then into another cell for expression.

Figure 2-3. Mobile plasmids.

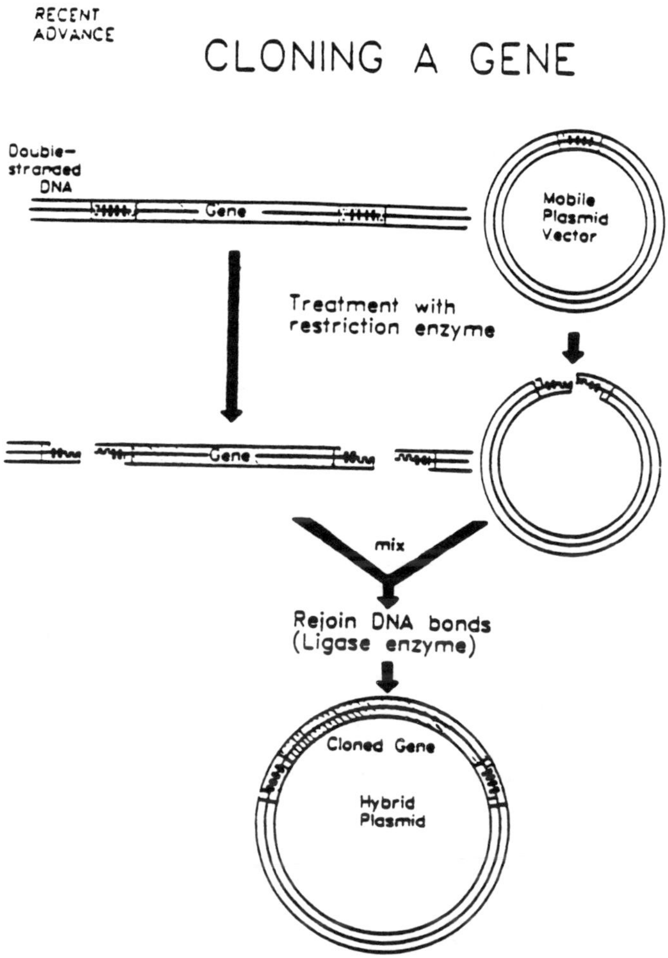


Figure 2-4. Cloning a gene.

**PROTEIN ENGINEERING**

The specific structural alteration of a protein (enzyme) by altering the information in the gene for that protein (enzyme) is referred to as protein engineering (see Figure 2-5). The topline of the figure shows a gene which encodes a protein product (enzyme). The stippled region in the gene depicts a structural alteration in the gene which was accomplished by the newly-developed in vitro mutagenesis techniques. This change of information in the gene results in a change in the messenger and, ultimately, a corresponding change in the amino acid sequence of the enzyme. The altered amino acid sequence may produce a structural change and potentially improve the function or stability of the engineered enzyme. Protein engineering techniques can be used for virtually any enzyme by first cloning the gene for the enzyme, and then applying the appropriate gene splicing and engineering procedures.

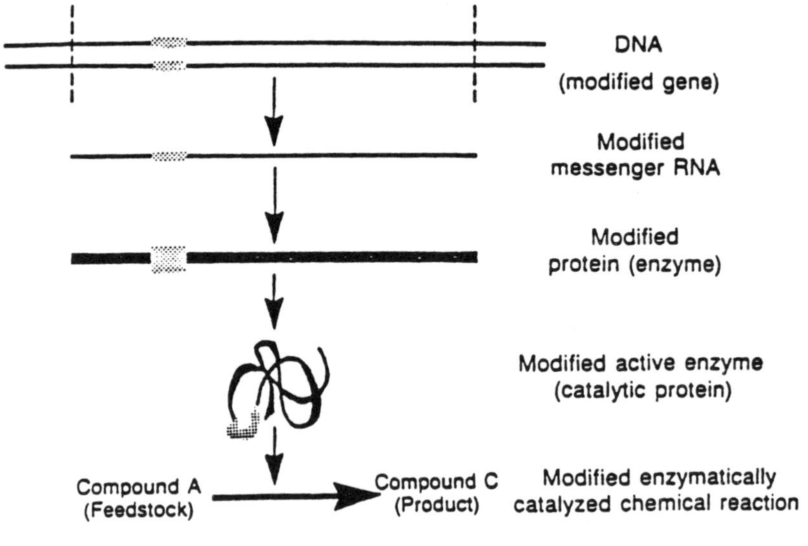


Figure 2-5. Mechanism of protein engineering.

# 3. Biochemical Process Technology in Japan (Bioreactors, Including Fermentation, Enzyme and Separation Technology)

Charles L. Cooney, Massachusetts Institute of Technology

## 3.1 Introduction

### 3.1.1 Rationale for Biotechnology in Japan

Japan is a country with limited natural resources and a population of about 120 million. As a consequence, it is not able to sustain itself in food and energy production or any other commodity product. Thus, its strategy for maintenance and growth is to add value through manufacturing and then, via export, generate the revenues to import all the materials necessary for its population. Japan's success in implementing this strategy is shown in its large excess in balance of trade. In recent years, there has been a general trend to use high technology to facilitate addition of value to feedstock. Biotechnology is very attractive because of its potential to add substantial value through high technology. Furthermore, Japan has a long and successful experience with biochemical manufacturing processes. At first glance, it appears as though it should be easy to build an advanced biochemical process industry on the old.

There are, however, several constraints in the development of new industry in Japan. These constraints include: a lack of venture capital, a high debt financing with low risk, a poor history of innovation and a strong public awareness and opinion on safety and environmental impact. As a consequence, any new industry must be "socially acceptable" and not be perceived as a dangerous or polluting industry. It will be shown later that this has had a major impact on the development of the biochemical process industry (BPI) in Japan (e.g. single-cell protein or SCP from hydrocarbons, approval of research on scale-up of recombinant DNA organisms). Japan also has been slow to develop an entrepreneurial infrastructure that enables it to rapidly transfer new science and technology into commercial realities. (In this context, the entrepreneurial infrastructure includes both the mechanism for generation of venture capital and availability of individuals to assume responsibilities and "get the ball rolling.") Furthermore, the educational system does not provide new highly skilled people quickly in response to new initiatives and it has a weak basic research structure, e.g. few Ph.D. programs. New biotechnology involves less capital per unit of product. This is attractive to an industry that tends not to finance risky ventures. Japan has a very sophisticated domestic health care market which is set up to service this industry well.

### 3.1.2 Historic Perspective on Biochemical Engineering in Japan

Fermentation technology has been important to the Japanese for millennia since many fermented foods and beverages are important to the Japanese diet. Modern fermentation technology, however, began in the post WWII years with the conscious development of process technology for the manufacture of higher valued goods such as amino acids and nucleotides. These products were for the food industry. The Japanese fermentation industry was able to establish a pre-eminent position in their manufacture by rapidly finding and exploiting high producing micro-organisms. This task was achieved at the time by using what was relatively low cost labor to play the tedious game of mutation and selection of large numbers of organisms. Japanese scientists did not play a major role in developing the science of microbial physiology; but rather positioned themselves to take rapid advantage of the discoveries in this area to seek even better overproducers of amino acids and nucleotides. It is interesting to note that the early success in the BPI in Japan came from microbiologists, whose training in "biotechnology" developed through microbiology and agricultural chemistry departments in the universities. This is in contrast to the U.S. where training became based in chemical engineering departments.

The second major stage of development of a modern BPI was in antibiotics manufacture. Here, the Japanese proceeded along a similar strategy and built an industry for the efficient production of known antibiotics. It was easy to attract foreign investment into this industry, both as capital for plant construction and to form joint ventures. Japan was quickly recognized as a major developing antibiotics market that could be best served by Japanese production. Furthermore, many of the joint ventures were set up to take advantage of the local capability to screen and improve new cultures for new products.

The third stage of BPI development was an attempt to move into more commodity product manufacture. Examples include: single-cell protein (SCP), citric acid, monosodium glutemate (MSG), and lysine. In the area of amino acid manufacture, where they had developed low-cost technology, the Japanese industry was quite successful. In SCP, however, they were unsuccessful. The primary reason for the lack of success was that the industry was trying to develop SCP technology being hydrocarbons as a feedstock at a time when our understanding of chemical carcinogensis was evolving, Japanese industry's image as a "polluter" was at a maximum, and it was realized that the substrate chosen for production of SCP, normal alkanes could contain heterocyclic hydrocarbons unless properly processed. Thus, public opinion in Japan forced the industry to abandon this business. It is interesting to note that later, the price of oil was to make this feedback too expensive for a product that would utilmately compete with low priced soy bean meal.

The fourth stage of the industry began in the early 1970's, when Japanese labor became increasingly expensive and there was a need to develop products that were more "home grown." The Japanese BPI became more broad in its products, and new therapeutic agents began to emerge from Japanese science (i.e., the work of Umezawa). Worldwide, biotechnology was beginning to evolve as a high tech area with potential far beyond earlier expectations. The area of enzyme technology began to develop, and the Japanese rapidly became proficient in seeking new methods for improving enzymes as catalysts, and in developing new applications. They developed this area in a "technology driven" manner and not a "market driven" manner; this led to some notable failures in terms of business development.

The Japanese were among the first to develop a technology for the isomerization of glucose to fructose in the manufacture of high fructose corn syrup (HFCS). Yet, they were unable to commercialize this technology and US industry led the way for establishment of this major (greater than $2 billion) biotech business. It was logical for the industry to be based in the U.S. since we have the feed stock, corn, however the Japanese killed the opportunity to provide the enzyme catalyst, glucose isomucose which is a $75 million dollar worldwide business. Japanese technologists led the way for development of whole cell immobilization technology but have not been able to generate a major business in this area except for specialty chemicals. They pioneered early research on novel air-driven fermentors, but it was ICI, Ltd. in the U.K. that was the first to really take advantage of this technology.

### 3.1.3 Expansion of Biotechnology

Japan's goal in biotechnology is to establish a pre-eminent position in the worldwide Biochemical Process Industry (BPI) by maximizing the value added to feedstocks through biochemical processing. In order to do this it seeks to expand its current base in the BPI and leverage this manufacturing base by development of an industry based utilizing the "new biotechnology." Industrial growth in Japan is dependent on the ability of its managers to leverage its resources effectively to maximize value added often on imported raw materials. One of the major resources is a well educated, hard-working, and aggressive labor force. Japan is well positioned to take advantage of high technology such as the microchip and genetic engineering as a means to generate value added. In the case of biotechnology, it is necessary for the Japanese to implement the following:

a. There must be an increase in the number of students prepared to develop and support the BPI.

b. There must be effective use of automation to maximize the return on invested capital, while minimizing labor requirements.

c. Japan must open and maintain a window to the US and Europe to follow the rapidly changing science of biotechnology.

d. MITI and other appropriate agencies must stimulate industrial development of biotechnology.

e. Industry needs to become more "market" rather than "technology driven" in its development.

### 3.2 Biochemical Process Industry in Japan and Strategy for Expansion

One of the important questions being asked in this study is, why does or how can the Japanese BPI maintain a competitive edge in biochemical process technology over the U.S.? One approach to this question is through an examination of the following equation:

$$P = VF_m(S_pS_a - C_m)$$

where P is profit, V is total market size (in mass), $F_m$ is the market share which a company can achieve, $S_p$ is the selling price (expressed as a value placed on utility) as \$/unit of activity, $S_a$ is the specific activity as units/mass, and $C_m$ is the manufacturing cost in \$/mass. Profits attract competition and thus this equation can be used to gain some insight into the ability of a competitor to be competitive. Often, the parameter of interest is <u>R</u>eturn <u>O</u>n <u>I</u>nvestment (ROI), which is the profit divided by the investment. The ROI provides the incentive to make an investment. The volume of the market is determined by the utility of the product and its price elasticity. The market share attainable by a company is determined by such factors as marketing capability, technical service, distribution costs, "brand name," time of market entry relative to competition, and proprietary protection such as patent protection. The driving force is the profit margin, which is $(S_pS_a - C_m)$ in the above equation. This margin can be increased by minimizing the manufacturing cost ($C_m$) and by maximizing the net selling price ($S_pS_a$). The selling price has two components - the specific activity and the utility value. Price is based on utility and there is a maximum value per unit of activity. The primary way to improve the profit margin is to develop a new product with an improved specific activity. Since manufacturing cost is approximately proportional to mass and not utility, it is often possible to introduce new products with increased activity and still sell them at the same price per unit of activity and thereby increase the profit margin.

Important questions are how well do, or can, the Japanese BPI compete in: (1) process improvement, reduction of $C_m$, (2) introduction of new, improved products with higher $S_a$, and (3) marketing to gain new worldwide market share.

Japanese industry has been successful in developing high productivity processes with low capital investment and labor requirements. However, there is difficulty in reducing the cost of the feedstock and energy costs because these are relatively expensive commodities in Japan. Introduction of new products with improved specific activity has not been a strength of Japanese industry; it, in fact, may become not only a problem in maintaining market share but a barrier to entering the business. Frequently, Japanese industry will go abroad to license new products. Marketing of a product overseas has not always been easy for Japanese industry despite their reputation as purvayors of electronic goods. In sum, the Japanese BPI has a difficult time in competing with many other producers. A strategy of distributing the manufacturing and establishing joint market agreements can have a major effect in reducing both the manufacturing and marketing costs. By distributing their manufacturing and marketing through joint ventures or by building companies in the U.S. they will establish the market presence necessary to be very competitive.

### 3.3 Academic Resources

In an assessment of the prospects for biotechnology in Japan, one should understand the educational system and the capabilities of that system to provide a continuous stream of professionals to work in biochemical engineering and applied microbiology. Recently, Drs. Wang and Cooney, at MIT's Chemical Engineering Department, completed a survey on worldwide biochemical engineering research and education. In the results of this survey, 13 organizations in nine universities were identified as major participants in training of biochemical engineers in Japan. These organizations and their staffing and space are listed in Table 3-1. In several universities, biochemical engineering, research, and training is divided between departments. Furthermore, the lead department providing this training is not a chemical engineering department, as is the case for the U.S., but agricultural chemistry, applied microbiology, or fermentation technology department.

Table 3-1. Summary of biochemical engineering and biotechnology activities in Japan.

| | | | Personnel Data | | | | | |
|---|---|---|---|---|---|---|---|---|
| Organization | Department Affiliation | Name of Contact | Total Faculty | Ph.D. Staff | Professional Staff | Ph.D. Students | B.S./M.S. Students | Space $M^2$ |
| Hiroshima Univ. | Ferm. Tech. | Nagai, S. | 24 | 5 | 22 | 3 | 40/18 | 1,800 |
| Inst. Phys. & Chem. Res. | Institute | Endo, I. | 7 | 3 | 7 | 1 | 10/ 4 | 400 |
| Kyoto Univ. | Chem. Eng. | Hashimoto, K. | 9 | 0 | 0 | 0 | 13 | 210 |
| Kyushu Univ. | Food Science | Ueda, S. | 2 | 3 | 0 | 2 | 8/ 6 | 145 |
| Nagoya Univ. | Chem. Eng. | Kobayashi, M. | 3 | 3 | 2 | 3 | 14 | 410 |
| Nagoya Univ. | Agriculture | Koga, S. | 4 | 0 | 1 | 1 | 6 | 250 |
| Nagoya Univ. | Food Science | Yamane, T. | 4 | 0 | 1 | 2 | 9 | 190 |
| Osaka Univ. | Ferm. Tech. | Aiba, S. | 12 | 24 | 9 | 7 | 40/20 | 4,000 |
| Osaka Univ. | I.C.M.E. | Taguchi, H. | 1 | 3 | 2 | 1 | 3 | 1,060 |
| Osaka Univ. | Pharm. Science | Miura, Y. | 4 | 0 | 1 | 4 | 12 | 270 |
| Tokyo Inst. Tech. | Lab. Resource Util. | Susuki, S. | 7 | 2 | 7 | 7 | 24 | 800 |
| Tokyo Univ. | R&D Microbiol. | Yano, K. | 3 | 1 | 0 | 4 | 8 | 240 |
| Tsukuba Univ. | Appl. Biochem. | Ueda, K. | 1 | 26 | 15 | 20 | 40/20 | 3,400 |
| TOTAL . . . . . . . . | | | 80 | 70 | 67 | 55 | 295 | 13,175 $M^2$ |

*From Wang and Cooney (1984)

The number of doctoral candidates relative to BS/MS candidates is relatively small. This proportionately low level of doctoral candidates suggests that relatively little "basic" research, is done in Japanese universities. It is also consistent with the observation by L. Smith (Fortune, Oct. 29, 1984, p. 22, "Creativity Starts to Blossom in Japan") that Ph.D.s in Japan are, in general, produced in lower proportion to the population (2,040 in science and technology in Japan versus 17,924 in the U.S. in 1983) than in the U.S. The reason for this is that industry prefers in hire BS/MS graduates early in their career in order to train them in "the ways of the company" since they will become life time employees. If a student requires further training, they may be sent abroad for postgraduate training. This is evidenced by the number of Japanese scientists who receive postgraduate training in US universities.

A summary and comparison of the biochemical engineering resources in the U.S., Japan, and other countries is provided in Table 3-2.

## 3.4 Strategies To Overcome Development Bottlenecks

The Japanese have had difficulty in marketing biochemical products in the U.S. and Western Europe. This is a consequence of difficulties in servicing some of these markets, e.g., diagnostic enzymes, pharmaceuticals, industrial enzymes. This is in contrast to products, such as amino acids, where the Japanese have completely dominated the business in the U.S. These products compete primarily on price and not availability of applications technology, or service. The approach that is being taken to overcome this "marketing" problem includes the following actions: distribution of manufacturing by building "foreign" plants to eliminate the distance/inventory factor, establishment of joint ventures with companies that have good marketing (e.g. sales, distribution, service) capability, and, more recently, establishing research contracts with smaller entrepreneurial companies to access new technology. They have chosen the areas of RDNA, mammalian cell culture, and bioreactors as important areas for both internal and U.S. development. They have not focused on down-stream processing and product recovery.

Table 3-2. Summary of estimated world-wide efforts in biochemical engineering and biotechnology research and education at universities (1983).

| Country | Total Faculty | Ph.D. Staff | Professional Staff | Ph.D. Students | B.S./M.S. Students | Facility/ Space (Sq. Meters) | Estimated Annual Equipment Budget for Personnel ($ Million) |
|---|---|---|---|---|---|---|---|
| W. Germany | 23 | 168 | 216 | 175 | 85 | 26,350 | 33 |
| Japan | 80 | 70 | 67 | 55 | 295 | 13,175 | 22 |
| U.K. | 45 | 62 | 69 | 56 | 82 | 8,030 | 22 |
| U.S.A. | 79 | 28 | 26 | 167 | 126 | 8,520 | 18 |
| Australia | 24 | 23 | 23 | 41 | 144 | 5,820 | 16 |
| France | 23 | 32 | 49 | 24 | 20 | 2,900 | 8.5 |
| Scandinavia | 13 | 16 | 26 | 63 | 94 | 2,750 | 6.8 |
| Switzerland | 8 | 26 | 9 | 34 | 63 | 3,300 | 5.1 |
| Netherlands | 7 | 6 | 16 | 19 | 64 | 2,450 | 3.3 |
| Canada | 6 | 12 | 10 | 14 | 19 | 1,190 | 2.7 |
| Austria | 6 | 0 | 14 | 7 | 72 | 1,600 | 2.3 |
| Israel | 8 | 0 | 9 | 5 | 8 | 800 | 1.6 |
| Italy | 5 | 6 | 2 | 0 | 6 | 160 | 1.2 |
| S. America | 22 | 2 | 7 | 1 | 110 | 1,520 | 4.1 |

**Annual Basis:** Faculty = $110,000; Ph.D. Staff = $80,000; Professional Staff = $50,000; Ph.D. Student = $30,000; B.S./M.S. Students = $10,000.

*From Wang and Cooney (1984)

## 3.5 Conclusions And Recommendations

Japanese industry has taken a major initiative in biotechnology which has the potential of making it a predominant force in the world's biochemical process industry in the 1990's. It is likely to achieve this predominance by a series of actions that include: an intensive effort to accumulate a large knowledge base in biotechnology, a trend to distribute its manufacturing and marketing capabilities throughout the world, a trend to learn and access Western research and development through joint and ventures, research contracts and equity deals with small biotech companies.

The Japanese government has catalyzed growth in the BPI by calling attention to the potential and funnelling money through organizations such as MITI to stimulate research. Much of the stimulation is aimed at manufacturing technology and is likely to assist industry in becoming a low cost producer.

The Japanese government has a history of making and keeping long term commitments to R&D development.

There is a strong academic commitment to fermentation technology and the published works of the following academic investigators in biochemical engineering should be followed:

| | |
|---|---|
| Nagai, S. | Hiroshima University |
| Endo, I. | Institute of Physics and Chemistry Research |
| Hashimoto, K. | Kyoto University |
| Kobayshi, M. | Nagoya University |
| Aiba, S. | Osaka University |
| Imanaka, T. | Osaka University |
| Taguchi, H. | Osaka University |
| Susuki, S. | Tokyo Institute of Technology |
| Yano, K. | Tokyo University |
| Ueda, K. | Tsukuba University |

# 4. Biosensors

John Wilson, Thomas Lord Research Center

## 4.1 Methodology

A preliminary list of companies and individuals active in the biosensor field was constructed through patent and publication searches. The results are shown in Table 4-1. The searches covered world patents through December 1981 and Japanese patents between April 1981 and September 1983. Several smaller separate searches were made for patents added during 1982, 1983, and 1984.

The patents retrieved by this strategy were limited to those which had been actually published in Japan. This may have excluded:

1. Patent applications by Japanese organizations which have been pursued via the world patent route with Japan as a designated state but which have not yet been published in Japan.

2. Patent applications by Japanese organizations in countries which publish specifications earlier than Japan.

Furthermore, the database contains only published patent specifications. Normally, Japan automatically publishes full patent specifications 18 months after application. Thus, patents filed during the past 18-20 months have not been retrieved.

A side benefit of this strategy was that a list of patents was generated of non-Japanese companies patenting in Japan. This gave a limited view of world-wide technical activity in this field.

Other sources of information suggested that several companies were seriously interested in biosensor technology but had been missed by the patent searches (Table 4-2). In particular, NEC has substantial activity, but

Table 4-1. Japanese commercial interests in biomedical sensors revealed by patenting and publishing activities 1977-1983.

| | Bio-element | | | Transducer | | |
|---|---|---|---|---|---|---|
| | Enz. | Microbe | Immuno | ISE & FET | Amp | Other |
| Asahi Medical | | | + | ++ | | |
| Anritsu Denki | | + | | | | |
| Ajinomoto | + | +++ | | | +++ | |
| Amano Seiyaku | + | | + | | + | |
| Fuji Electric | + | | | | + | |
| Hitachi | +++ | | | | +++ | |
| Horiba | | | | + | | |
| Japan Tobacco & Salt | + | + | | | + | |
| Kankyo Bunseki Centre Co. | | + | | | + | |
| Kuraray | + | | + | +++ | | |
| Kurita Kogyo | | + | | | | |
| Kyoto Daiichi Kagaku Kogyo | + | + | | | + | |
| Kyowa Hakko Kogyo | | ++ | | | + | |
| Matsushita Electric | +++ | | | | +++ | |
| Mitsubishi Electric | + | + | | | + | |
| Mitsubishi Rayon | + | + | | | + | |
| Nikki-so | + | | | | + | |
| Olympus Optical | | | | +++ | | |
| Omron Tateisi | +++ | | | +++ | | |
| Sankyo | | | + | | | |
| Shimadzu | | | | ++ | | |
| Shindengen Kogyo | | | | + | | |
| Showa Denko | | | | + | | |
| Toshiba | +++ | | | ++ | + | |
| Toyobo | + | | | | + | |
| Toyo Jozo | ++ | | | | ++ | |
| Yokogawa Electric | ++ | | | | ++ | |

Key:
+ 1-4 patents or publications
++ 5-9 patents or publications
+++ 10+ patents or publications

Table 4-2. Other Japanese companies with some interests in the biomedical sensor field.

| | Preliminary Information |
|---|---|
| Denki Kagaku Keiki | Attendees at Fukuoka Conference, 1983 |
| Fuji Film | Dry reagent chemistry |
| Fujisawa Pharmaceuticals | Attendee of Fukuoka Conference, 1983. Developing insulin closed-loop delivery system |
| Japan Scientific Instrument Co | Exhibitors at Fukuoka 1983; Marketing only? |
| Mitsubishi Chemical | Have set up the Mitsubishi-Kasei Life Sciences Institute; with some biosensor work. |
| Nippon Electric (NEC) | Very recent announcement of micro enz-FET. |
| Nissin Electric | Working with Ajinomoto on biosensors |
| Sumitomo Electric | Interests in trans-cutaneous blood gas monitoring; attendee at Fukuoka, 1983. |
| Suntory | 1976 paper concerned with glucose oxidase, published from Suntory central Research Institute. |
| Teijin | Attendee at Fukuoka, 1983. Interests in monoclonal antibodies and medical diagnostics. |

appears to have avoided revelation of their involvement through patenting. Other companies that have adopted a similar strategy may have been missed by our search.

A similar procedure was used to generate a list of academic groups active in biomedical sensor research (Table 4-3). In this case, both patents and publication indices were searched, and it is expected that coverage is fairly comprehensive, given the favourable attitude to publication in Japanese academic life.

Further information was obtained from a report prepared by the Comet Research Institute in Japan. This report summarizes Japanese patents and technical newspaper articles written in Japanese over the past 2-3 years, supplemented by information from conversations with technical journalists. Finally, a number of interviews were conducted by. Those interviewed included both commercial and academic people and three persons with an overview of the requirements of Japanese medical markets for biosensor technology. Those persons interviewed directly are listed in Table 4-4. Wherever possible inconsistencies in information were resolved and every attempt was made to obtain as accurate a picture as possible of current activities in Japan.

## 4.2 Overview of Sensor Technologies

At least four different configurations of sensors are under development for both biomedical and process applications. Most current research activity is directed at biomedical products for which the need is thought to be considerable and the requirements less demanding of the developer.

The four most common sensor configurations are:

1. Potentiometric or Amperometric Electrodes.

   In this case, typically, a surface reaction or other process produces a potential difference or current as the result of a

Table 4-3. Japanese academic groups with biomedical sensor interests.

| Group | Institutions(s) | Area of Interest |
|---|---|---|
| M. Aizawa | Tsukuba University, Ibaraki | Enzyme Immunosensors, luminescence $O_2$ sensor |
| I. Chibata | Institute of Applied Biochemistry | Enzyme Immobilisation |
| M. Esashi, T. Matsuo | Tohoku University, Sendai | Micromechanical fabrication; ISFETs |
| K. Hiiro | Government Industrial Research Institute, Osaka | Ion Selective Electrodes |
| K.Imaeda | Hoshi College of Pharmacy | luminescent cell $O_2$ sensor |
| H. Imai | Horishima University | glucose oxidase electrode |
| Ito | Nagoya Institute of Technology | glucose sensor |
| H. Inokuchi, T. Yagi | Institute of molecular Science, Okazaki; Shizuoka University, Oya | cytochrome c and organic conductors |
| H. Kaneko | Tokai University | Magnetic bacteria sensor |
| I. Karube | Tokyo Institute of Technology | All biosensors |
| T. Kawashima | Kagoshima University | uricase membranes |
| H. Kitano | Kyoto University | enzyme electrodes |
| Kobayashi | Nagoya University | enzyme/luciferin electrodes |
| B. Kurihara | Jikei University | glucose sensor |
| Y. Nakamoto | Kanazawa University, | composite urease membranes |
| S. Nakamura | Kitasato University, Kanagawa | glucose sensors |
| M. Shichiri | Osaka University | *in vivo* micro-glucose sensors |
| S. Shimizu | Magoya University | luminescent cell $H_2O_2$ sensor |
| N. Shinohara | Tokyo University | urease electrode |
| H. Tsubomura, N. Yamamoto | Osaka University | Immunoelectrodes |
| Y. Umezawa | University of Tokyo | ion-selective immunoelectrode |

Table 4-4. Interviews in Japan.

| | | | |
|---|---|---|---|
| 1. | Commet Research Institute | Information | Mr. Kodera |
| 2. | Tokyo Institute of Technology | Academic | Prof. Karube |
| 3. | Tsukuba University | Academic | Prof. Aizawa |
| 4. | MITI | Government | Mr. Seki |
| 5. | Toshiba | Commercial | Mr. Wada |
| 6. | Ajinomoto | Commercial | Dr. Hikuma |
| 7. | Osaka University Hospital | Medical | Prof. Hayashi |
| 8. | Osaka University | Academic | Dr. Tsubomura |
| 9. | National Cardiovascular Centre | Medical | Mr. Katayama |
| 10. | Omron Tateisi | Commercial | Mr. Yoshida |
| 11. | Nissin Electric | Commercial | Mr. Ohtani |
| 12. | Fuji Electric | Commercial | Mr. Nakamura |
| 13. | Mitsubishi Chemical | Commercial | Mr. Sugai |
| 14. | Nikkei Biotech | Technology Journal | Mr. Kawata |
| 15. | Toyo Jozo | Commercial | Mr. Izumi |
| 16. | NEC | Commercial | Dr. Tsuji |
| 17. | Showa Denko | Commercial | Mr. Kimura |
| 18. | Defence Force Hospital | Medical/Academic | Prof. Sekiguchi |
| 19. | Tohoku University | Academic | Prof. Matsuo |
| 20. | Mitsubishi Electric | Commercial | Mr. Maeda |
| 21. | Matsushita Electric | Commercial | Mr. Date |

shift of electrode potential or in electrode polarization. Such devices may be direct, where the effect that is measured actually takes place on the electrode of a cell, or indirect where the electrochemical effect is caused by a shift in bulk concentration of the product of a reaction which is measured by a secondary electrode system. The latter type is particularly common and is exemplified by the many sensors for such analytes as glucose, which can be converted enzymatically to oxygen or hydrogen peroxide. These, in turn, can be sensed by a conventional Clarke oxygen electrode.

2. Chemically-Sensitive Electronic Devices (CSED's).

The electrical characteristics of a solid state device, typically a field effect transistor (FET), are modified by a surface reaction or other process often occurring on or in a membrane attached to the surface of the device.

3. Enzyme Thermistors.

These are devices whose resistance is usually sensitive to the thermal effect caused by a reaction. They have seen limited commercial application.

4. Opto-electronic Devices.

These vary widely, but are usually sensitive to changes in fluorescence or other properties (such as light transmissivity) that may result from biological processes.

### 4.3 Indirect Electrochemical Devices

Historically, most development work has focussed on this type of sensor. Typically it consists of an enclosed cell containing an electrolyte and a secondary detector. One end of the cell is closed by a selectively permeable membrane on which, or in which, is a catalyst - typically an enzyme - that is designed to produce a product that is detectable by the secondary detector. The product diffuses electrolyte into the cell that results in a shift in cell potential that is related to the concentration of the measured species in solution external to the membrane. In a common commercial example of this type of device. The enzyme is glucose oxidase or glucose dehydrogenase. The secondary detector is a Clarke oxygen electrode which is used to determine the concentration of glucose in blood or urine samples. Such devices tend to be relatively bulky and, because they depend upon the chemical integrity of the enzymatic protein attached to the membrane, are likely to be unstable at elevated temperatures (e.g., during steam sterilization) or at low or high pH.

Any improvement in robustness will depend upon improving the inherent stability of the enzyme - through the use of hermophilic enzymes or replacing the enzyme with a stable, biochemically-specific synthetic biocatalyst (a step that is still far in the future). A reasonable degree of chemical stability and some improvement in thermal stability, can be achieved by encapsulation of the enzyme either within the membrane or separately.

Similar concerns apply to many of the other types of devices that are discussed here. Any device that depends upon the use of biologically-selective material, whether enzyme, whole cell, or monoclonal antibody, suffers from the same limitation in application that requires exposure to high temperature or relatively aggressive conditions. Fortunately, these problems are primarily limited to process applications; however, some biomedical applications require long-term stability, which is also difficult to achieve.

Table 4-5 suggests some sensor enzymes which may be useful (and in some cases have been applied) in device applications to detect or analyse the substrates noted. In some applications the immobilized enzyme can be replaced with a simple permselective membrane that passes (without modification in this case) only the species to be measured. This approach is limited to those species for which permselective membranes and suitable secondary detectors are available. However, such devices are far more inherently robust than enzyme-based systems.

Table 4-5. Sensor enzymes.

| SUBSTANCE DETECTED | USE | ENZYME |
|---|---|---|
| Lactate | Clinical Research, Sports Medicine, Fermentation | Lactate Oxidase, Lactate Dehygrogenase, Lactate Dehygrogenase |
| Pyruvate | Clinical Research, Sports Medicine, Fermentation | Pyruvate Oxidase |
| Cholesterol | Clinical, Food Processing | Cholesterol Oxidase |
| TNT | Explosives Detection | Trinitrotoluene Oxidoreductase |
| Carbon Monoxide | Combustion Control, Fire Alarms, Pollution Monitor, Clinical | Carbon Monoxide Oxidoreductase |
| Glucose | Clinical, Fermentation, Food Processing | Glucose Oxidase, Glucose Dehydrogenase |
| Alcohol | Fermentation, Biodeterioration Alarms, Law Enforcement | Methanol Dehydrogenase Alcohol Oxidase |
| Formaldehyde | Biocide and Pollution Monitor | Methanol Dehydrogenase |

Prototypical indirect sensors of the type described here have now been made with a variety of secondary detectors. For example, a variety of ion-selective electrodes and even semi-conductor devices have been used to detect the secondary effect of the enzymatic reaction on the primary membrane. The response of such devices is often surprisingly fast and sensitivity is good. However, most devices are non-linear with respect to substrate concentration except at very low concentrations and, in most cases, difficulty has been experienced in maintaining calibration. This difficulty may be associated with enzyme instability with changes in membrane characteristics due to mishandling or misuse.

## 4.4 Direct Electrochemical Sensors

Much of the recent sensor development effort has concentrated on a variety of electrochemical cells in which one or more electrodes of the cell are modified by treating the surface with a biological material such as an enzyme or antibody. Often this is done in combination with a "mediator" that serves as a transfer agent for electrons between the biological material and the electrode surface. Early work in this area used low-molecular-weight mediators (e.g. benzoquinone) in homogeneous solution. More recently, other more complex heterogeneous mediators attached to the electrode surface have been used. Most of these are redox compounds such as ferrocene and its insoluble derivatives such as dimethyl ferrocene. Soluble ferrocene (e.g. carboxy ferrocene) may also be used if it is desirable to have the mediator in homogeneous solution, rather than localized on the electrode. Recent work on such sensors has been reported by Higgins et al. in the U.K.

Wingard and co-workers at the University of Pittsburgh have also made excellent progress in the development of mediated enzyme-based electrochemical systems. These have been developed primarily for biomedical sensor applications including in-vivo measurement of biologically important substrates such as glucose. While such devices are applicable to the measurement of substrate concentrations during biological processing, they suffer the same instability of the biological material as their indirect counterparts.

Even at body temperatures, instability of enzyme-based sensors such as enzyme electrodes remains a serious problem. Recent efforts have concentrated on the use of amperometric electrochemical devices that can measure substrate concentration through oxidation of the substrate at the electrode surface without the need for a biological intermediary of any kind. In cells of this kind, the electrical current passing through the cell is proportional to the amount of substrate available to be oxidized. In the potentiometric case, the potential generated by the cell is related to the amount of oxygen detected by the electrode. This is related to the amount of substrate "seen" by the enzyme catalyst.

In principle, amperometric devices that avoid the use of enzyme or other biological material should offer both selectivity and stability. In practice, this is seldom the case in complex biological systems -- more than one electrode process, including the desired one, may contribute to the observed current. Such devices are still far from being commercial successes.

Serious problems remain regarding enzyme stability, even for commonly available and much-researched enzymes such as glucose oxidase. Any electrochemical device in a complex environment, such as is encountered in the human body or during fermentation, is prone to side effects particularly to parasitic electrochemical reactions which may generate spurious signals from the cell. In the case of the indirect electrochemical sensors discussed earlier, unless enzyme stability can be greatly increased (and enzyme activity maintained) through the use of thermophilic species, or unless sensors can be devised which do not require biological materials for their action, we are unlikely to have a successful robust sensor for either biomedical or bioprocess applications.

Similar concerns apply to the use of sensors of this kind in other areas such as the detection of industrial or other environmental pollutants. Unless sensors of this kind can retain calibration and sensitivity and remain robust for extended periods of time, their industrial application will be extremely limited.

If the issues of robustness and stability can be resolved for sensors of the kinds discussed here, few other problems appear to remain. The sensitivity of such sensors is remarkably high and response times are generally good.

## 4.5 Enzyme Thermistors

A number of conductometric devices, of which the enzyme thermistor is the most sensitive and also the most technically successful, have been devised to measure concentrations of substrates in biological solutions. In the enzyme thermistor an enzyme appropriate to the substrate to be determined is immobilized on or close to the surface of a thermistor, a device whose resistance is highly temperature-sensitive. When exposed to the substrate, the enzyme-catalysed reaction that takes place on the surface of the thermistor generates a thermal effect owing to the reaction enthalpy. As a result, the resistance of the thermistor changes. This change can be measured in a normal Wheatstone bridge configuration. The resistance can be related directly, but not linearly, to the substrate concentration.

Variations on the enzyme thermistor involve the use of immunosorbents such as monoclonal antibodies in place of or in conjunction with enzymes. The enzyme may also be used in conjunction with a cofactor, but, in this case, no mediator is required since no electron transfer is involved. Enzyme thermistors have been used to analyse for a wide variety of substrates.

A major advantage of the enzyme thermistor is its simplicity and its avoidance of any need to make measurements with complex electronic, electrochemical, electro-optical, or spectrophotometric equipment. As a result, the device is not prone to the interferences that plague electrochemical sensors but, like most other biosensors, it is dependent upon the stability and continued catalytic activity of the enzyme or other biological material.

Thus, the enzyme thermistor is subject to the same limitations for long-term applications as the other devices discussed here (the robustness of the sensor depends upon robustness of the enzyme used in its construction). A further limitation is that it must be used in a system that is otherwise thermally stable since it measures very small thermal effects. This requires the use of precise temperature control equipment, often in a slipstream away from the main body of fluid on which the determination is being made, or a large system with high thermal inertia. These are major limitations and are likely to prevent the successful commercialisation of the enzyme thermistor.

### 4.6 Enzyme Transistors

Interest has been growing in devices which combine semiconductor technology with enzymes or other biological entities. Most studies have focussed on the use of field-effect transistors (FET's). Appropriate modification of the surface of an FET can make it sensitive to ionic species in solution (ion-sensitive field-effect transistors (ISFET's) or other chemical agents (CHEMFET's). FET's modified with enzymes are commonly referred to as ENFET's or ENZFET's.

A particularly attractive feature of semiconductor sensors is their potential for miniaturization and for direct integration with signal processing micro-electronics. The sensing portion of the semiconductor can be very small which makes these devices especially interesting for biomedical applications. Most early work on FET-based sensors was focussed on the development of very small scale ISFET's. Pioneering studies by Caras and Janata led to the development of small scale ENFET's for penicillin determination. Danielson et al. have pointed out that a simple hydrogen sensor may be made if the metal gate of a MOS-structure FET is made of catalytic metal such as palladium instead of gold. According to the inventors, the sensitivity of such a device can be greatly enhanced by associating it with a co-enzyme regeneration system. Such devices may also be used for ammonia determinations. Again, the sensitivity of the device can be increased by immobilization of a suitable enzyme on the device surface, thus introducing the same stability problems as other devices discussed earlier.

ENFET's are part of a larger family of chemically-sensitive electronic devices (CSED's) which can be made to sense virtually any species, ionic or non-ionic, electroactive or non-electroactive, chemically reactive or inert by the appropriate selection of reagents coupled to the sensors. The advent of CSED's has added a new dimension in chemical detection which, when augmented by the more conventional electrochemical measurement approaches, greatly broadens the scope and fields of application.

Most of these devices have been developed to monitor metabolites in patient care. Many consist of FET's coupled with an appropriate permselective membrane, the selectivity of which may itself be enhanced by modifying the membrane through attachment of ion-specific ligands such as valinomycin or crown ethers. Such devices are more stable than those involving the use of biological species such as enzymes for reasons already discussed. One prospect for future advances in this field is the development of membrane systems that are biochemically selective through appropriate modification by non-natural organic species.

Finally, an inherent advantage of CSEDs is the potential for producing multiple sensors. These may consist of multiple sensors of the same kind mounted on a single chip, or the establishment of several sensors of different kinds on a single chip (a more complex technical problem). In the first case, a simple statistical analysis of signals from individual sensors can be used to eliminate readings from failed or off-calibration sensors, thus improving the quality of the average result. The analyis can be performed in circuitry built into the sensor. In the second case, multiple analytical data can be obtained from a single sensor, thus minimizing the number of probes inserted into the patient. Pioneer work in the area of multiple sensors has been done by Pace and by Cheung et al. Commercial devices have been produced by Toshiba, and devices developed by Covington in the U.K. have been used to monitor blood chemistry during surgery. It is not clear that sensors of this kind will be of particular value in bioprocess control since the space available for the sensor is seldom limited and separation of individual sensors may actually enhance reliability. However, the use of multiple sensors of the same kind with statistical analysis may be a useful way of prolonging effective sensor life.

## 4.7 Optoelectronic Devices

Recent developments in fibre optic sensor technology and fluorometry has led to the development of a number of inherently simple yet highly sensitive and robust devices that may be of critical importance in bioprocess control in the near term.

Most fibre optic chemical sensors, whether of single- or double-fibre design, follow the same principle. Light passing down the fibre is used to "excite" a fluorescence, either directly in the fermentation broth or other biological fluid or in a susceptor which is sensitive to changes in the chemistry of the fluid, located at the end of the fibre. The resulting emitted radiation is collected and provides a measure of the concentration of the fluorescing substrate. Several such devices have been proposed by Hirschfeld. Others are under development in a number of laboratories. While those devices that include a susceptor depend upon the stability and robustness of the susceptor itself, the inherent stability of a fibre optic device presents a major advantage over a sensor involving the enzymes or other biological materials. A number of prototype instruments are in use, and commercial versions can be expected soon. Devices exist that can measure cell fluorescence (NAD/NADH equilibrium), pH, and dissolved oxygen. Other sensors await the evolution of appropriate susceptors.

One American company (Biochem Technology) has announced the commercial introduction of a new fermentation sensor based on fluorometric technology. While fluorometry has been used for a number of years in enzyme essays and other analytical procedures, no device suitable for continuous fermentation use has been developed. In the present fluorometer, a collimated light beam of controlled wavelength is used to excite fluorescent radiation from the cell mass in the fermenter. The intensity of the fluorescent signal is measured photometrically and provides a very sensitive indication of the cell status during fermentation. While it is presumed that the fluorometer "sees" primarily the NAD/NADH equilibrium in the cell during fermentation, an equilibrium that is highly sensitive to the cell status as determined by oxygen or nutrient supply, the exact relationship between the fluorescent signal and cell chemistry is not clearly understood.

At least one other company is known to be developing a fluorometer. A wide range of novel applications of this technology can be expected during the next few years.

The fluorometer can also be used as an indirect device, using a susceptor at the fluorometer tip, whose contents are separated from the biological process by an appropriately permeable membrane. The fluorometer now measures variations in fluorescence in the susceptor as a function of changes in pH or oxygen level. In this case, the source of fluorescence is better understood, and more easily related quantitatively to the status of the process.

Both fibre optic measurement techniques and those based upon fluorometry appear to have a significant future since both offer a robust, sensitive and apparently versatile method of measuring a number of biochemical variables in real time during the course of fermentation. However, in at least some cases, the signal obtained is only empirically related to the biological process. While such signals can be used as the basis for automatic process control, particularly if adaptive control methods are used, the absence of an analytical relationship between the measured parameter and process chemistry precludes an analytical approach to process control.

## 4.8 Markets in Japan

Interviews with senior staff at Japanese hospital analytical laboratories revealed a widespread reliance on high throughput multichannel analysers and of emergency use single function analysers for the bulk of clinical analyses. These facilities are typically found in regional hospitals and in commercial service laboratories. Hitachi, Shimadzu and Toshiba supply the bulk of the multi-analyzers with a number of companies supplying the single function unit such as those incorporating the widely-used "indirect" glucose oxidase electrodes.

The introduction of home testing was considered highly likely and made more probable by the national preoccupation with gadgets and a general concern about health and quality of life. Acceptance of devices for use by patients themselves is likely to be rapid in Japan. A strong home market is available to support and test a fledgling disposable biosensor industry.

## 4.9 Government Activity in Japan

The level of government financial support for this field, whether for applications in the biomedical or the process industry, is quite small. Support is largely confined to encouragement of general inter-company communication. No specific development programs have been undertaken by MITI or, as far as we can determine, by any other government agency. MITI is, however, supporting other forms of biotechnology research. MITI believes that the current industrial activity in Japan is sufficient to ensure progress in biosensor development without further government intervention.

## 4.10 Academic Activity in Japan

In general, those academic groups in Japan that appear to be most productive in developing biosensor technology are closely linked with industry and are carrying out research with a strong developmental flavour. These Japanese academic groups (listed below) appear to have achieved a high level of expertise with particular groups being perceived as leaders of individual subdisciplines.

| | |
|---|---|
| Karube | Biosensor Design |
| Chibata | Enzyme Immobilization |
| Inokchi/Yagi | Electron Transduction |
| Aizawa | Immunosystems |
| Matsuo/Esashi | Microfabrication |

In many respects, these groups correspond to those found in the U.K. and elsewhere. For example, Covington's group at Newcastle (leaders in biosensor design) works with the Wolfson Institute in Edinburgh (microfabrication experts).

These groups have good to excellent international academic contacts and international reputations for excellence that, at least in one or two cases, are built on what appears to be a carefully managed publicity campaign. The groups provide up-to-date information and technology to their commercial colleagues. Close liaison between academic and commercial research does not appear to have led to philosophical conflicts, since both partners evidently find satisfaction in their joint contribution to Japanese society.

The technical scope of the academic work uncovered in Japan is apparently adequate for current biosensor development, but there are only a small number of groups involved and these are focussed on relatively short term commercial research. The majority of concepts appear to have originated from Karube or Aizawa. There are several groups studying longer range prospects such as the use of enzymes in sensors and the use of luminescent bacteria. A narrow focus to most work is confirmed by the paucity of serious work on opto-electronic, calorimetric, piezoelectric, or optical transducer mechanisms. No patents on these other transducer designs were detected in our searches.

More detail of academic and industrial sensor activity is provided in the following section.

## 4.11 Technical Overview of Japanese Research

For many analytes, the most effective way of achieving selectivity in measurement is by using an enzyme specific for the analyte. No matter which transduction system is used, (amperometric, potentiometric, etc. as outlined in Section 4-2) the effective immobilization of the enzyme component is crucial to the stability and function of the sensor. To effectively immobilize an enzyme to a transducing system, the following criteria must be met:

| | |
|---|---|
| long shelf life | stability in operation |
| good reproducibility | ease of manufacture |

The weightings of these criteria will change as sensors come into their second and third generation of development. Requirements for a successful biosensor are shown in Table 4-6. Present and future stages of biosensor development are indicated in Table 4-7. Presently-available sensors are defined as being at stage 0 with considerable development work being focussed on stages 1 and 2.

A number of companies have manufactured first generation enzyme sensors (stage 0 sensors) using a variety of immobilization techniques. Table 4-8 shows the immobilization system used by these companies.

Table 4-6. Requirements for a successful biosensor.

WHAT IS NEEDED?

SENSORS THAT ARE:

- FOOLPROOF
- RUGGED
- STABLE
- REPRODUCIBLE
- PRE- OR AUTO-CALIBRATED
- LIGHTWEIGHT/SMALL
- INEXPENSIVE
- MINIMAL OR NO SAMPLE PRETREATMENT

Table 4-7. Present and future states of biosensor development.

| | |
|---|---|
| STAGES OF MEDICAL SENSOR DEVELOPMENT | |
| STAGE 0: | CURRENT ANALYTICAL SYSTEMS |
| STAGE 1: | SENSORS FOR USE WITH REAGENTS IN ANALYSERS |
| STAGE 2: | SENSORS WHICH CAN BE MANUFACTURED CHEAPLY, RELIABLY, AND IN QUANTITY. PROGRESSIVELY REDUCING NEED FOR CALIBRATION (FINALLY REQUIRING NONE). SYNDROME RELATED MULTIFUNCTIONALITY |
| STAGE 3: | NONINVASIVE: QUASI NONINVASIVE SENSING TECHNOLOGIES, INCLUDING IMAGING. PERMANENT AND SEMIPERMANENT IN VIVO SENSORS |

Table 4-8. Enzyme membranes.

IMMOBILIZATION

FIRST GENERATION:

- COLLAGEN/ALBUMIN
- COVALENT LINKS (GLUTARALDEHYDE)

IMPROVEMENTS:

- PROTECTIVE MEMBRANE (E.G POLY-CARBONATE)

KEY DEVELOPMENTS

TOSHIBA

- ASYMMETRIC ULTRAFILTRATION MEMBRANE

NEC

- COVALENT LINKAGE OF ENZYME TO GATE OF FET

BOTH COMPANIES USING IC FABRICATION TECHNOLOGY

In addition to commercial work, a number of techniques are being developed by academic groups. Professor Karube believes that for small sensors, Langmuir-Blodgett and vapor deposition technologies will be used to form submicron layers of biomolecules, protected by cellulose acetate membranes. Professor Matsuo uses a cross-linked urea-albumin membrane for his work on enzyme-FETS. Although the majority of immobilization work being carried out in Japan such as the cross linking of enzyme with albumin or collagen has been pioneered elsewhere, there are some significant exceptions. The formation of a composite membrane with a protective membrane covering the enzyme layer leads to greater stability and prevents blocking of the membrane by sample constituents. Fuji Electric and Matsushita use a polycarbonate membrane. Omron Tateisi and Toyobo have jointly developed an enzyme sandwich protected on the outer layer with porous PVC and attached to the electrode via cellulose. The developments by NEC and Toshiba appear to be the most significant.

NEC has developed a system which can use conventional defense-quality semiconductor fabrication technology to manufacture a urease enzyme FET. They use spin coating to form the enzyme membrane followed by photo-masking and UV irradiation to inactivate one sector of the urease membrane to act as a reference. Once they have improved the stability of the electrode, they will be well placed to mass produce disposable sensors. Toshiba also appears to be using their skills in semiconductor fabrication technology. During discussions, they commented on the use of "photolithographic techniques for embedding enzymes." This almost certainly relates to the manufacture of a multi-enzyme membrane by selective photo-cross-linking of sections of the membrane and removal of unpolymerized material. Thus a membrane can be built up which contains a number of different enzyme sectors.

In forecasting the immobilization techniques which will be used for second generation medical sensors, it seems likely that the most successful technologies will be those which are compatible with standard micro-fabrication techniques. The development of effective protecting layers to cover the enzyme will also increase stability and biocompatibility.

## 4.12 Transducer Mechanisms

### 4.12.1 Amperometric Systems

The development of amperometric enzyme electrodes dominates biosensor development work in Japan, particularly glucose measurement. The principle of amperometry relies on the production of electrons either directly or indirectly during a chemical reaction; the selectivity is provided by the enzyme of choice. The majority of these amperometric systems use oxidase enzymes which oxidize their substrates using molecular oxygen, yielding hydrogen peroxide which is measured by a secondary detector. The developments by companies are summarized in Table 4-9.

A number of academic groups are also working on amperometric systems. Karube, in cooperation with NEC, has pioneered a micro-oxygen electrode using gold and silver electrodes. This development paves the way for miniaturized amperometric enzyme sensors. Professors Moriizumi and Karube have manufactured a glucose sensor measuring only some 4mm x 2mm x 0.6mm on a single silicon crystal wafer, and several groups have been involved with the development of glucose sensors especially for in vivo use. However, it appears that the fundamental problems of stability and biocompatibility required for in vivo applications have not yet been overcome.

Assuming that the appropriate enzyme can be immobilized stably, (while retaining sufficient activity) the current generation of amperometric sensors are adequate for incorporation into analysers. For second generation sensors, however, a number of improvements will be required. The most significant developments in overcoming these problems have been made by Karube (working with NEC) and by Matsushita.

The micro-oxygen sensor fabricated by NEC using conventional microfabrication technology will lead to inexpensive multiple-enzyme sensors. With continued development, this technology should satisfy the very stringent demands of second generation sensors.

Table 4-9. Japanese companies developing amperometric systems.

| Company | Analyte | Transducing System |
|---|---|---|
| Fuji H202 | Uric Acid<br>Glucose<br>Amylase | Measurement of<br>by Pt:Ag |
| Hitachi | Various:<br>No details available | |
| Matsushita | Glucose | Thin film sensor<br>based on Pt<br>electrode |
| direct | | Sensor based on<br>electron transfer<br>via electron<br>mediators |
| NEC<br>electrode | Oxygen | Micro-oxygen<br>using gold & silver<br>electrodes<br>(manufactured for<br>Karube) |
| Omron Tateisi<br>$H_2O_2$ | Glucose<br>lactic acid | Measurement of<br>Pt:Ag |
| Toyo Jozo | Triglycerides<br>Cholesterol<br>P.choline<br>Glucose | Measurement of<br>$O_2$ by Pt/Ag:AgCl |
| Toshiba | Glucose<br>Uric acid<br>Lactate<br>GPT | Measurement of<br>$O_2$ by<br>Pt/Pb:KOH<br>Pt/Ag:AgCl (most<br>recent work) |

Matsushita has made two interesting developments: (1) a redox system which transfers electrons directly from glucose oxidase and (2) the thin film electrode. In the latter they chose to use oxygen as the electron carrier with a platinum electrode rather than the redox system. Their redox mediator system appears to be less efficient than oxygen as an electron carrier due to slower movement within the membrane.

The thin film electrode, though not as sensitive at present as other electrodes, will be inexpensive and represents a novel approach and a significant leap forward; the technology will be rapidly adaptable to a number of other oxidase enzymes.

### 4.12.2 Potentiometric and FET Mechanisms

For the detection of ions, ion selective electrodes have provided the first generation of successful sensors in Japan by the marketing of single- and multi-ion sensors by both Hitachi and Toshiba (Table 4-10). Some development problems must remain since Hitachi has temporarily withdrawn their 9-function device, but it is likely that these relate to the enzyme-ISE functions (see below). Future development of ion-selective charge-sensitive devices will concentrate on size reduction, cost reduction and performance increases. ISFET's hold promise for the second generation of smaller ion-sensitive sensors and are being actively pursued for this purpose by Hitachi, Toshiba, and Kuraray with academic input from Professor Matsuo (Table 4-11). The technology for mass producing single- and multi-ion micro FET sensors appears to be very close to fruition in Japan. In itself, a multi-ion sensor cannot satisfy a large proportion of the market need; quantification of glucose, urea, etc. are required simultaneously.

Table 4-10. Potentiometric and Ise-based sensors.

| COMPANY OR GROUP | STATE OF WORK | ANALYTES |
|---|---|---|
| Prof. Tsubomura | Immuno-electrode work proceeding very slowly | Trypsin |
| Hitachi | Multi ISE-based devices nearing full production | $H^+$,$Na^+$,$K^+$,$Cl^-$ |
| Showa Denko | Developing ISE for HPLC monitoring | $Cl^-$ |
| Toshiba | Triple ion sensor on market, 5 enzyme ISE's developed, 1 year from market launch | $Na^+$,$K^+$,$Cl^-$ various metabolites |

Table 4-11. FET-based sensors.

| COMPANY OR GROUP | STATE OF WORK | ANALYTES |
|---|---|---|
| Prof. Karube | enz-FET's discontinued, urease FET initiative to NEC. Affinity-FET's underway | glucose, antigens |
| Prof. Matsuo | Several micro-FET's constructed. Development and membrane attachment methods | pH, $pCO_2$, $Na^+$ |
| Hitachi | Some research on ISFET | $Cl^-$ |
| Kuraray | Manufacture of dual micro recently delayed(?) Collab. with Matsuo | $pCO_2$, pH |
| Mitsubishi | Small dual sensor fabricated; used as development tool. Collab. with Matsuo | urea, glucose antigens in future |
| NEC | SOS FET fabricated using IC technology; used as development and publicity tool. Collab with Karube | urea, multi-enzyme in future |
| Toshiba | New developments | ions and antigens in future |

Enzyme-ISE's and enzyme FET's are, therefore, under development (Tables 4-10 and 4-11) for first-generation sensors. Both Hitachi and Toshiba have enzyme-ISE devices in an advanced state of development. For single function applications, the relatively large-sized enzyme-ISEs appear to be commercially adaptable. However, in the context of multi-function enzyme-ISEs, it is apparently hoped that adequate freedom from cross-talk and buffer sensitivity will be provided simply by the localization of the enzyme and its ionic product close to the ISE surface. It is likely that significant flow rates over adjacent enzyme-ISE's are required to completely control cross-talk and that problems in this area have beset final production development of the Hitachi multi-function device. Toshiba has followed the alternative route of developing numerous single-function enzyme-IJSEs which they expect to market within one year.

It is evident that Toshiba and Hitachi enzyme-ISE's will be among the competitors in the first generation of biosensors.

Since miniaturization of potentiometric enzyme electrodes generates a similar group of problems as development of enzyme-FET devices, without the inherent advantages of the low impedance signals provided by the latter, work on micro enzyme ISE's was not expected. In contrast, two companies (Mitsubishi Electric and NEC) are developing enzyme FET's, although neither Hitachi nor Toshiba revealed any active interest in enzyme-FET development despite their work on enzyme-ISE's. Mitsubishi Electric is collaborating with Prof. Matsuo in enzyme-FET developments. Prof. Matsuo believes that the biggest remaining fabrication problems stem from the need to form thin and stable enzyme membranes over the gate regions. However, this difficulty appears to have been successfully addressed by NEC in a very short time by making use of IC fabrication concepts. Encapsulation difficulties are also reduced by their use of SOS technology. Mitsubishi Electric and Prof. Matsuo both claim that cross-talk problems in multi-enzyme FET arrays can be overcome either by high sample flow rates or by positioning of each enzyme FET adjacent to others which operate through production and detection of different charged species.

The major hurdle remaining for all enzyme FET technology is the sensitivity of the devices to buffer strength; no solutions to the problem were offered. A real contribution by enzyme FET's to the second generation of medical sensors will depend on some means of overcoming this fundamental difficulty.

In general, the advantage of micro-enzyme FET devices over micro-amperometric enzyme electrodes is expected to be their low impedance signal which will not be reduced in proportion to reduced gate size. Against this, amperometric devices are inherently less susceptible to interference from buffers, nonselectively bound charged species, and cross-talk. For this reason, Professor Karube is not actively developing enzyme-FET devices, but his group has recently put effort into affinity-FET systems including antibody-coated FET's and FET's coated with a glucose-binding protein from yeast.

Since free-moving charged species are not generated by such devices, cross-talk is not a problem. As they rely entirely on the selective binding component (antibody or glucose receptor) to generate different amounts of binding between analyte and irrelevant molecules, specificity is likely to be their weak point. Nevertheless, if selectivity is adequate, and given the fabrication technology developed at NEC, it seems that affinity-FET devices might well be commercially feasible during the second stage of sensor development. Both Mitsubishi Electric and Toshiba stated their interest in this approach, although NEC technology appears to be ahead at present.

### 4.12.3 Luminescence Mechanisms

While two Japanese academic groups, those of Professors Imaeda and Shimizu, have made use of immobilised luminescent bacteria for the detection of $O_2$ and $H_2O_2$ respectively, we judged that this technology is unlikely to impact medical sensors and it was not subjected to further evaluation.

Perhaps of greater potential medical importance is the work of Kobayashi et al., who have immobilized cholesterol oxidase separately with a Cypridina luciferin analogue. The finding of a linear relationship between light output and substrate concentrations suggest this may form the basis of a practicable "oxidase electrode." Similar development by Professor Aizawa employs luminol. No commercial interest in these research programs was revealed.

While functional, it is unlikely that these devices will be developed rapidly enough in Japan to have an impact during the first generation of enzyme-based sensors. In the second generation, they will compete directly with amperometric devices, and will probably contain direct electron mediators between enzyme and electrode. As the energy pathway of such systems is considerably more direct than the "electron to light to electron" path involved in luminescent systems, it is unlikely that the latter will be competitive.

The attractions of antibody-based sensors for quantification of a very large number important analytes were clear to all our discussants. Active development work was revealed only by Professors Aizawa, Karube, and Tsubomura. The latter was disappointing and, even if the work is valid, is likely to proceed too slowly to have a commercial impact. Of Professor Aizawa's several immuno sensor-related projects, most are not strictly sensors but require added fluid reagents. The studies on antigen-quenching of electro-luminescence might be of future importance.

Professor Karube had conducted limited studies on immuno-FET performance and felt that such a system might be effective in the future. This view was shared by Prof. Matsuo, Mitsubishi Electric, NEC, and Toshiba. Both Mitsubishi Electric and Toshiba appeared to be considering immuno-FET development.

With active involvement in immuno-diagnostics through Mitsubishi Chemical and some FET expertise at Mitsubishi Electric, a joint Mitsubishi effort in this direction could be fruitful if the principle is valid. Nevertheless, progress is likely to be limited by bio-FET fabrication technology rather than immuno-technology.

From our study of commercial activity in Japan, a pattern emerges as to which companies are likely to be successful in pushing developments forward to second generation sensors, and those which are unlikely to proceed beyond the first generation. As highlighted above, one of the key skills which will be required for marketing second generation sensors is mass production. It is, therefore, the electronic companies whose business is in high volume manufacture rather than the biotechnology companies who will be better placed to develop and manufacture the second generation sensors. Companies such as Toshiba and Hitachi already have well-respected names in the medical products area. Others such as NEC and Matsushita will be penetrating these markets.

Table 4-12 summarizes the writer's views on those companies as to: (1) their ability to produce effective second generation sensors, (2) those which will stop at the first generation, and (3) the "dabblers" who are unlikely to produce effective medical sensors. While this table represents an objective view of the companies we visited or collected sufficient data on to make a judgement, there may well be developments we are unaware of which lead to additions to this list. The companies likely to be most active in second generation medical sensor developments come from the electronics field.

The development of dry-reagent chemistry-based test strips is a technology in direct competition with sensors. Only Fuji-Film company is known to be active in Japan, although Canon Cameras is reputed to be interested in this area. They are currently well behind Kodak who have their Ektachem system developed.

Table 4-12. Ability of Japanese companies to manufacture and develop medical sensors.

| Likely to produce second generation medical sensors | Likely to produce first generation sensors | "Dabblers" |
|---|---|---|
| Hitachi | Hitachi | Ajinomoto |
| Matsushita | Fuji Electric | Nissan Elec |
| Mitsubishi Chemical | Kuraray | Showa Denko |
| Mitsubishi Electric | Toshiba | |
| NEC | Omron Tateisi | |
| Toshiba | Toyobo | |
| Kuraray(?) | | |

## 4.13 Conclusions On Japan

The Japanese have a substantial home-market of 120 million people, with some 153,600 GP's. They can develop and test their products in the domestic market before exporting them.

Concepts for medical sensor development come from the literature both in and outside of Japan. The academics in Japan have good international contacts and an excellent relationship with industry. There is therefore a free flow of information to Japanese industry from academics around the world. The group run by Karube at the Tokyo Institute of Technology is a major driving force behind biomedical sensor development. Karube and his colleagues will continue to feed industry with concepts and developments, and to interest new companies in entering this area.

The innovation we have seen within companies has concentrated at finding solutions to manufacturing problems, not in the generation of new concepts. Historically, Japanese industry (and most notably the electronics industry), has been effective in taking developments forward to a point where they can manufacture goods economically. It appears that medical sensors for some analytes are now at this stage and will receive the full might of Japanese industrial development capabilities.

Their focus has been on those techniques which are compatible with current semiconductor fabrication methods. Although this may not be the ideal system for medical sensors, the ability to manufacture with proven technologies will be of major benefit in bringing these sensors to the market. As a result, the Japanese developments have a high chance of commercial success.

The technologies in which the semiconductor companies are weak are those related to biochemistry. For immobilisation of enzymes, where only limited biochemical knowledge is required, the companies have recruited one or two biochemists and maintained close links with university departments. For techniques which call for greater biochemical experience such as immunosensors, the semiconductor companies will need a more significant biological input. We therefore predict that developments in Japan on second generation biosensors will concentrate initially on micro amperometric enzyme electrodes, ISFET's, and enzyme FET's. The development of immunosensors based on antibody/antigen reactions will lag behind these other technologies. The companies will establish products for the analysis of simple metabolites (glucose, uric acid, urea), ions and trace metals ($Na^+$, $L^+$, $Ca^+$), and enzymes

(GPT, amylase, GOT) -- and then broaden into the technically more demanding area of immunosensors. This view is supported by the fact that the current research into immunosensors within companies has been given a lower priority than the development of other types of medical sensors.

All the companies we interviewed judged the dry reagent chemistry-based tests as major competition to the developing medical sensors. In the near future, dry reagent chemistries have the advantage of more advanced technical development and the benefits of proven user acceptability. Although not dismissing dry reagent strips lightly, the majority of companies were confident that multifunctional micro sensors would be cheaper to manufacture and offer significant advantages in the future to justify their development.

In summary, Japan has high calibre academics working in close liaison with companies who are well placed to develop and manufacture medical sensors. Japanese companies represent serious competition in the development of medical sensors.

# 5. Large Scale Tissue Culture Activity in Japan

Gordon Sato, W. Alton Jones Cell Science Center

## 5.1 Introduction

### 5.1.1 Newly Emerging Technology

Much of the past activity devoted to the industrial production of biotechnology products involves transferring specific genes into microorganisms, such as bacteria, yeast, and fungi, and then growing large quantities of these organisms in a fermenter. The microorganisms are harvested, and the desired product is then separated and purified.

Significant interest is now developing for the use of cultured animal cells for production of speciality high market value products. The use of animal cell cultures in industry is a newly-emerging activity which is growing rapidly. As such, it is difficult to assess the current situation or predict the future. Currently, the main cell culture products that various companies throughout the world have targeted are lymphokines, interferons, tissue plasminogen activators, urokinase, and many monoclonal antibodies for therapeutic and diagnostic use. However, no products of any commercial significance are being produced by large scale tissue cultures at the moment. While there is a doubt about potential commercial value of any of the currently targeted products, there is no doubt that the production of useful products by large scale tissue cultures will one day become a reality.

A few words should be said about a competing technology, the introduction of animal or human genes into microbial organisms. Bacterial cells can divide every twenty minutes or so, and can be grown in large fermentors on relatively inexpensive defined nutrients. On the other hand, animal cells take 15 to 20 hours to divide and usually require expensive components such as growth factors or serum. These considerations provide microorganisms a cost advantage over the more expensive animal cell culture methods. This advantage can be offset by several considerations. It is likely that protein modification by glycosylation will be important for the activity and metabolic stability of many peptides and proteins. Only animal cells will be able to carry out these modification reactions in the foreseeable future. Many companies are now developing systems of getting cells to secrete the desired protein into the medium. A complex series of reactions are needed to properly translate and process a protein for secretion, and these processes are different for microbial and animal cells. Large scale fermentation reactors are already highly developed.

### 5.1.2 Technology Development Needed

The main approach to making animal cell cultures more competitive is to develop efficient bioreactors and tissue culture strains which make a greater quantity of a given product. Many research groups are working on large scale cultures of animal cells and are using technologies such as bead cultures, encapsulated cultures, and hollow fiber culture vessels. There is no doubt that this technology will become better and better and eventually approach the efficiency of mass cultures of microbes. The main theoretical consideration is the much larger size and fragile nature of animal cells, which does not permit the energetic agitation used in microbial cultures for efficient energy and material transfer. All of the technologies mentioned previously are aimed at offsetting these disadvantages.

### 5.1.3 Gene Amplification Needed

An important approach to increasing the production of a single product by an animal cell is to amplify the gene in question. Amplification of the gene will lead to increased levels of proteins which are the gene products. At the present time, gene amplification has only been achieved in systems where a drug kills the host cell unless a particular gene is amplified. Many research groups are working to achieve gene amplification for genes which do not, themselves, provide a selective advantage to the cells by being present in many copies; therefore, the amplification must be achieved by an indirect means. This will be a crucial development of great commercial importance and will, if patented, provide the patent holder with an overwhelming advantage.

## 5.2 Historical Developments

Until the middle 1950's, the best scientists were working on problems such as genetics, molecular biology, virology, metabolic pathways, and enzyme mechanisms. Tissue culture research was carried out by a small group of specialists whose chief concerns were the techniques of cultivation. These groups were typified by the Tissue Culture Association of the United States, and the Japanese Tissue Culture Association. These groups maintained the point of view that the culture of human and animal cells was an occult art which was too difficult for all except the dedicated specialists. This monopoly was broken in the 1950's when Drs. Eagle, Puck, and Dulbecco entered the field and helped identify the required nutrients so that the cells could be grown on a more defined nutrient medium. Today, the culture of animal cells is becoming an exact science, and procedures can be transmitted by detailed receipes. The realization that tissue culture is an exact science is new and is a reason why the commercial use of cultures is in such an undeveloped state.

## 5.3 Assessments

To assess the commercialization of cell cultures is difficult. The situation is changing rapidly, no commercially significant products are on the market, the clinical utility of the currently targeted products is not yet established, and much of the information is proprietary. In Japanese Universities, basic research ranges from mediocre to excellent. The better laboratories in Japan are not as good or as numerous as the better laboratories in the universities in the United States. The fundamental discoveries have been made in the United States, or in western Europe. The general lack of true innovative research in Japan is the result of culture, education, hierarchical organization of research groups, mechanisms of funding, and government policy. Many creative Japanese scientists are established outside of Japan, mainly in the United States. Very little effort has been made to induce these highly-productive scientists to return to work in Japan.

While basic research in the United States is much better than basic research in Japan, Japanese industrial research is carried out at a level at least as good as that in the United States. In my judgment, the competence and industriousness of the Japanese industrial research worker is probably higher than that of his American counterpart. For the present, Japan can maintain its competitive edge by buying or adapting innovative technology developed elsewhere and using it to increase the value of a product. This edge results from the high efficiency of the industrial sector.

The Japanese have not entered biotechnology through the creation of new genetic engineering firms supported by entrepreneur-venture capitalists as in the United States. Instead, these new activities are carried out as new ventures within large established companies with the encouragement of government agencies, such as MITI and the Ministry of Education.

## 5.4 Commercial Development of Cell Culture Technology

Large scale cell culture technology is being developed mainly in the larger petroleum and chemical companies in order to provide diversification from petrochemical products. Some of the major brewery companies (Asahi, Kirin, Sapporo) have extensive fermentation facilities, but have not, as yet, made significant commitments to the development of cell culture technology.

## 5.5 Some Industries Committed to Cell Culture Technology

**AJINOMOTO** is a large company currently making amino acids, with branches in the United States, and has great potential for using cell cultures. They are producing interleukin 2.

**HAYASHIBARA** is a small but active company. They are producing interferon from tumor cells propagated in hamsters.

**KYOWA HAKKO** has attempted to develop more of an academic atmosphere, and encourages publication.

**MITSUI PETROCHEMICAL INDUSTRIES** is developing plant cell culture techniques, and already markets pigments used in cosmetics and foods. The cosmetic company, **KANEBO LTD.** markets the pigments as "Biocosmetics."

**MITSUBISHI CHEMICAL** is the largest chemical company in Japan and has developed an early interest in biotechnology. They formed Mitsubishi Life Science Research Institute to carry out a more basic style of research.

**MORINAGEN** is a small company that custom produces monoclonal antibodies.

**SUMITANO CHEMICAL** is a large company with United States connections. They currently have the production of interferon and tissue plasminogen activator under development.

**SUNTORY** is a brewing company that is rapidly changing to biotechnology development. They have established connections with Rockefeller University.

**TAKEDA** Chemical Industries is the largest pharmaceutical company in Japan and is active in many areas of biotechnology using cell culture.

**TEIJEN LTD.** is a large textile company that is developing the application of monoclonal antibodies as drug delivery systems.

**TORAY INDUSTRIES** is a large textile company that early developed biotechnology interests, mainly focussed on interferon.

**TOYO SODA MFG.** is an example of one of the largest petrochemical companies developing biotechnology. They plan to develop monoclonal antibodies for clinical use.

## 5.6 Biotechnology Development at the Universities

The seven major universities (Tokyo, Osada, Kyoto, Kyushu, Nagoya, Tokoku and Hokkaido) have biotechnology developments. With the help of government funding, their early interest in biotechnology was mainly directed to the immobilized enzyme technology.

The major university centers for cell culture development which have been supported by funds from the Ministry of Education are:

**Osaka University** has a Research Center for Cell Culture in the medical school which specializes in animal cell culture. The primary scientists at Osaka are Professors Okada and Matsubara.

**Kyoto University** has a Research Institute in its Agricultural School for plant cell culture techniques, including the use of cell fusion to improve the quality of plant proteins. Dr. Sendra at the Institute developed an electrofusion apparatus. Dr. Yamada is a plant biochemist there from the Agriculture school.

**Nagoya University** also has a Research Institute in its School of Sciences developing biotechnology.

## 5.7 Additional Japanese Scientists That Have Expertise in Gene Expression in Animal Cells

There are two important Japanese text books that concern the molecular biology of animal cells. The authors who have contributed chapters to these texts are listed below and can be taken as representing Japanese scientific expertise in the development of biotechnology involving large-scale cell culture.

Somatic Cell Genetics, edited by Tsutomu Yamane (Tohoku Univ.), Yoshio Okada (Osaka Univ.), Masaaki Horikawa (Kawasawa Univ.), Tosio Koroki (Tokyo Univ.)

Somatic Cell Genetics (continued)

Akira Asano, Research Laboratory of Protein, Osaka University

Yozo Igawa, Society for the Study of Cencer, Research Laboratory of Cancer

Masahide Ishibashi, Cancer Research Institute, Aichi Prefecture

Naomichi Kan, Biological Experimental Center, Central Laboratory, Japan Government

Yoshiki Oba, Department of Pharmacy, Kanazawa Univ.

Yoshio Okada, Research Laboratory of Microbiology, Osaka Univ.

Takeo Kadonaga, National Cancer Laboratory of America

Yoshihisa Kusano, Department of Domestic Science, Hiroshima Girl's College

Kiichi Kuraka, Department of Pharmacy, Kanazawa University

Toshio Kuroku, Research Institute of Medicine, Tokyo University

Masao Sasaki, Research Laboratory of Radiobiology, Kyoto University

Koki Sato, General Research Laboratory of Radiology

Fumio Suzuki, Department of Pharmacy, Kanazawa University

Toyozo Sekiguchi, National Cancer Research Institute

Hideo Tsuji, General Research Laboratory of Radiology

Takeo Toban, General Research Laboratory of Radiology

Yoshisuke Uishi, Biological Experiment Center, Central Laboratory, Japan Government

Fumio Hanaoka, Faculty of Pharmacy, Tokyo University

Yoshisado Fujiwara, Department of Medicine, Kobe University

Masakatsu Horikawa (deceased), Department of Pharmacy, Kanazawa University

Yutaka Matsutani, Laboratory of Acid-Fast Bacterium, Tohoku University

Masamo Muramatsu, Society for the Study of Cancer, Research Laboratory of Cancer

<u>Somatic Cell Genetics</u> (continued)

Hiroshi Yasue, Cancer Research Institute, Aichi Prefecture

Katsu Yamaizumi, Research Laboratory of Microbiology, Osaka University

Masaatsu Yamada, Faculty of Pharmacy, Tokyo University

Seki Yamane, Laboratory of Acid-Fast Bacterium, Tohoku University

Makoto Watabe, General Laboratory of Isotope, Tokyo

Masaki Watanabe, Department of Pharmacy, Kanazawa University

<u>Experiment Methods of Cultured Cell Genetics</u>, edited by Yukiaki Kuroda

Tatsuro Ikeuchi, Obstinate Disease Research Laboratory, Tokyo Institute of Medicine and Dentistry

Naomichi Inui, Central Laboratory, Japanese Government

Shiro Kato, Research Laboratory of Microbiology, Osaka University

Akiyoshi Kawamura, Emeritus Professor, Tokyo University

Masayoshi Kume, Department of Dentistry, Josei Institute of Dentistry

Atsuhiro Sakagami, Research Laboratory of Microbiology, Osaka University

Jiro Sato, Department of Medicine, Okayama University

Koki Sato, General Research Laboratory of Radiology

Yutaka Shimada, Department of Medicine, Chiba University

Toyozo Sekiguchi, National Cancer Research Institute

Toyozo Terashima, General Research Laboratory of Radiology

Takayoshi Tokiwa, Department of Medicine, Okayama University

Ten Nakamura, Faculty of Medicine, Tokyo University

Masayoshi Naniwa, Department of Medicine, Toyama Institute of Medicine and Dentistry

Masakatsu Horikawa (deceased), Kanazawa University

Experiment Methods of Cultured Cell Genetics, (continued)

Geshiharu Matsumura, Research Laboratory of Dentistry, Tokyo University

Jun Mihashi, Research Laboratory of Agricultural Technology

Toshio Mori, Department of Pharmacy, Kanazawa University

Takeshi Yasuda, Department of Agriculture, Kobe University

Masaatsu Yamada, Faculty of Pharmacy, Tokyo University

Seki Yamane, Laboratory of Acid-Fast Bacterium, Tohoku University

Michihiro Yoshida, Faculty of Science, Hokkaido University

Masaki Watanabe, Department of Pharmacy, Kanazawa University

# 6. Protein Engineering

David Jackson, GENEX Corporation
Dale L. Oxender, University of Michigan

## 6.1 Interest and Committment of Japan to Protein Engineering

The July 10th, 1985 issue of Biotechnology News reports that Japan's Ministry of International Trade and Industry (MITI) is seriously considering spending $122 million over the next six years beginning in April 1986 in the field of protein engineering. The plans are aimed at designing new proteins, artifically producing enzymes and antibodies, and developing proteins with new functions. MITI has plans to set up a research organization with cooperation from industry, universities and the national research organizations to share funding, personnel and research results. Eventually, cooperation from foreign "enterprises" will be sought for the project.

The project will cover both basic research involving the determination of structure-function relationships in proteins and applied research in the form of actually synthesizing new proteins with specific functions for use in biotechnology and computer-aided design. Research in this area is currently supported on a small scale in Japan by the Education Ministry, The Science and Technology Agency, and the Ministry of Agriculture, Forestry and Fisheries at a few Japanese universities.

Clearly there is significant interest in Japan in protein engineering, and they have up until now been willing to enter into off-shore research programs in this area with a conviction that this field will become very important for the development of biotechnology in the future.

## 6.2 What is Protein Engineering?

In an issue of Science devoted to "Biotechnology", Philip H. Abelson, the Editor, notes that "We are now in the beginning phase of exploitation of the ability to engineer proteins. ... Through the use of synthetic genes and recombinant DNA techniques, it is now possible to begin to improve on the evolutionary process .... Objectives are to create superior enzymes for use as catalysts in the production of high-value specialty chemicals ... and to produce enzymes for large-scale use in the chemical industry."

Clearly, protein engineering is a newly emerging area of biotechnology. It concerns the systematic altering of the structure of a protein to produce a new and improved protein or hormone. As shown in Figure 6-1, the top line represents a gene which encodes a protein product (enzyme or hormone). The stippled region in the gene depicts a structural alteration introduced into the gene by newly-developed directed mutagensis techniques. This change of information in the gene results in a change in the messenger and, ultimately, a corresponding change in the amino acid sequence of the protein product of the gene. The altered amino acid sequence may produce a structural change and potentially improve the function or stability of the engineered enzyme. Protein engineering techniques can be used for virtually any enzyme or protein hormone by first cloning the gene for the protein and then applying the appropriate gene splicing and engineering procedures.

Recent advances in genetic engineering methodology, which are the subject of Chapter 7 of this report, together with improvements in the techniques for structural analysis of proteins have enabled scientists to begin to develop strategies for protein engineering.

## 6.3 Applications and Products

Genetic engineering companies as well as pharmaceutical companies are particularly interested in the possibility of designing new hormones, artificially producing enzymes, and antibodies that either have improved properties such as thermostability or resistance to oxidation. One area of immediate interest is the recruitment of an existing enzyme or hormone by structural alteration to carry out a new reaction of commercial interest.

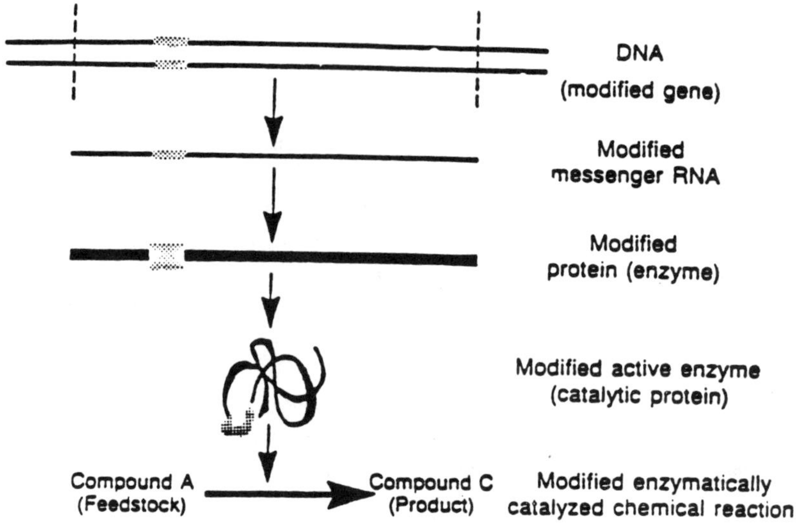


Figure 6-1. Mechanism of protein engineering.

## 6.4 Required Technologies

### 6.4.1 Recombinant DNA Technology

The basis for much of the current excitement over protein engineering results from recent advances in gene splicing and genetic engineering technologies. As described above, the gene for a given protein or hormone can easily be removed from its tissue source and spliced into a suitable vector for efficient expression in a host microorganism. One of the key capabilities for protein engineering concerns the new techniques for directed mutagensis of the cloned gene. There are 5 or 6 companies in the United States and one (Nippon Zeon, Ltd.) in Japan that now offer fully automated DNA synthesizers which can be used to synthetically produce all or part of a gene, thus greatly facilitating the design of synthetic enzymes. Once a cloned gene for a protein has been suitably altered, it can be used to produce an "engineered" protein.

### 6.4.2 Protein Structural Analysis

In order to fully exploit protein engineering, it will be necessary to gain a better understanding of the relationship between the structure of a protein and its function. A protein is made up of a linear sequence or chain of individual amino acids that constitute its primary structure. The chain of amino acids is folded into a unique three-dimensional structure that represents the functional form of the protein and determines whether it can serve as an enzyme, a hormone, or has some other function. The three-dimensional structure of a protein can be determined using x-ray analysis of crystals or by spectroscopic analysis of the protein in solution.

The techniques for determining the structure of proteins have been greatly improved during the last few years, primarily in the United States. It is now often possible to collect enough x-ray data with a computer coupled to a modern two-dimensional area detector to solve the structure of a protein in two weeks, while a few years ago it would take two years to obtain the same information.

### 6.4.3 Computer Graphics

The protein engineering field provides an important application for computer graphics. The data for protein structural analysis which are collected by the computer can be displayed on a screen using computer graphics technologies. The use of the computer for molecular modeling and computer-aided-design will play a key role in the future of the protein engineering field. At the present time, there is a rapid development of new hardware and software for computer-aided structural analysis and design in most industrialized countries of the world.

## 6.5 Interdisciplinary Nature of Protein Engineering

An examination of the variety of techniques described above clearly establishes the interdisciplinary nature of this new field. It is important for universities and industries to develop interdisciplinary research and training efforts in this area in order to bring together scientists and engineers from the fields of molecular genetics, protein structural analysis, and computer sciences.

## 6.6 Assessment of Protein Engineering Technology in Japan

The following projections of Japanese capabilities are based on interviews with senior officials of eighteen major Japanese companies (pharmaceutical, chemical, food processing, and electronic). In addition, key personnel from the universities of Tokyo and Osaka and a major Japanese trading company were interviewed. The interviews were conducted during the latter part of 1984.

Much of the technology base was developed in the United States, Great Britain and several countries in Western Europe. Japan seems more content at this point in acquiring the technology for protein engineering from outside the country. Of the three technologies required, the genetic engineering and gene splicing technologies are the easiest to acquire, and Japan has as good a capability in these areas as the United States. In fact, many of the Japanese scientists received their training in good overseas laboratories.

The developments in protein structural analysis require sophisticated instrumentation and highly trained crystallographers. Japan has a relatively small pool of trained experts in protein structural analysis, and it is not easy to rapidly expand this base. Only 32 out of 1000 members of the Japanese section of the International Crystallographers Union identify protein structural determination as a major interest. This list includes postdoctoral and graduate students. Most of the crystallographers in Japan are in industry working in the field of material sciences.

There are only two major centers for protein structural analysis in Japan. One is at the Institute for Protein Research in Osaka, where eleven professionals headed by Dr. Kutsube are listed. They operate with a $3.5 million per year budget but do not seem to have significant industrial contacts. In Tokyo, five professionals are listed on the faculty of Pharmaceutical Sciences at the University of Tokyo. They are directed by Professor Yukio Mitsui, but they do not have state-of-the-art equipment for protein structural analysis. Dr. Mitsui and his group have submitted the crystal data for the enzyme, subtilisn, to the protein data base in the United States at Brookhaven National Laboratory. Neither director knew of any significant effort in protein engineering anywhere in Japan as of late 1984. There are fewer than five students per year being graduated. The demand is small and so is the supply. This area of technology obviously represents one of the most significant bottlenecks in developing the protein engineering field in Japan.

The computer graphics equipment and programs are being imported into Japan and are usually a generation behind that now being used in the United States. A major problem with computer graphics applications has been the incompatability of various hardware and software developments. The Rikei company is the sole Japanese distributor of the Evans and Sutherland equipment which is state-of-the-art computer graphics equipment for molecular modeling. Generally, the most recent versions of software are not available, and since the crystallographic software is often written for a VAX mainframe computer, the Japanese must re-write the programs to run on the government supported Japanese computers.

## 6.7 Conclusions and Assertions

1) There is significant interest in the field of protein engineering in Japan for the development of industrial enzymes, pharmaceuticals, diagnostics, biosensors, and microelectronics.

2) Japan currently ranks fourth to the United States, United Kingdom, and Western Europe in protein engineering. There is not much activity at the present time at either the universities or industries.

3) Japan has a small pool of trained scientists for basic structural analysis, but significant abilities to acquire technologies developed outside of Japan.

4) The government agency, MITI, plans to channel significant resources, $122 million, over the next six years into a protein engineering project. The agency will coordinate Japan's efforts in this area between universities, industries, and the national research institutes and provide a format for sharing the basic technology developments among these organization.

5) Japan seems willing to acquire protein engineering expertise from other companies and universities outside Japan by cooperative arrangements.

6) Japan appears willing to take a relatively long range view of the commercialization of protein engineering technology.

# 7. Recombinant DNA Applications

Reed B. Wickner, National Institutes of Health

## 7.1 Introduction

Classical genetics - the interbreeding of strains selecting for desirable traits among the offspring - has long been used for strain improvement in industrial microbiology as well as in plant and animal breeding. The recent discoveries in molecular biology have made possible a more deliberate construction of organisms with desirable traits. The basic method that allows this to be done is the ability to connect particular genes (the units of inheritance) to carrier molecules in the laboratory. The genes on the carrier molecules can now be introduced into a cell - either a microorganism, like a bacterium or a yeast, an animal or plant cell - or a whole animal or plant. Once inside the cell or organism, the carrier maintains the gene and allows the gene's information to be expressed. Classical genetics allows one to make favorable combinations of the genes of one species. These new "recombinant DNA" methods allow one to construct useful combinations of genes from several different species together in one cell or one organism. In addition, a gene whose product (a protein) may normally be made in only small amounts can, using the new methods, be coaxed into producing enormous amounts of a useful protein (e.g., a hormone or an enzyme).

Among the realized and potential applications of recombinant DNA are the following:

- Production of useful proteins in large amounts, generally in microorganisms:
  - hormones - growth hormone, insulin, thyrotropin, etc.
  - enzymes - both those used for research and those used for commercial products
    - enzyme replacement therapy for patients lacking specific enzymes

- use of thermostable enzymes in chemical processes to replace certain less efficient chemical methods

interferons - potential treatments for viral diseases.

- Gene replacement therapy for patients with inherited diseases.
- Development of vaccines by producing key viral antigens without the viral genome.
- Construction of plants with more favorable characteristics for agriculture.
- Improvement of animal stocks for agriculture.
- Research in biomedical and agricultural areas is itself an application and the one which has already been most realized. By making possible better understanding of medical, biological, and agricultural problems, solutions can more readily be found, even if these solutions do not themselves involve recombinant DNA.
- Design of new genes and proteins for the above purposes. This subject is discussed in a separate chapter.
- Diagnosis of inherited disorders before or after birth.
- Construction of modified microorganisms with higher efficiencies in certain fermentation processes, e.g., yeasts able to use starch or cellulose to make alcohol, microorganisms more efficient at producing amino acids or nucleotides.

These new methods have made certain previously laborious tasks easy and have made some previously impossible tasks approachable. While recombinant DNA technology may not be quite the revolutionary answer to all of man's problems, it has, together with the other techniques described in this report, stimulated both research and industry in recent years.

As will be seen in detail in this chapter, Japan's situation in recombinant DNA technology is not substantially different from our own.

## 7.2 Technologies

The gene is the fundamental unit of heredity, as the cell is the basic unit of which the body is composed. Genes are made of DNA. Each gene determines the structure of a particular protein in the cell. These proteins include, (1) structural components (collagen is the main protein of which tendons are made); (2) the substances that provide the force when muscles contract (the proteins actin and myosin); (3) the proteins (called enzymes) that carry out the body's multitude of chemical reactions; (4) proteins that regulate the activity of genes; (5) cell surface proteins that are involved in communication between cells and in regulating the flow of materials into and out of the cell; and (6) hormones (such as insulin, growth hormone, etc.) and interferons - protein messengers involved in regulating metabolism and growth, defending against virus infection, etc.

The production of a particular protein is determined by several factors which are roughly multiplicative (within limits):

(1) The number of copies of the gene in the cell. This can be increased by attaching the gene to a molecule which the cell replicates to produce many copies, called a plasmid. The regulation of plasmid replication can be artificially adjusted to amplify the number of plasmid molecules per cell, thus increasing the amount of a particular protein made per cell.

(2) The stability of the maintenance of the gene in the cell. This usually boils down to the stability of the plasmid molecule on which the gene is located. If the plasmid, and thus the gene it is carrying, are lost from the cell, none of the desired protein will be made by that cell. This is a particular problem in yeast, where the high copy number plasmids are unstable and the stable plasmids are low copy number.

(3) The regulation of the expression of the gene. The transfer of information from gene to protein is a multistep process. The rate at which this process takes place varies from one gene to another. For some genes, many protein molecules are made each second from a single copy of the gene. For others, only one is made every few minutes. For yet others, none are

made. This "efficiency of expression" is determined by two factors: (a) the structure of the gene and (b) the structure of the cell. Both of these factors can be manipulated to optimize the production of the desired protein. (a) includes the strength of the promoter and upstream activating sequence or "enhancer." These factors determine the efficiency of transcription (production of an RNA copy of the gene). The efficiency with which the RNA copy is converted into a protein ("translation") is determined, at least partly, by the strength of the ribosome binding site and the use of codons for abundant tRNAs rather than for rare tRNAs.

(4) Secretion of the protein from the cell. Industrial production of a protein generally requires that it be relatively pure. There are thousands of proteins in even the simplest of cells (bacteria), and most often it is desirable that these proteins not be present in the final preparation. Although there are many powerful methods known for separating proteins differing only slightly from each other, one simple, economical solution to this problem is to have the cell selectively release the desired protein into the medium in which the cells are growing. This, too, can be manipulated by altering the same two factors as in (3) above: (a) the structure of the gene and (b) the structure of the cell. (a) is generally altered by recombinant DNA methods, while (b) is altered using the methods of classical genetics or manipulation of the environment.

Other methods are vital if the full range of benefits of recombinant DNA technology are to be realized. Among these are:

(5) Methods to introduce genes into different types of cells. A wide variety of methods exist for this purpose, but the most useful here is called "transformation" and consists of mixing the gene - usually on a plasmid molecule - with cells and some special treatments of the mixture to coax the cells into swallowing the gene. Here it is important to develop plasmids with the various special properties that make them most suited to this task. These are called vectors. The "special treatments" that induce the cells to take up the gene on the plasmid are not always so easy to guess and often require extensive empirical study.

(6) The entire range of methods to analyze the structure of genes, including "sequencing," "blotting," and many others.

## 7.3 Current Japanese Capabilities

In order to assess the Japanese capabilities in the area of recombinant DNA, we have combined several approaches. First, the author has visited Japan four times in the last four years, including twice in the last year, for a total of about six months. This time was mostly spent working in the laboratory, visiting other laboratories, and meeting other scientists at various meetings. Second, we read the literature on recombinant DNA since this is necessary to our own research.

The third approach is a more deliberate survey of the current activity in recombinant DNA in Japan. Abstracts (in the Japanese language) from the following meetings in Japan during the last 12 to 15 months were obtained and scanned: Japan Agricultural Chemistry Society, Japan Biochemical Society, Japan Genetics Society, Japan Society of Fermentation Technology, and Japan Society of Molecular Biology.

A large sample of those abstracts involving recombinant DNA were read, and the principal findings of those read which were considered to report substantial results are tabulated at the end of this chapter. This presumably represents a substantial sample of the work in this area going on in academic institutions. A great deal of research in Japan is carried out in company laboratories, and it is safe to assume that only a fraction of this work is published.

This survey of abstracts published in the Japanese language has a second purpose. While most work of sufficiently high caliber carried out in Japan is eventually published in English language journals, there is often a substantial time lag of a year or more between the completion of the work and this publication. This results, in part, from the task of translating everything into the English language and, to a lesser extent, because of the distance manuscripts must travel back and forth. The questions which might be considered, using the information tabulated in this appendix, are: (1) whether scientists not familiar with the Japanese language would find this information sufficiently useful to justify the work of obtaining prompt translations, and (2) whether all of the important and useful information presented at such meetings does eventually appear in the English literature.

I believe the answer to the second question is that it generally does, that Japanese university scientists are anxious that their work be internationally recognized, and that this means publishing in the English language. The answer to the first question will be more subjective.

The state of Japanese technology in each of the areas outlined in the section entitled "Technologies" is as follows:

(1) Plasmid copy number. Plasmids were first discovered in Japan (Watanabe), the first cloning of a plasmid replication origin was done in Japan, and work has been done by Yura, Hiraga, Hirota, Toh-e, and others in Japan on control of plasmid copy number. The vectors in current general use in the United States are the same as those used in Japan. The Japanese scientists show equal ability to design new vectors to allow high expression (e.g., recent papers by Matsubara et al.), transformation of new industrially important hosts (see Table 2), cloning of promoters, etc.

(2) Stability of plasmid maintenance in the cell. Here again work from Japanese laboratories is among the best being done. For yeast vectors using the $2^2$ DNA plasmid as a replicon, Kikuchi (Keio University) has recently shown that its instability is due to a segregation problem rather than replication as was previously thought based on work done in the United States. Hiraga did early work on the segregation of mini F plasmids. As a general problem for production, this has not been completely solved, but Japanese workers are certainly at the same level as the United States in the quality of their work.

(3) The regulation of expression of the gene. This usually involves manipulation of the "promoter sequences", or the "upstream activating sequences", or the "enhancer sequences," depending on the organism. But

manipulation of the host genotype is important as well. Evidence of a substantial amount of this type of work can also be found in the table in the appendix. For example, several comparisons of the strength of various promoter sequences have been published (e.g., T. Oshima et al. and Matsubara et al.). At least two groups are chemically synthesizing genes in order to use codons optimum for the production organism rather than those suitable for the original organism. By using codons whose tRNAs are abundant in the organism used for production (e.g., yeast or Escherichia coli), translation efficiency can be increased. Vectors suitable for cloning promoters by a shotgun approach have also been designed. Studies of control of gene expression by regulatory genes in yeast were pioneered by Y. Oshima's laboratory. The generally excellent background in Japan in microbiology and genetics has resulted in there being no barrier to reconstructing host strains or picking for study the hosts suitable for particular industrial processes.

(4) Secretion of proteins from the cell. The study of this area is still at a relatively early stage with many mechanisms apparently utilized by cells to secrete proteins and the signals for such excretion not fully clarified, although much has been done. The Japanese are active in this area as well, but they are not really among the leaders.

(5) Transformation methods and vector design. Here Japanese workers have made many important contributions, such as the lithium acetate transformation method for yeast [Kimura (Kyoto University)] and the development of host vector systems in a number of industrially important microorganisms.

(6) General "sequencing," "blotting," and other methods. Familiarity with these methods is rapidly becoming widespread in Japan, as reflected by the results shown in the table of work using these methods. Advances in this area have generally been published in the open literature and are available to all.

In summary, it is quite clear that not only are the Japanese able to use any technology developed in the United States or elsewhere, but in many of the areas central to applications of recombinant DNA, the Japanese have been among the leaders, often making seminal discoveries.

## 7.4 Special Factors Influencing The Development of Biotechnology In Japan

A brief summary of factors influencing the relative position of Japan and the United States is listed in Table 7-1. These factors are discussed in detail below.

Table 7-1. Japanese factors affecting competitive ability.

| |
|---|
| Positive |
| Strong background in microbiology, genetics, and molecular biology. |
| Strong pharmaceutical and ferementation industries. |
| Uniformity of education means high literacy and competence level. |
| Low crime and drug use frequency. |
| High cultural pressure to work hard. |
| Negative |
| Overstringent guidelines. |
| Much lower level of funding for basic research. |
| Uniformity of education stunts creativity. |
| Independence in research comes very late. |
| Other |
| Absence of natural resources -- Focus on applied aspects |
| Company collaboration with university laboratories<br>Company workers studying in university laboratories |
| Facilitates transfer of technology to companies |

### 7.4.1 History

Japan has a long history of scholarship and accomplishment in the biological sciences. Among their many important discoveries is the discovery by Watanabe of plasmids, the molecules whose use is so central to the molecular cloning methods we are discussing here, and the early work of Matsubara on λdv in which the regulation of replication of a plasmid was studied for perhaps the first time. Hiraga was the first to clone an origin of replication of a plasmid. This was carried out by genetic methods and precedes the recombinant DNA era. Okazaki discovered the discontinuous nature of DNA replication. Hirota was the first to isolate DNA replication-defective mutants, a key contribution which allowed the unraveling of the physiology and enzymology of DNA replication by others. Tomizawa has done outstanding work on T4 phage replication and recombination and, more recently, on the detailed mechanism of the Col E1 replicon, perhaps the most widely used replicon in research and applications. Takanami discovered several of the early restriction enzymes and was among the first to apply them to molecular biological analysis of M13 transcription.

More important is the broad base in Japan of work and trained people in molecular biology, genetics, and microbiology, which were in place before the beginning of the recombinant DNA era. This gave Japan the potential to quickly become competitive in this highly technical area.

Japan has also traditionally been very strong in the fermentation-based industries, such as production of alcoholic beverages, soy sauce, natto (a fermentation product of soy beans using a Bacillus strain), and others. There is a large Japanese pharmaceutical and chemical industry as well. These industries could be expected to provide both a source of interest in the applications of recombinant DNA and a source of expertise in large-scale production methods of biologicals and marketing.

Soon after the invention of cloning methods in the United States, it was widely speculated here that the new DNA combinations produced by the in vitro methods might have unforeseen dangerous effects if allowed to escape into the environment. This idea was widely debated in the United States and resulted in the "NIH Guidelines" for the conduct of such studies. The "NIH Guidelines" classified experiments based on the level of danger thought to be present in the type of DNA used, the vector used to propagate the DNA, and the ability of the host to survive in a natural environment outside the laboratory. The supposed risks were at first purely speculative, but some time later "risk assessment" experiments were carried out to determine if, for example, an oncogene cloned in bacteriophage lambda could be cajoled into producing cancer in mice. Such experiments generally showed that the risks of the recombinant DNA were orders of magnitude less than the risks of the parent DNA molecules from which the recombinants were constructed. Likewise, recombinant (constructed) plasmids were generally less stable in nature and less widely propagated than their natural ancestors.

Nevertheless, these experiments were necessarily few compared with the virtually infinite possible combinations of molecules that could be made in the laboratory, and it is only with years of experience and the absence of any known damage to humans or the environment from such experiments that the general attitude in the United States is that the dangers of these experiments is probably much less than the dangers of the DNA recombinations constantly being carried out in far greater numbers in nature. As a result, the "NIH Guidelines" have become substantially less restrictive, while still limiting or even prohibiting certain special experiments.

In Japan, the concern expressed by scientists in the United States was seen as a cause for even greater fear than was generally expressed here. The dangers of recombinant DNA were felt there to be comparable to those of radioisotopes and comparably restrictive guidelines were established. The Japanese exposure to nuclear war has produced a "nuclear allergy" among the Japanese people. As a result, it was politically necessary that the controls on the use of radioisotopes be much more stringent than they are in the United States. All experiments with radioisotopes are done in a separate room set aside for that purpose, requiring a change of shoes and various other inconveniences, such that many investigators simply avoid using radioisotopes.

Thus, when "guidelines" were drawn up in Japan for recombinant DNA experiments, the (speculative) risk assigned to a given experiment was higher than was assigned in the United States, and the precautions to be taken were higher for a given level of risk assignment. As a consequence, doing what would have been a P2 experiment in the United States required a laminar flow hood (a P3 precaution in the United States) costing $20,000, as well as the import of restriction enzymes, $^{32}$P-labeled nucleoside triphosphates, etc. Since the general level of funding of biological sciences in Japan is far below that in the United States, many even relatively well-funded laboratories simply could not afford to enter this area. This resulted in a "clone gap" between Japan and the United States. There were laboratories in Japan that were fully entered into the cloning business (Honjo, Saito, Nakanishi, Takanami, and Matsubara, among many others).

As the "NIH Guidelines" were gradually made less restrictive, so were those in Japan. The barrier to recombinant DNA work in Japan was decreasing at a time when the potential practical benefits were increasingly close to being realized in countries with less restrictive policies, and the realized benefits in research in these countries were everywhere to be seen. This combination gave rise to the "cloning fever" in Japan in the early 1980's. A sample of the fruits of this effort can be seen in the tables and will be discussed further below.

## 7.5 Current and Projected State-of-the-Art

The large number of junior faculty Japanese scientists (Jushu) doing postdoctoral work in laboratories in the United States and Europe, and the rapid dissemination of new information in both the usual English language journals and in Japanese language review journals, has meant that there was never any real knowledge gap. While there are variations from one laboratory to another, the general level of technology in the recombinant DNA field is not much different between Japan and the United States. Japan is said to have the lead in the general area of fermentation, and since the large-scale production of engineered gene products will require application of general fermentation methods, this could eventually translate into a Japanese advantage. However, while there are some spectacularly productive Japanese laboratories, there are a number of factors outside of the restrictiveness of the guidelines, which make research in Japan more difficult. The most outstanding such factor is the low level of funding of biological research. Except for a few laboratories (e.g., Mitsubishi Kasei Seimei Kenkyo-jo), there is no such thing as disposable glassware (e.g., test tubes) or disposable plasticware (petri dishes, pipets, etc.). In spite of our image of Japan as having abundant "cheap labor," technicians are relatively few compared to university laboratories in the United States. Students and junior faculty even have to clean the floors in many cases. Equipment varies from excellent to poor, with the average much below standards in the United States. Average commuting times of Japanese scientists to and from work may be an hour each way, and longer times are not unusual. Factors such as these are very variable from one institution to another. Moreover, there are factors that affect the ability to do any biological research, not just recombinant DNA work.

One factor which is particular to recombinant DNA work was the initial necessity of importing nearly all of the restriction enzymes, other enzymes (e.g., ligase, polymerase), radioactive nucleoside triphosphates, and other chemicals peculiar to this type of work. At first this resulted in either long delays in obtaining materials or the necessity for preparing some of the enzymes or other reagents. Recently, the production of most enzymes by Japanese companies (e.g., Takara Shuzo) and the establishment of branches in Japan of the Western companies making these materials has eliminated this problem.

In conclusion, there is no real technology gap, only a lag between the time it was practical for workers in the United States and their Japanese counterparts to begin recombinant DNA research. Also, a shortage of funding in Japan (compared to the United States) makes all biological research comparatively difficult.

## 7.6 Creativity In Japanese Research

In Japan, as in the United States, creativity in research is variable. It is often said, both here and by the Japanese themselves, that

Japanese research is much less creative and original than that in the United States. Indeed, specific efforts to stimulate creativity are now underway in Japan. The basis of this difference, if there is indeed a difference, may be in the severe pressure toward conformity in Japanese society (an old Japanese proverb says "the nail that sticks up gets hammered down"), in the uniformity of the educational system (every 3rd grader throughout Japan learns the same thing from the same textbook on the same day of the year), or in the cultural uniformity reinforced by high population density. The Japanese academic system does not allow total independence of faculty members until they are about age 40-45, and then only if they can get one of the relatively scarce professorships. Until then, most junior faculty are in positions called Joshu (literally, "helping hand" or assistant) or Jokyōju (assistant professor) and are at least formally under the direction of the professor. The resulting relationship can be one of collaborator or one of master and slave or anything in between, but cooperation and consensus are more often seen than simple direction from top to bottom. It is perhaps significant that in the rare cases where Joshu and Jokyōju are completely independent (e.g., Professor Yura's laboratory in the Virus Institute of Kyoto University), some highly creative work has been done by the junior faculty (Hiraga, Ishihama, Itoh, et al.).

In spite of these factors, it is not clear that there is a real difference in creativity between Japan and the United States when allowance is made for the various factors discussed in the previous section. Creativity is difficult to measure, but, in the author's experience, Japanese research <u>is</u> very creative. As just one example, the control of gene expression by gene rearrangement - one of the really revolutionary concepts to come out of recent molecular biology - was, except for McClintock's work on corn, first discovered in bacteria (by Iino), in yeast (by Oshima), and in immunoglobulins (by Tonegawa and Honjo), all Japanese.

The author is inclined to agree with Herman Lewis's suggestion [<u>Biotechnology in Japan</u> (1984)] that Japan is not intrinsically deficient in science or technology, but that the emphasis has been on acquisition and efficient application of established knowledge rather than discovery of new knowledge because this is a more economical use of limited resources and a more effective way to compete in the world market. It is, to use Dr. Lewis's word, a "strategy."

One index of the practical difficulty of doing basic research in Japan is the relatively high proportion of the really famous Japanese researchers who have done their most widely known work abroad, largely in the United States. Such names as Tomizawa (plasmid replication), Nomura (ribosomes), Tonegawa (immunoglobulin gene rearrangement), Hirota (DNA replication), Itakura (oligonucleotide synthesis), and Hanafusa (RNA tumor viruses) are giants in their fields, and many other examples could be listed. All found work in the United States, France, or Switzerland easier for various reasons. Few Japanese are drawn to the United States for cultural or political reasons, and it is common for distinguished researchers to return to Japan later (e.g., Hirota). It is likely that these excellent researchers come here to do research. Of course, one can list many equally famous Japanese who have spent most of their careers in Japan, such as Okazaki (DNA replication), Sugimura (cancer research), Honjo (immunoglobulin genes), Oshima (yeast genetics), Matsubara (plasmid and phage replication), Takanami (phase development, restriction enzymes, etc.), and Ogawa (recombination, etc.). Both these groups of scientists (and many others of equal excellence) are products of the Japanese education system and Japanese society. One does not suddenly become creative at age 30 or 35.

## 7.7 Basic Versus Applied Research

Japan is a series of islands whose total land area is less than that of California, and whose population is half that of the United States. Of this area, about 80% is mountainous and essentially uninhabitable. Japan has no oil, no natural gas, only a tiny amount of coal, no iron deposits, and, except for a fair amount of water power usable for hydroelectric generation, very little of any other natural resources. Perhaps as important, every Japanese knows these facts intimately and knows that they imply that Japan can only survive economically by importing raw materials and exporting manufactured (or processed) goods. In order to successfully export manufactured or processed goods in a competitive world market, all the details of production must be optimized.

The importance of hard work is constantly emphasized. It is a central feature of Japanese culture. While we glorify the stylish, brilliant individualist who, apparently effortlessly, accomplishes something, the Japanese ideal is someone who accomplishes something as a result of years of hard work. Hard work is said to have a sort of mystical quality.

These two factors result in an orientation of scientists toward applications and a willingness to pay enough attention to detail to produce a high quality product at optimum efficiency.

Japan has much first-rate basic research and some researchers have no interest in applications. However, it is my impression that this group is a relatively smaller proportion of researchers than in the United States. Often, important basic findings in Japan are made in the course of a search for something of applied significance. Also, the direction of analysis of findings is usually strongly influenced by potential applications. In the past, the best researchers in the United States were invariably in basic research, and they were even scornful of applied aspects. This has not been true in Japan, and the situation in the United States appears to have changed dramatically in this regard in the last 5 years, perhaps in part because of the sudden realization that the frontier area of recombinant DNA was naturally leading to potentially exciting new applications, perhaps in part because of the general conservative philosophical shift of the country.

Japanese companies, willing to establish or improve their own recombinant DNA research and development units, send employees to various laboratories in Japan and in the United States for 1 to 2 years to learn the basic methods and approaches all at company expense. While companies in the United States hire people from academic laboratories to fill their research and development positions, they do not generally farm them out as in Japan. This may be because there is no formal postdoctoral system, as such, in Japan, so that a person with a Master's or Ph.D. degree may be hired by a company, then later sent for a postdoctoral-like experience. In the United States there are plenty of Ph.D.s with postdoctoral experience already in the job market. It seems possible, however, that the Japanese system may result in a more efficient transfer of information from academic or government laboratories into the companies interested in pursuing applications.

Most of the Japanese national universities carrying on extensive research programs are not only government-supported but are a part of the government itself. The Japanese government's policy with regard to transfer of technology from such laboratories to private companies may differ somewhat from ours. Our heterogeneous society focuses on fairness and not giving any company an advantage over others at public expense. The Japanese may be more practical, with closer connections resulting in part from the absence of a prohibition on collaboration between government scientists and industry and in part from the frequent presence of company employees in the government laboratories as postdoctoral and predoctoral trainees.

Another factor which may give Japan a competitive advantage in applying RDNA technology to production of saleable products is the Japanese education system. The Japanese education system consists not just of schools, teachers, and books, but of parents and a whole structure of society that tells children that they must study hard. It is estimated that as many as 20% of adults in the United States are essentially illiterate, whereas the figure in Japan is probably less than 1%. Note that literacy in Japan means being able to read and write 1,800 complex characters and their combinations, not just 26. While our brightest, best educated people are as well or better educated than their Japanese equivalents, our standard deviation of degree of education is higher.

When RDNA-based products reach the point of production, the Japanese production workers may be more uniformly well-educated. The emphasis at all levels on hard work and the roughly ten-fold lower levels of crime and drug use may mean a more uniformly high quality product produced at higher efficiency than in the United States, as it apparently has in the auto industry, in cameras, electronics, etc. This may prove the most difficult challenge to us of all.

## 7.8 Predictions of Future Trends

Predicting the future is always a risky business, but it is accurate to say that while 5 years ago Japan was well behind in recombinant DNA work, they are now approximately even. Whether this recent surge in activity will translate into a lead in this area in the future is difficult to know.

Japanese study English, and Japanese researchers pay close attention to work done in the United States. Perhaps more Americans should study Japanese to facilitate the flow of information from Japan to the United States.

The U.S. will continue to maintain a lead in basic research as a result of higher levels of government funding for this area, and as a result of greater interest and value placed on basic research in the United States than in Japan. However, I see no reason why the Japanese should not be able to pass us in the commercial applications areas. I think it is likely that they will be able to produce higher quality products for less money. The difference may be one of social and cultural factors rather than one of governmental policy, but it will, be useful for us to follow the scientific and technological developments in this area closely.

**Addendum to Section 7: Sample of Current Work In Recombinant DNA In Japan**

The following table presents a sampling of current research activities in university and company laboratories presented at four recent scientific meetings in Japan. The abstracts for these meetings are published in Japanese, and the author has briefly summarized the results of a number of the more interesting papers. The work presented at the meetings of the Japan Society for Fermentation Technology and the Japan Agricultural Chemistry Society tends to be applied in orientation, while that presented at the meetings of the Japan Biochemical Society and the Japan Molecular Biology Society tends to be more basic. The table is arranged by the organism whose DNA (or RNA) was cloned.

Most quality work in basic areas that is first published in Japanese (as an abstract or a paper) will later appear in the English language literature. However, the same is probably not true for applied work. There is less motive to publish applied work widely, and it is more difficult to get an applied paper published in English language journals.

This suggests that continued study of the Japanese language literature will be profitable by allowing an earlier look at work which is to appear later in English and by bringing out some work which would not be published in English at all.

References noted in the table are as follows:

(1) Japan Society for Fermentation Technology Abstracts, Meeting of November 7-9, 1983, Osaka, Japan.

(2) Japan Biochemical Society Abstracts, 56th Meeting, September 29-October 2, 1983, Fukoka City, Japan.

(3) Japan Agricultural Chemistry Society Abstracts, Meeting of April 1-4, 1984.

(4) Japan Molecular Biology Society Abstracts, 6th Meeting, August 22-25, 1983.

RECOMBINANT DNA PROJECTS IN JAPAN

| | Organism | Work | P.I. | Institution | Reference |
|---|---|---|---|---|---|
| (1) | Human | Chemically synthesized human growth hormone gene expressed in E. coli under control of the trp promoter. | Kenichi Matsubara | Osaka University<br>Cell Engineering Center | (4) 2B5 |
| (2) | Human | Human β-interferon gene synthesized chemically using high-use codons of E. coli and cloned by cDNA method. | Ei Mochida | Mochida Pharmaceutical Co. | (4) 2B6 |
| (3) | Human | Human γ-interferon gene and β-interferon genes produced using E. coli lipoprotein promoter. $5 \times 10^9$ U/l of γ-interferon obtained. | Tetsuo Oka | Cancer Research Center<br>Dept. of Biochemistry<br>Kyowa Fermentation Company<br>Tokyo Research Dept. | (4) 2B7 |
| (4) | Human stomach cancer cells | Amplification of c-myc found; not of c-mos (same chromosome). Normal restriction pattern, however. | Jun Yokota | Tokyo University<br>Medical Sciences Institute | (4) 1A12 |
| (5) | Human | Gastrin cDNA clone made from poly A RNA from surgically removed human stomach. This used to isolate human genomic clone. | Kenichi Matsubara | Osaka University Medical School<br>Cell Engineering Center | (4) 3A33 |
| (6) | Human | Using previously isolated interleukin-2 cDNA clone, a genomic clone was isolated. The IL-2 gene has one intron. | Koreaki Taniguchi | Cancer Center (Tokyo)<br>Dept. of Biochemistry | (4) 3A34 |
| (7) | Human | Analysis of pseudo V genes formed in human B cells. | Tasuku Honjo | Osaka University Medical School<br>Dept. of Genetics | (4) 3A18 |

| | Organism | Work | P.I. | Institution | Reference |
|---|---|---|---|---|---|
| (8) | Human | A secretion vector using the E. coli ompF outer membrane protein gene was made, and the β-endorphin gene was attached. All β-endorphin was secreted into the medium. | Shoji Mizushima | Mitsubishi Chemical Co., and Nagoya University, Dept. of Agrictural Chemistry | (3) 3R-2 (p. 569) (2) 1 Q-a8 (p. 680) |
| (9) | Human | A clone of nuclear DNA with 77% homology to mitochondrial rRNA isolated. | Kazunori Shimada | Kumamoto University Medical School First Dept. of Biochemistry | (4) 4B2 |
| (10) | Humans and Old World monkeys | Comparison of sequences of δ-globins. Evolutionary conclusions. | Yasutaka Takagi | Kyushu University Medical School Dept. of Biochemistry | (4) 3A21 |
| (11) | Human | Gastrin coding sequence synthesized as a tetramer for production purposes. | Hirosuke Okada | Osaka University Dept. of Fermentation Technology | (3) 1G-24 |
| (12) | Influenza virus | Inhibitors of the endonuclease that cuts a cap-containing primer RNA off of mRNAs studied. ApG inhibits, as does poly U. | Kiyoshi Kawakami | Jichi Medical School | (4) 2A2 |
| (13) | Human adult T cell leukemia virus (ATLV) | Cloning of viral RNA. Several pieces cloned and sequenced. Cloning of proviral DNA. Seven λ charon 28 clones obtained. | Masakazu Hatanaka | Kyoto University Virus Institute | (4) 1A15 (4) 1A16 |

| | Organism | Work | P.I. | Institution | Reference |
|---|---|---|---|---|---|
| (14) | Human adult T cell leukemia virus | Study of provirus structure. Previously reported whole sequence of ATLV. Leukemia cells from patient have one provirus, but cell lines *in vitro* have eight, of which six have deletions in 5' end. Some patients lack gag-pol env; have only pX gene. | Mitsuaki Yoshida | Cancer Institute<br>Tumor Virus Division | (4) 1A18<br>(4) 1A19 |
| (15) | Adenoviruses of various mammals | Sequencing of the terminal inverted repeats and comparison. | Gihei Sato | Obihiro Animal Husbandry University | (4) 1A20 |
| (16) | Adenovirus | First 36 nucleotides synthesized. *In vitro* priming protein-dependent DNA synthesis demonstrated. Starts at G at position 30. Primase recognition sequence is TTATTTT. | Joei Ikeda<br>Hideaki Tanaka | Plant Virus Institute<br>Chemical Technology Institute | (4) 1A21 |
| (17) | Adenovirus-transformed cells | Sequencing of cellular DNA in which adeno inserts. The adeno and cellular sequences form a repeating unit. | Masahide Ishibashi | Aichiken Cancer Center | (4) 1A23 |
| (18) | Adenovirus-transformed cells | *Hind*III G fragment carrying E1, a gene used to transform cells. Transformed cell DNA carrying these sequences cloned. Various insertion places; E1a gene transcribed. | Kaoru Fujinaga | Sapporo Medical School<br>Cancer Institute<br>Dept. of Molecular Biology | (4) 1A24<br>(4) 1A25 |
| (19) | Adenovirus type 12 | Mutants of E1b region lacking either the 19K protein or 54K protein constructed *in vitro*. 19K protein needed for transformation, but not for vegetative growth. 54K protein opposite. | Yasuhisa Fukui | Tokyo University<br>Medical Sciences Institute | (4) 1A27 |

| | Organism | Work | P.I. | Institution | Reference |
|---|---|---|---|---|---|
| (20) | EBV-infected cells | Pseudovirions isolated. Specific regions of cellular DNA found in pseudovirions. | Toero Ozato | Hokkaido University Medical School<br>Cancer Institute | (4) 1A30 |
| (21) | Varicella Zoster virus | $tk^-$ L cells made $tk^+$ by infection with VZV. VZV genome present in cells, but not the ends. Homology of cellular repeat DNA with the ends of VZV detected. | Takahashi Rimei | Osaka University<br>Microbiology Institute | (4) 1A31 |
| (22) | HSV-TK gene on a plasmid transformed into L cells | λ charon 4A used to clone the integrated plasmid sequences. The plasmid integrated at a site of partial homology with the chromosomal DNA. | Masuo Taito | Tokyo University<br>Pharmacy School<br>Dept. of Microbiology | (4) 1A32 |
| (23) | Hepatitis B virus | Study of surface antigen gene promoter location. Found within 1 kb of start of transcription. | Kenichi Matsubara | Osaka University Medical School<br>Cell Technology Center | (4) 1A33 |
| (24) | Hepatitis B virus | All four genes expressed in <u>S. cerevisiae</u>. | Kenichi Matsubara | Osaka University Medical School<br>Cell Technology Center | (4) 1A34 |
| (25) | Influenza virus (types A, B) | Sequences of various segments compared; evolutionary analysis. | Takashi Miyata | Kyushu University<br>Faculty of Science<br>Dept. of Biology | (4) 3C11 |

| | Organism | Work | P.I. | Institution | Reference |
|---|---|---|---|---|---|
| (26) | Retrovirus, hepatitis B virus, cauli-flower mosaic virus | These three types of viruses may all involve reverse transcription. The authors point out sequence homologies among them. | Takashi Miyata | Kyushu University<br>Faculty of Science<br>Dept. of Biology | (4) 3C12 |
| (27) | Sendai virus | 4000 bp from 3' end cloned as cDNA and sequenced. Two genes found. | Hiroshi Shibuta | Tokyo University<br>Medical Research Institute | (4) 2A1 |
| (28) | Human JC virus | This papova virus causes progressive multifocal leukoencephalopathy. These workers sequenced the whole virus. It has large T, small T, VP1, VP2, and VP3. | Eiichi Soeda | National Genetics Research Institute | (3) 1G-29 |
| (29) | Mouse | Isolation from L cells of 5.1 and 5.3 kb plasmids, both partly homologous to polyoma virus. Approximately 5,000 copies/cell. *E. coli* - L cell shuttle vector constructed. | Michio Oishi | Tokyo University<br>Dept. of Applied Microbiology | (4) 2B1 |
| (30) | Rat | β-Globin cloned and sequenced. | Toyoharu Hozumi | Wakunaga Pharmaceutical Company<br>Central Research Institute | (4) 3A22 |
| (31) | Golden hamster | The long terminal repeat sequence of intracisternal A particles was sequenced. Different chromosomal copies differ in their 5' ends. A 6 bp duplication of host sequences is generated on insertion. | Masaya Kawakami | Kitazato University Medical School<br>Dept. of Molecular Biology | (4) 1A6 |
| (32) | Rat | Preangiotensinogen cDNA clone isolated and sequenced. 24 amino acid signal, three glycosylation sites. | Shigetada Nakanishi | Kyoto University Medical School<br>Dept. of Immunology | (4) 3A29 |

| | Organism | Work | P.I. | Institution | Reference |
|---|---|---|---|---|---|
| (33) | Cow | Bradykininogen cDNA clones made from bovine liver mRNA. There is a large and a small form. | Shigetada Nakanishi | Kyoto University Medical School<br>Dept. of Immunology | (4) 3A30 |
| (34) | Sheep | Corticotrophin releasing factor cDNA clone made from hypothalamic poly A RNA. Sequence shows 41 amino acids before CRF sequence. Used as probe to isolate human genomic CRF gene. It has one intron. Seven amino acids differ between human and sheep CRF. | Shosaku Numa | Kyoto University Medical School<br>Dept. of Medicinal Chemistry | (4) 3A31 |
| (35) | Morris murine hepatomas | vHa-ras and v-myc probes used in Northern and Southern blot analyses. Amplification of both oncogenes in Morris hepatomas was seen, as was increased transcription | Takashi Sugimura | National Cancer Center<br>Dept. of Biochemistry | (4) 1A10 |
| (36) | Wistar rats | cDNA library of fetal mRNA was made. Two clones hybridized with B cell tumor mRNA, but not with normal B cell mRNA. These clones are not known oncogenes. | Toshio Ando | Meiji Pharmaceutical University<br>Dept. of Hygienic Chemistry | (4) 1A11 |
| (37) | Rat | Cloned U1 RNA-hybridizing genomic DNA. Also chemically synthesized presumed splicing substrate RNA and put (as DNA) into pBR322 for production. Also purifying RNP containing U1, U2, U4, and U5 from rat liver and HeLa cells. U1-Containing RNP fraction binds well to synthetic splice site as do other fractions. | Yasumi Oshima | Tsukuba University<br>Dept. of Biological Sciences | (4) 3B30<br>(4) 3B33<br>(4) 3B34<br>(4) 3B35 |
| (38) | Friend leukemia virus | The LTR and env regions, needed for leukemia induction, were sequenced. Their detailed relationship to F-SFFV and Mo Mu LV are described. | Yoji Ikawa | Physical Chemical Research Institute<br>Cancer Research Institute | (4) 1A7 |
| (39) | Kirsten murine sarcoma virus | End regions of wild type and an unrescuable mutant were sequenced. A defect in the mutant in a region involved in packaging was found. | Nobuo Tsuchida | Yakult Company<br>Central Research Institute | (4) 1A8 |

| | Organism | Work | P.I. | Institution | Reference |
|---|---|---|---|---|---|
| (40) | Rat-Harvey sarcoma virus | cHa-ras transcription levels in rat liver cancer and in cells treated with 3'-methyl DAB (a carcinogen) were elevated. | Takashi Sugimura | National Cancer Center<br>Dept. of Biochemistry | (4) 1A9<br>(4) 1A10 |
| (41) | Chicken | α-Crystalline gene cloned and sequenced. Two δ-crystalline genes cloned and partially sequenced. δ-Crystalline gene attached to various promoters and microinjected into cultured animal cells. With its own promoter, only expressed in lens-covering cells. | Setsuto Okada | Kyoto University<br>Faculty of Science<br>Dept. of Biophysics | (4) 3A14<br>(4) 3A15<br>(4) 3A16<br>(4) 3A17 |
| (42) | Chicken | Myosin alkali light chain gene cloned. The genes for the two distinct alkaline light chains overlap and have some exons in common and some specific for one or the other alkaline light chain. | Masami Muramatsu | Tokyo University Medical School<br>Dept. of Biochemistry | (4) 3A25 |
| (43) | Chicken | Discovery of enzyme in bursa cells that makes a double-strand break specifically in J-region DNA. | Tasuku Honjo | Osaka University Medical School<br>Dept. of Genetics | (4) 3A19 |
| (44) | *Xenopus laevis* | $V_H$ gene cloned using mouse probe. Sequence shows similar V-J joining sequences. | Tasuku Honjo | Osaka University Medical School<br>Dept. of Genetics | (4) 3A20 |
| (45) | Rous sarcoma virus | A packaging-defective mutant, TK15, was studied by sequencing of the genome and provirus. A 250 bp deletion in the 5' leader region was found. | Masashi Yamamoto<br>Kumao Toyoshima | Tokyo University<br>Medical Sciences Institute | (4) 1A2 |
| (46) | Chicken erythroblastosis virus | The *erbA* protooncogene was sequenced. Also sequenced was the *erbB* gene of a mutant virus which was still oncogenic but had lost the ability to produce erythroblastosis. A 169 bp deletion was found. | Masashi Yamamoto<br>Kumao Toyoshima | Tokyo University<br>Medical Sciences Institute | (4) 1A4<br>(4) 1A5 |

| | Organism | Work | P.I. | Institution | Reference |
|---|---|---|---|---|---|
| (47) | New avian sarcoma virus 2635H | Genome cloned into *E. coli* and analyzed. Sequence is 5'-(Δgag)-src-related sequence-3'. No *pol* or *enV* found. | Masashi Yamamoto<br>Kumao Toyoshima | Tokyo University<br>Medical Sciences Institute | (4) 1A1 |
| (48) | Silkworm | Fibrin gene upstream activating sequence identified using cloned gene and *in vitro* transcription system from silk gland cell extracts. Activating protein factor isolated from silk gland cell extracts. | Yoshiaki Suzuki | Basic Biology Research Institute<br>Dept. of Cell Differentiation | (4) 3A12 |
| (49) | *Drosophila* | Myosin L chain cDNA cloned. | Gaiju Hotta | Tokyo University<br>Faculty of Science | (4) 3A24 |
| (50) | Various metazoa | Looked for genes with homology to the oncogenes *yes* and *src*. Multiple homologous genes found. Gene families. | Masashi Yamamoto<br>Kumao Toyoshima | Tokyo University<br>Medical Sciences Institute | (4) 1A3 |
| (51) | *Torpedo californiа* | Cloned and sequenced cDNAs of the four subunits of the acetylcholine receptor. They show homology with each other. Speculation about function included. | Shosaku Numa | Kyoto University<br>Medical School | (4) 1C18 |
| (52) | *Bombyx mori* | Sericin (silk glue) gene family - cloned one 11 kb mRNA. Expression of gene family during life cycle studied; silk fibroin gene promoter identified by deletions; site-directed mutagenesis. | Yoshiaki Suzuki | Basic Biology Research Institute<br>Dept. of Cell Differentiation | (4) 2A9<br>(4) 2A10<br>(4) 2A11<br>(4) 2A12 |
| (53) | *Drosophila* | Retrovirus-like particles contain 4.5S RNA that is nearly identical to previously described U6 small nuclear RNA. | Tadayoshi Shiba | Mitsubishi Institute<br>of Life Sciences | (4) 3B28 |

| | Organism | Work | P.I. | Institution | Reference |
|---|---|---|---|---|---|
| (54) | Silkworm-cytoplasmic polyhedrosis virus | Sequenced 5' ends of mRNA of segment 8 (high efficiency translation) and segment 7 (low efficiency). Eight had more homology with 18S rRNA than 7. | Kinichiro Miura | National Institute of Genetics | (4) 3A3 |
| (55) | Various plants | A promoter cloning vector based on URA3 expression in yeast was constructed. Plan is to clone a strong plant promoter. | Kenji Sakaguchi | Mitsubishi Institute for Life Sciences | (1) 137 |
| (56) | Coin moss = liverwort | The ribulose bisphosphate carboxylase large subunit promoter and chloroplast ARS sequences cloned to make a candidate vector for plant cell transformation. | Kanji Oyama | Kyoto University<br>Dept. of Agricultural Chemistry | (1) 261<br>(3) S IV-9<br>(p. 696) |
| (57) | Wheat | Histone H4 gene sequenced. 200-250 copies/genome. | Masaki Iwabuchi | Hokkaido University<br>Faculty of Science<br>Dept. of Botany | (4) 3A26 |
| (58) | Moss (Zenigoke) | Chloroplast DNA (121 kb circle) cloned into *E. coli*. An inverted repeat present. Sites of transcription starts and translation starts identified. | Tohru Komano | Kyoto University<br>Dept. of Agricultural Chemistry | (3) 1G-23<br>(3) 1H-6 |
| (59) | Cauliflower mosaic virus | An 856 bp deletion occurs at high frequency in CaMV-S. Its ends have the donor-acceptor sequences typical of splicing sites. So does CaMV-M strain where this does not occur. | Joei Ikeda | Plant Virus Research Institute | (4) 1C24 |
| (60) | Tobacco mosaic virus | Complete cDNA made and plasmid carrying c DNA constructed. No rearrangements. Introduction into plant cells did not lead to TMV-RNA or protein production. Infection of tobacco protoplasts with TMV-RNA yields transient production of the 30K protein. | Yoshimi Okada | Tokyo University<br>Faculty of Science | (4) 1C25<br>(4) 1C26<br>(4) 1C27 |

| | Organism | Work | P.I. | Institution | Reference |
|---|---|---|---|---|---|
| (61) | Hops stunt viroid and cucumber pale fruit viroid | HSV cloned and sequenced. HSV clone hybridizes with CPFV-RNA. | Yoshimi Okada | Tokyo University<br>Faculty of Science | (4) 1C28<br>(4) 1C29 |
| (62) | Saccharomyces cerevisiae | Analysis of PHO5 (repressible acid phosphatase) control region by serial deletions using BAL31 exonuclease. | Akio Toh-e | Hiroshima University<br>Faculty of Engineering<br>Dept. of Fermentation Technology | (4) 2A5 |
| (63) | S. cerevisiae | GAL7 control region sequenced and BAL31 deletions made; transcript start sites determined. Three super-repressor alleles of GAL80 cloned. Study of transcription of GAL80. | Toshio Fukasawa | Keio University School of Medicine<br>Dept. of Molecular Genetics | (4) 2A6<br>(4) 2A7<br>(4) 2A8 |
| (64) | S. cerevisiae | For hepatitis B antigen production, an expression vector was constructed using the PHO5 promoter. | Akio Toh-e<br><br>Kenichi Matsubara | Hiroshima University<br>Dept. of Fermentation Technology<br>Osaka University Medical School<br>Cell Engineering Center | (3) S IV-7<br>(p. 694)<br>PNAS 80: 1<br>(1983) |
| (65) | S. cerevisiae | To find the best promoter for protein production, promoter strengths were compared for TRP1, glyceraldehyde 3-P-dehydrogenase, phosphoglycerate kinase, and repressible acid phosphatase. PGK was the strongest. | Takehiro Oshima | Suntory<br>Biomedical Research | (1) 106 |

| | Organism | Work | P.I. | Institution | Reference |
|---|---|---|---|---|---|
| (66) | Saccharomyces, S. bisporus, S. bailli | Plasmids from these organisms cloned, analyzed, and one (from S. rouxii) sequenced. All are nonhomologous but have similar size and inverted repeats. Each replicates in all Saccharomyces tested. Used to make vectors compatible with 2 μ DNA. Can be used in soy sauce production organism. | Akio Toh-e | Hiroshima University<br>Dept. of Fermentation Technology | (1) 105 |
| (67) | S. cerevisiae | Mitochondrial EF-Tu cloned (a nuclear gene) using E. coli EF-Tu gene as probe. Sequences similar, but the mitochondrial enzyme has a signal sequence. | Yoshito Kamishiro | Tokyo University<br>Medical Science Institute | (4) 4B1 |
| (68) | Saccharomyces rouxii pSR1 | 2 μ DNA-like plasmid sequence. Two inverted repeats of 959 bp. ARS's very close to inverted repeats. | Yasuji Oshima | Osaka University<br>Faculty of Engineering<br>Dept. of Fermentation Technology | (4) 3B5 |
| (69) | S. cerevisiae | Restriction endonuclease from yeast isolated and purified. 4-Base sticky end produced. Recognition sequence not yet completely clear. The enzyme (SceI) cuts its own cells DNA in vitro. | Takehiko Shibata | Physical Chemical Research Institute<br>Dept. of Microbiology | (4) 3C29 |
| (70) | S. cerevisiae | Made a 2 μ DNA-HIS3-LEU2 plasmid and studied if polyploidy aids enzyme production. It does not. | Yasuji Oshima | Osaka University<br>Dept. of Fermentation Technology | (1) 420 |

| | Organism | Work | P.I. | Institution | Reference |
|---|---|---|---|---|---|
| (71) | S. cerevisiae | An improved simple fast transformation method developed. All transformants are haploid. Very useful. But rate is several-fold lower than protoplast method. | Akira Kimura | Kyoto University<br>Food Science Research Institute | (1) 103<br>J. Bact. 153, 163 (1983)<br>Agric. Biol. Chem. 47, 1691 (1983) |
| (72) | Saccharomyces diastaticus | The secreted glucoamylase gene cloned and inserted into S. cerevisiae to allow the latter to metabolize starch. To stabilize it, the repeated δ sequence of S. cerevisiae used to integrate the gene at several places in the genome. | Fukui Sakuzo | Hiroshima University<br>Faculty of Engineering<br>Dept. of Biophysical Chemistry | (3) 1F-18 |
| (73) | Schizosaccharomyces pombe | Two genes for α-tubulin cloned, one of which was shown to be the same as the nda2 gene affecting nuclear division. | Mitsuhiro Yanagida | Kyoto University<br>Faculty of Sciences<br>Dept. of Biophysics | (4) 3A23 |
| (74) | Schizosaccharomyces pombe | Two ARS sequences isolated, one from rRNA, one from the α-tubulin gene. The later was was sequenced. | Mitsuhiro Yanagida | Kyoto University<br>Faculty of Science<br>Dept. of Biophysics | (4) 3B3 |

| | Organism | Work | P.I. | Institution | Reference |
|---|---|---|---|---|---|
| (75) | *Schizosaccharomyces pombe* | Transformation with nonreplicating (in *S. pombe*) plasmid occurs at high frequency if the host carries a replicating pBR322-containing plasmid already. Due to recombination. Also the *ARS* sequence near *ura1* sequenced. | Masayuki Yamamoto | Tokyo University Institute of Medical Sciences | (4) 3B1 (4) 3B2 |
| (76) | *Schizosaccharomyces pombe* | By integrating cloned piece of rRNA, the repeat cluster was located on the long arm of chromosome III near *ade5*. Isolation of topoisomerase 1 mutants and of endonuclease mutants by screening extracts. | Mitsuhiro Yanagida | Kyoto University Dept. of Biophysics | (4) 2C8 (4) 2C9 |
| (77) | *Schizosaccharomyces pombe* | Histone genes cloned and H2A gene sequenced. | Mitsuhiro Yanagida | Kyoto University Faculty of Science Dept. of Biophysics | (4) 3A28 |
| (78) | *Kluyveromyces lactis* | *ARS* sequences for *S. cerevisiae* isolated from *K. lactis*. Linear DNA killer plasmid pGKl1. pGKl1 sequenced. 202 bp terminal inverted repeats and ORFs for 129K, 69K, 51K, 45K, and 29K proteins found. | Norio Gunge | Mitsubishi Institute for Life Sciences | (4) 3B4 (4) 3B6 |
| (79) | *Kluyveromyces lactis* | Linear DNA plasmids determining killer trait (toxin). Potentially useful as vector with secreted protein toxin and *ARS* sequences for *S. cerevisiae*. | Norio Gunge | Mitsubishi Institute for Life Sciences | (1) 104 |
| (80) | *Tetrahymena* | Macronuclear and micronuclear histone genes compared by Southern blotting using sea urchin histone probe. Each was different. | Takashi Mita | Industrial Medical School (Kushu) Dept. of Molecular Biology | (4) 3A27 |
| (81) | *Candida pelliculosa* | β-Glucosidase of *C. pelliculosa* cloned into *S. cerevisiae* to make a strain that can use cellulose to make alcohol. | Akio Toh-e | Hiroshima University Dept. of Fermentation Technology | (1) 136 |

| | Organism | Work | P.I. | Institution | Reference |
|---|---|---|---|---|---|
| (82) | *Aspergillus oryzae* | ARS sequences cloned into *S. cerevisiae*. Plan to make *Aspergillus* vectors. | Kohki Horikoshi | Chemical and Physical Research Institute | (3) 1H-8 |
| (83) | *Bacillus licheniformis* | Extracellular protease cloned into *B. subtilis* in order to produce large amounts of protease and to develop a secretion vector. | Akira Kimura | Kyoto University<br>Food Science Research Institute | (3) 3R-9<br>(p. 573) |
| (84) | *Bacillus amyloliquefaciens* | To increase the production of secreted enzymes, a gene promoting the secretion of extracellular protease was cloned into *B. subtilis*. The same gene promotes secretion of several other enzymes. | Akiko Sawakura | Mitsui Toatsu, Bioengineering Institute | (3) 3R-10<br>(p. 573) |
| (85) | *Bacillus subtilis* | α-Amylase gene promoter and signal used to make secretion vector. This was effective when used with *E. coli* β-lactamase. | Gakuzo Tamura | Tokyo University<br>Dept. of Agricultural Chemistry | (4) 2B3 |
| (86) | *Bacillus circulans* | Cloning of *B. circulans* amylase into *B. subtilis*. | Fumio Hishinuma | Mitsubisishi Institute of Life Sciences | (4) 2B4 |
| (87) | *Bacillus stearothermophilus* and *Bacillus licheniformis* | A host-vector system set up using part of a natural *B. stearothermophilus* plasmid. Replicates in *B. stearothermophilus* or in *B. subtilis*. Natural copy number = 43. Transforms *B. stearothermophilus* at $2 \times 10^7$ per μg DNA. Used to clone pencillinase from *B. licheniformis* and a protease and α-amylase from *B. stearothermophilus*. | Shiichi Aiba | Osaka University<br>Dept. of Fermentation Technology | (3) S IV-6<br>(p. 693)<br>(1) 138<br>(1) 151 |
| (88) | *Bacillus amyloliquefaciens* | Using a temperate phage, α-amylase of *B. amyloliquefaciens* was cloned into *B. subtilis*. | Huiga Saito | Tokyo University<br>Applied Microbiology Institute | (1) 139 |

| | Organism | Work | P.I. | Institution | Reference |
|---|---|---|---|---|---|
| (89) | *Bacillus stearothermophilus* and *B. licheniformis* | To maximize production of penicillinase, strains with the gene on plasmid pLP11 were tested at various temperatures and in various media. | Shuichi Aiba | Osaka University<br>Dept. of Fermentation Technology | (1) 231 |
| (90) | *Bacillus stearothermophilus* | To find a strong promoter to use in expressing heat-resistant proteins in *B. stearothermophilus*, promoters were cloned into *E. coli* using an erythromycin-resistance-based vector. | Yoshiaki Nomune | Tokyo Engineering University<br>Faculty of Science<br>Dept. of Natural Products | (2) 4S-a8 |
| (91) | *Bacillus pumilus* | Xylanase and β-xylulosidase cloned into *E. coli* and *B. subtilis*. | Hirosuke Okada | Osaka University<br>Dept. of Fermentation Technology | (1) 140, 141 |
| (92) | *Bacillus stearothermophilus* | To convert the lactose in whey (from cheese-making) into glucose and galactose and to make low-lactose milk for people with lactose intolerance, β-galactosidase cloned into *E. coli*. This heat-stable enzyme should survive pasteurization. | Hirosuke Okada | Osaka University<br>Dept. of Fermentation Technology | (1) 142 |
| (93) | *Bacillus megaterium*, *B. licheniformis*, *B. stearothermophilus* | To obtain high α-amylase production, the gene was cloned from these *Bacillus* species. | Juzo Utaka | Nagoya University<br>Dept. of Food Engineering Chemistry | (3) 3R-5 (p. 571)<br>(3) 3R-6 (p. 571) |

| | Organism | Work | P.I. | Institution | Reference |
|---|---|---|---|---|---|
| (94) | Bacillus stearothermophilus | α-Amylase cloned into E. coli. Production of stable α-amylase to degrade starch for ethanol production or other use is the goal. Secretion, high production level, and heat stability observed. The structure of the signal peptide determined in order to make a secretion vector. | Kumio Yamane | Tokyo Engineering University Faculty of Science and Tsukuba University Dept. of Biological Sciences | (3) 3R-7 (p. 572) (3) 3R-8 (p. 572) J. Bact. 156, 327 (1983) J. Biochem. 95, 87 (1984) |
| (95) | B. subtilis | Cloning and sequencing of two rRNA operons (rrnO and rrnA) near the origin of replication. | Hiroshi Yoshikawa | Kanazawa University Cancer Institute | (4) 1B10 |
| (96) | B. subtilis | SpoOF clone prevents sporulation if present in high copy number, but is not due to overproduction of spoOF product. | Hiuga Saito | Tokyo University Applied Microbiology Institute | (4) 1B11 |
| (97) | B. subtilis | spoOC and spoOA cloning. | Kiyoshi Kurahashi | Osaka University Protein Institute | (4) 1B12 |
| (98) | Bacillus thuringiensis | Cloning of the structural gene for the crystalline insect toxin produced by the spores of this bacteria. The toxin is produced in E. coli. The 2.8 kb carrying the gene was sequenced. | Yuji Shibano | Suntory Applied Biology Research Institute | (4) 3B8 (3) 1F-8 |

| | Organism | Work | P.I. | Institution | Reference |
|---|---|---|---|---|---|
| (99) | B. subtilis | Constructed expression vector for B. subtilis using erythromycin-resistance promoter. | Teruhiko Beppu | Tokyo University<br>Dept. of Agricultural Chemistry | (3) 1F-2 |
| (100) | B. subtilis | spoIVC gene cloned. | Yasuo Kobayashi | Hiroshima University<br>Dept. of Biological Production | (3) 1H-11 |
| (101) | B. subtilis | Tunicamycin resistance gene tmrA cloned into B. subtilis and E. coli. tmrA ia just upstream of amyE, the structural gene for amylase, and mutations in tmrA result in increased amylase production. tmrB and aroI, just downstream of amyE, were also cloned. | Gakuzo Tamura | Tokyo University<br>Dept. of Agricultural Chemistry | (3) 1H-13<br>(3) 1H-14<br>(3) 1H-15<br>(3) 1H-16 |
| (102) | B. stearo-thermophilus | A host-vector system for this organism developed in order to clone thermostable enzymes for applications. α-Amylase was cloned using this system. | Shuichi Aiba | Osaka University<br>Dept. of Fermentation Technology | (1) 138 |
| (103) | B. subtilis phages Ø29, M2 | The 5' ends of the phage DNAs each have a bound 30K protein that acts as a primer for DNA synthesis. An in vitro DNA replication system was established. The gene 2 and 3 products are essential for binding dAMP to the primer protein. The DNA elongation is aphidicoline-sensitive. | Hideo Hirokawa | Jochi University<br>Life Sciences Institute | (4) 3C32<br>(4) 3C33 |
| (104) | Escherichia coli | Efficiency of transcription-translation of galK attached to various strong λ and E. coli promoters compared in various bacterial hosts. The best combination was a combined $P_L$-$P_R$ promoter in a Klebsiella aerogenes MK9000 host. | _____ Muroooka and Isao Mitsuya | Hiroshima University<br>Faculty of Engineering<br>Dept. of _____________ | (4) 2B2<br>(3) 1F-5 |

| | Organism | Work | P.I. | Institution | Reference |
|---|---|---|---|---|---|
| (105) | E. coli | Isolated secretion-defective temperature-sensitive mutant, secY, in one of the ribosomal protein operons. A cold-sensitive mutant suppressing the secY temperature-sensitive mutant maps at 53-54 minutes. | Takashi Yura | Kyoto University<br>Virus Institute | (4) 2C6 |
| (106) | E. coli | Signal peptidase gene is essential for E. coli growth. | Takayasu Date | Kanazawa University Medical School<br>Dept. of Biochemistry | (4) 2C5 |
| (107) | E. coli | Constructed plasmid to use E. coli DHFR gene to measure promoter strength by degree of resistance to trimethoprim. | Keishiro Tsuda | Industrial Technology Institute<br>Fiber Polymer Research Institute | (4) 1B1 |
| (108) | E. coli | Sequencing of dnaQ-rnh (RNAse H) genes. Genes oriented oppositely; promoters overlap. | Mutsuo Sekiguchi | Kyushu University<br>Faculty of Medicine<br>Dept. of Biology | (4) 1B2 |
| (109) | E. coli | Cloning and sequencing of ftsI, a gene of cell division; codes for penicillin binding proteins. Amino acid sequence has homology with other pencillin binding proteins. | Yukitaka Hirota | National Institute of Genetics | (4) 1B3<br>(4) 1B4 |
| (110) | E. coli | RNase III (rnc) clone and sequence. | Hisao Uchida | Tokyo University<br>Medical Sciences Institute | (4) 1B5 |
| (111) | E. coli | Sequence of the RNA portion of RNase P - the gene of a temperature-sensitive mutant whose RNA portion of RNase P is defective. The RNA part of RNAse P is itself processed by removal of part of the 3' end of the primary transcript. | Toshio Shimura<br>Dept. of Biophysics | Kyoto University<br>Faculty of Science | (4) 1B6<br>(4) 1B7 |

| | Organism | Work | P.I. | Institution | Reference |
|---|---|---|---|---|---|
| (112) | *E. coli* | Alkaline phosphatase cloned. | Keishiro Tsuda | Industrial Technology Institute | (4) 1B8 |
| (113) | *E. coli* | Phosphoenol pyruvate carboxylase gene promoter; sequencing, *in vitro* and *in vivo* transcription, and regulation by ppGpp. | Hirohiko Mazuki | Kyoto University<br>Faculty of Science<br>Dept. of Chemistry | (4) 1B17 |
| (114) | *E. coli* | *ompF* promoter - delimitation and study of effects of osmotic pressure and *ompB*, *ompR*, and *envZ* mutations. Recognition site of *ompB* located. Transcription start site located. | Shyoji Mizushima | Nagoya University<br>Dept. of Agricultural Chemistry | (4) 1B20<br>(3) 1F-4 |
| (115) | *E. coli* | *cya* (adenylate cyclase) promoter region defined and sequenced. CRP-cAMP binding sequence present. CRP binds - footprinting experiments. | Koji Kyoba | Kyoto University<br>Medical School | (4) 1B23 |
| (116) | *E. coli* | Cloned the "stringent starvation protein" using oligonucleotides based on amino acid sequence. SSP binds to RNA polymerase and alters its *in vitro* activity. | Akira Ishihama | Kyoto University<br>Virus Institute | (4) 1B25 |
| (117) | *E. coli* | *htpR*, a regulatory gene for heat shock proteins, is necessary for cell growth only at 42°C. Based on amber mutants. | Takashi Yura | Kyoro University<br>Virus Institute | (4) 1B26 |
| (118) | *E. coli* | *lexA* transcription analysis. Regulated by attenuator mechanism. | Jun Nakazawa | Yamaguchi University<br>Medical School<br>Second Dept. of Biochemistry | (4) 1B27 |

| | Organism | Work | P.I. | Institution | Reference |
|---|---|---|---|---|---|
| (119) | *E. coli* | rho has an attenuator and regulates its own expression. Promoter and attenuator sequenced and S1 mapping of transcripts done. There are two promoters - 760 bp and 260 bp upstream - and two attenuators, and termination at them depends on the rho protein. | Mutsuo Inai | Kyoto University<br>Virus Institute | (4) 1B28 |
| (120) | *E. coli* | nusA is a termination factor. Suppressors of a nusA temperature-sensitive mutant were mapped (snaA at 68 minutes; snaB at 89 minutes, and snaA was cloned. The snaA mutant grows slowly; all sna mutants grow λ poorly. | Yoshikazu Nakamura<br>Hisao Uchida | Tokyo University<br>Medical Sciences | (4) 1B32 |
| (121) | *E. coli* | nusB gene cloned and sequenced. Purified labeled protein from maxicells. | Toshio Maegawa | Tokyo University<br>Medical Sciences | (4) 1B32 |
| (122) | *E. coli* | $tRNA^{MET}_{f2}$, a 15K protein of unknown function, nusA, and infB (protein synthesis initiation factor $IF_{2\alpha}$) form an operator with that order.<br>NusA sequenced and compared with sigma (it can replace sigma in vitro). Considerable homology and structural similarity. | Fumio Imamoto | Physico-Chemical Research Institute<br>Dept. of Molecular Genetics | (4) 1B33<br>(4) 1B34 |
| (123) | *E. coli*<br>ColE1 | In the ColE1 plasmid there is a mitomycin-induced transcript, upstream from the colicin gene, coding for a 17,000 dalton protein. | Jun Nakazawa | Yamaguchi University Medical School<br>Dept. of Biochemistry | (4) 3B9 |
| (124) | *E. coli*<br>ColE2,<br>ColE3 | In vitro system replicating ColE2 and ColE3 DNA requires RNA and protein synthesis for a trans-acting template-specific plasmid-specified factor. The coding region of the factor and the origin have been located by mutation and sequenced. | Tateo Itoh | Osaka University<br>Faculty of Science<br>Dept. of Biology | (4) 3B10<br>(4) 3B11 |
| (125) | *E. coli*<br>pSC101 | pSC101 origin of replication and a gene (ped) essential for replication sequenced.<br>Origin compared to oriC; both require dnaA for replication. | Kazuo Yamaguchi | Kanazawa University<br>Cancer Research Institute | (4) 3B12 |

| | Organism | Work | P.I. | Institution | Reference |
|---|---|---|---|---|---|
| (126) | E. coli plasmid R100 | Sequenced the origin of replication, a positive protein regulator (repA1), and a negative RNA regulator (RNAI, 91 bases) all within the 1.8 kb replication region. | Eiichi Otsubo | Tokyo University<br>Applied Microbiology Institute | (4) 3B13 |
| (127) | E. coli plasmid ColIb | Origin sequenced. | Kiyoshi Mizobuchi | Tokyo University<br>Faculty of Science<br>Dept. of Biochemistry | (4) 3B15 |
| (128) | E. coli R6K plasmid | A protein binding specifically to the plasmid oriγ origin of replication was labeled and its binding site defined and sequenced. | Kazumi Nakano | Fukui Medical School<br>Dept. of Biochemistry | (4) 3B21 |
| (129) | E. coli mini F plasmid | F3 protein is needed for replication from origin I, but not from origin II. | Tetsuo Iino | Tokyo University<br>Faculty of Science | (4) 3B23 |
| (130) | E. coli RNA phage GA | Cloned into pBR322. | Akikazu Hirashima | Keio University Medical School<br>Dept. of Molecular Biology | (4) 3C9 |
| (131) | E. coli | Cloned E. coli phosphofructokinase, triose phosphate isomerase, and glucokinase. Plan is to make an ATP-generating bioreactor from AMP and glucose. | Akira Kimura | Kyoto University<br>Food Research Institute | (1) 444 |
| (132) | E. coli | Xylose isomerase of E. coli put into S. cerevisiae in order to allow yeast to make alcohol from xylose, but no activity was observed. | Hiuga Saito | Tokyo University<br>Applied Microbiology Institute | (1) 133 |

| | Organism | Work | P.I. | Institution | Reference |
|---|---|---|---|---|---|
| (133) | E. coli | Analysis of cloned phoR gene. This 50,000 dalton protein acts as an activator of alkaline phosphatase at high levels and a repressor at low levels. The phoS gene also sequenced. phoT = pstA, and the order of pho genes near 83 minutes is phoU, pstB, phoT, phoS. | Atsuo Nakada | Osaka University<br>Institute of Microbiology | (4) 2C2<br>(4) 2C3 |
| (134) | E. coli | Glycerol kinase cloned and λ "sleeper vector" used to overproduce the enzyme. Cells with 25% of protein as glycerol kinase obtained. | Eiichi Nakano | Kikkoman, Inc. | (1) 147<br>Agric. Biol. Chem. 46, 313(1982) |
| (135) | E. coli | To maximize protein production, a comparison was made of various E. coli promoter strengths using lacZ fusions. recA was strongest. | Kenichi Matsubara | Osaka University Medical School<br>Cell Technology Center | (4) 1B18 |
| (136) | E. coli | To improve the production of aspartate from fumarate, aspartase was cloned. A 17-fold increase in aspartase activity was observed. | Tomoyasu Taniguchi | Tanabe Pharmaceutical Co. | (1) 260 |
| (137) | E. coli | The alkB gene, concerned with repair of methylated DNA, was cloned. Its product is a 27 kdal protein. The alkA gene was also cloned; it codes the 28 kdal 3-methyl adenine DNA glucosylase. | Mutsuo Sekiguchi | Kyushu University<br>Faculty of Science<br>Dept. of Biology | (4) 3C13<br>(4) 3C14 |
| (138) | E. coli | The uvrD gene (helicase II) cloned and sequenced. Its regulation also studied. | Hideyuki Ogawa | Osaka University<br>Faculty of Science | (4) 3C15 |

| | Organism | Work | P.I. | Institution | Reference |
|---|---|---|---|---|---|
| (139) | E. coli | Using specially constructed λ and plasmids, a DNA gyrase-dependent in vitro illegitimate recombination system was developed and analyzed; stimulated by oxolinic acid. | Hideo Ikeda | Tokyo University<br>Medical Sciences Institute | (4) 3C20<br>(4) 3C21 |
| (140) | E. coli | RecA-promoted recombination is stimulated in the neighborhood of Tn 3. | Haruo Ozeki | Kyoto University<br>Dept. of Biophysics | (4) 3C23 |
| (141) | E. coli | Tn 2603 resolution (res) site sequenced. | Tetsuo Sawai | Chiba University<br>Dept. of Microbiology | (4) 3C24 |
| (142) | E. coli | Transposition frequency of IS1 is repressed by readthrough transcription from outside the IS1 element. | Eiichi Otsubo | Tokyo University<br>Applied Microbiology Institute | (4) 3C25 |
| (143) | E. coli | 70 kb of DNA around the chromosomal replication termination site cloned. | Yukitaka Hirota | National Genetics Research Institute | (4) 4C9 |
| (144) | E. coli | Sequenced the region near oriC where the RNA primer-DNA switchover occurs. Several distinct sites on each strand located. | Tsuneko Okazaki | Nagoya University<br>Dept. of Molecular Biology | (4) 4C2 |
| (145) | E. coli | Hydrogenase of E. coli cloned to combine with other hydrogenases (such as that of C. butyricum) to make a better one. | Shuichi Suzuki<br>Akihiko Kikuchi | Tokyo Engineering University<br>Natural Resources Institute<br>Mitsubishi Institute for Life Sciences | (1) 146 |
| (146) | E. coli | Sequenced pbpA and rodA genes (concerned with cell shape) and studied their expression using lacZ fusions. | Takahisa Ohta<br>Sadamitsu Asoh | Tokyo University<br>Dept. of Agricultural Chemistry | (3) 1F-2 |

| | Organism | Work | P.I. | Institution | Reference |
|---|---|---|---|---|---|
| (147) | E. coli | The two enzymes of glutathione synthesis cloned and several copies of each inserted into pBR325. Strain carrying this plasmid makes 60-fold increased GSH. | Akira Kimura | Kyoto University<br>Food Science Research Institute | (3) 1F-12 |
| (148) | Ø80 phage | Sequencing of control region (including cI, cII, gene 30, and part of gene 15) and identification by footprinting of cI binding site. The site at which the Ø80 cI repressor is cut by the recA protease during induction was determined - right in the middle. | Hideyuki Ogawa | Osaka University<br>Faculty of Science<br>Dept. of Biology | (4) 1C20<br>(4) 1C21 |
| (149) | Ø80 phage | Sequence of N-like gene, $P_L$ and $t_{L1}$. | Aizo Matsushiro | Osaka University<br>Microbiology Institute | (4) 1C22 |
| (150) | T4 | T4 dC infecting plasmid-carrying strain transduces the plasmid at 10- to 100-fold increased frequency if the plasmid has homology with T4. | Hideo Takahashi | Tokyo University<br>Applied Microbiology Institute | (4) 1C23<br>Mol. Gen. Genet. 186 (1982)<br>Plasmid 8, 29 (1982) |
| (151) | λ phage | 80U gene E amber mutants found at 47 sites in the gene by sequencing. Suppression pattern studied. | Isao Katsura | Tokyo University<br>Faculty of Science | (4) 1C15 |
| (152) | T3, T7 | Packaging in vitro uses ATP, gp18, gp19, concatemer DNA. phage. | Sadakazu Minagawa | Kyoto University<br>Faculty of Science<br>Dept. of Botany | (4) 1C16 |

| | Organism | Work | P.I. | Institution | Reference |
|---|---|---|---|---|---|
| (153) | λ phage | Amplification of a UGA suppressor blocks phage replication. | Haruo Ozeki | Kyoto University<br>Faculty of Science | (4) 1C19 |
| (154) | T7 phage | A study of RNA priming of DNA synthesis near the origin of replication shows most of this priming is due to T7 RNA polymerase, not T7 primase. | Tsuneko Okazaki | Nagoya University<br>Dept. of Molecular Biology | (4) 3C31 |
| (155) | E. coli phage G4 | In vitro modifications of cloned G4 replication origin correlated with activity in vivo. | Tohru Komano | Kyoto University<br>Dept. of Agricultural Chemistry | (3) 1G-26 |
| (156) | E. coli phage ØX174 | The heat-stable mutant ØXam3trD was found to be due to a 3 bp insertion in the G gene adding a methionine residue. | Tohru Komano | Kyoto University<br>Dept. of Agricultural Chemistry | (3) 1G-28 |
| (157) | Streptomyces kasugaensis | To improve kasugamycin production in this organism, a host-vector system was developed using an endogenous S. kasugaensis plasmid and pIJ702, a plasmid from S. lividus. A vector able to multiply in either was developed. | Masanori Okanishi | Tokyo University<br>Dept. of Preventive Medicine | (1) 157<br>J. Antibiotics 36, 99 (1983) |
| (158) | Streptomyces griseus<br>Streptomyces rimosus | To make a host-vector system for S. griseus, the organism that makes streptomycin, an oxytetracycline-resistant plasmid from S. rimosus, and a "pock formation" plasmid from S. griseus, were isolated. | Shuichi Aiba | Osaka University<br>Dept. of Fermentation Technology | (1) 158<br>(1) 156 |

| | Organism | Work | P.I. | Institution | Reference |
|---|---|---|---|---|---|
| (159) | Streptomyces kanamyceticus | The kanamycin-resistant gene, resulting in $kan^r$ ribosomes in vitro, was cloned into S. lividus, S. parvulus, S. lavendulae. | Hiroshi Ogawara | Meiji Pharmaceutical University Second Dept. of Biochemistry | (2) 4S-a6 (p. 1022) (2) 4S-a7 (p. 1022) |
| (160) | Streptomyces lividans | For cloning in the antibiotic-producing organisms of the Streptomyces group, a phage R4-based vector was constructed, and transfection frequencies of $10^6$ plaques/μg of DNA were obtained. | Hiuga Saito | Tokyo University Applied Microbiology Institute | (3) S IV-3 (p. 690) |
| (161) | Streptomyces hygroscopicus | S. hygroscopicus α-amylase was cloned into E. coli via a cosmid library and an oligonucleotide probe based on the protein sequence. In order to find strong promoters and secretion signals for Streptomyces, this gene was re-introduced into S. lividans and S. griseus. | Yukuzo Nagaoka | Meiji Baking Co. Pharmaceutical Research Institute | (3) 3R-1 (p. 569) |
| (162) | Streptomyces coelicolor bikiniensis lividans | The afsB gene that controls "A factor" (2-isocapryloyl-3-R hydroxymethyl-r-butyrolactone), a positive hormone-like regulator of spore formation, streptomycin biosynthesis, streptomycin resistance, and actinorhodin and prodigiosin synthesis, was cloned and sequenced. The promoter region was also defined. Gene cloned from S. coelicolor and from S. bikiniensis. | Teruhiko Beppu | Tokyo University Dept. of Agricultural Chemistry | (3) 1G-21 (3) 1G-22 |
| (163) | Pseudomonas | Pyocin gene cloning and sequencing of wild type and mutants. | Makoto Kageyama | Mitsubishi Kasei Life Sciences Research Institute | (4) 1B13 |

| | Organism | Work | P.I. | Institution | Reference |
|---|---|---|---|---|---|
| (164) | Salmonella typhimurium, Serratia marcescens, Klebsiella pneuomonia, Proteus mirabilis | RNA polymerase sigma subunits cloned from these organisms using E. coli mutant with an amber mutant in sigma and a temperature-sensitive suppressor. | Takashi Yura | Kyoto University Virus Institute | (4) 1B30 |
| (165) | Sulfolobus acidocal-darius | tRNA genes cloned; one $tRNA^{thr}$ gene sequenced. | Susumu Nishimura | National Cancer Center Dept. of Biology | (4) 3C6 |
| (166) | Brevibac-terium lacto-fermentum | Host-vector system established and homoserine dehydrogenase cloned. Used in threonine production. | Shigeru Nakamori | Ajinomoto Company | (1) 446, 445 (3) S IV-2 (p. 689) |
| (167) | Flavobac-terium sp. | Genes from two different organisms, each coding for hydrolases of the nylon-like oligomer of 6-aminohexanoic acid had similar restriction maps. Hybrid enzymes were made and examined for activity, but no improvement was observed. | Hirosuke Okada | Osaka University Dept. of Fermentation Technology | (1) 134 |

| | Organism | Work | P.I. | Institution | Reference |
|---|---|---|---|---|---|
| (168) | Clostridium butyricum | Hydrogenase cloned into E. coli for use in making $H_2$ gas. Energy source. | Shuichi Suzuki<br>Kenji Sakaguchi | Tokyo Engineering University<br>Natural Resources Institute<br>Mitsubishi Institute for Life Sciences | (1) 143 |
| (169) | Clostridium butyricum | β-Isopropylmalate dehydrogenase cloned into B. subtilis in order to make C. butyricum-B. subtilis shuttle vector. | Shuichi Suzuki | Tokyo Engineering University<br>Natural Resources Institute | (1) 144 |
| (170) | Pseudomonas paucimobilis | For potential use in biodegradation of industrial wastes, plasmids were isolated carrying the ability to degrade biphenyl, toluene, and salicylate. | Noboru Tomizuka | Engineering Technique Institute | (1) 152<br>J. Bact. 154, 1356 (1983) |
| (171) | Alcaligenes faecalis<br>Pseudomonas | Plasmids were isolated from these organisms encoding enzymes able to degrade phenanthrolene, maphthalene, and toluene. | Norizo Kiyohara | Okayama Science University | (1) 154 |
| (172) | Micromonospora olivasterospora | In this astromycin-producing organism, three steps in the biosynthesis are plasmid-encoded. They are located on an acridine-curable 20 megadalton plasmid. | Kiyoshi Nakayama | Kyowa Hakko | (1) 159 |

| | Organism | Work | P.I. | Institution | Reference |
|---|---|---|---|---|---|
| (173) | Pseudomonas | TOL plasmid, coding for toluene- and xylene-degrading enzymes, analyzed to construct Pseudomonas expression vector. Operator-promoter locations and transcript starts of xyl ABC and xyl DEGF operons detemined. Operators sequenced. xyl R and S (control genes) cloned. | Jun Nakazawa and Akiko Nakazawa | Yamaguchi University Medical School Second Dept. of Biochemistry | (2) 4S-a3 (p. 1022) (3) S IV-5 (p. 692) (4) 1B16 |
| (174) | Acetobacter aceti and Gluconobacter suboxydans | Endogenous plasmids from each of these organisms used with antibiotic-resistance genes from E. coli to make vectors. Low level transformation of each achieved. | Teruhiko Beppu | Tokyo University Dept. of Agricultural Chemistry | (3) S IV-4 (p. 691) |
| (175) | Corynebacterium, C. herculis, Brevibacterium flavus, Microbacterium ammoniaphilum (all glutamic acid producers) | Host-vector system established using pCG4 plasmid from C. glutamicum T250. Protoplast transformation method established. Purpose: to improve glutamate production | Akira Komuro | Kywa Hakko | (3) S IV-1 (p. 688) J. Bact. 159, 306 (1984) |
| (176) | Sulfolobus | From this extremely thermophilic oerganism, an enzyme-inducing positive supercoils in DNA was isolated. | Akihiko Kikuchi | Mitsubishi Institute for Life Sciences | (4) 3C26 |

| | Organism | Work | P.I. | Institution | Reference |
|---|---|---|---|---|---|
| (177) | Thermus thermophilus | Isopropylmalic acid dehydrogenase cloned and sequenced. | Toshio Oshima | Jichi Medical University Dept. of Biochemistry | (4) 4C13 |
| (178) | Pseudomonas, E. coli | A vector constructed from RSF1010 and various antibiotic resistance genes that can replicate and be expressed in P. putida and E. coli. | Keiji Yano | Tokyo University Dept. of Agricultural Chemistry | (3) 1F-1 |
| (179) | Thermus flavus | Malate dehydrogenase gene cloned and sequenced. It is very heat-stable. | Teruhiko Beppu | Tokyo University Dept. of Agricultural Chemistry | (3) 1F-10 |
| (180) | Serratia marcescens | A proline-producing strain constructed by classical genetic methods and found to produce 50 mg proline per liter. Cloning the proline genes (proA and proB) and introducing them on a stable plasmid improved output to 75 mg per liter. | Tsutomu Takagi | Tanabe Pharmaceutical Co. Biochemistry Research Institute | (3) 1H-1 |
| (181) | Pseudomonas | Vectors constructed for P. aeruginosa, P. fluorescens, P. putida, P. acidovorans, P. stutzer, P. cepacia, P. syringae, and P. medicocina. | Yoshibumi Itoh | Shinshu University Medical School Dept. of Bacteriology | (3) 1H-3 |
| (182) | Klebsiella oxytoca | nif genes (nitrogen fixation) cloned, and nifA subcloned and made constitutive. | Teruhiko Beppu | Tokyo University Dept. of Agricultural Chemistry | (3) 1H-10 |
| (183) | Azotobacter vinelandii | Optimizing conditions for transformation. | Keiji Yano | Tokyo University Dept. of Agricultural Chemistry | (3) 1H-9 |

| | Organism | Work | P.I. | Institution | Reference |
|---|---|---|---|---|---|
| (177) | *Thermus thermophilus* | Isopropylmalic acid dehydrogenase cloned and sequenced. | Toshio Oshima | Jichi Medical University, Dept. of Biochemistry | (4) 4C13 |
| (178) | *Pseudomonas*, *E. coli* | A vector constructed from RSF1010 and various antibiotic resistance genes that can replicate and be expressed in *P. putida* and *E. coli*. | Keiji Yano | Tokyo University, Dept. of Agricultural Chemistry | (3) 1F-1 |
| (179) | *Thermus flavus* | Malate dehydrogenase gene cloned and sequenced. It is very heat-stable. | Teruhiko Beppu | Tokyo University, Dept. of Agricultural Chemistry | (3) 1F-10 |
| (180) | *Serratia marcescens* | A proline-producing strain constructed by classical genetic methods and found to produce 50 mg proline per liter. Cloning the proline genes (*proA* and *proB*) and introducing them on a stable plasmid improved output to 75 mg per liter. | Tsutomu Takagi | Tanabe Pharmaceutical Co., Biochemistry Research Institute | (3) 1H-1 |
| (181) | *Pseudomonas* | Vectors constructed for *P. aeruginosa*, *P. fluorescens*, *P. putida*, *P. acidovorans*, *P. stutzer*, *P. cepacia*, *P. syringae*, and *P. medicocina*. | Yoshibumi Itoh | Shinshu University Medical School, Dept. of Bacteriology | (3) 1H-3 |
| (182) | *Klebsiella oxytoca* | *nif* genes (nitrogen fixation) cloned, and *nifA* subcloned and made constitutive. | Teruhiko Beppu | Tokyo University, Dept. of Agricultural Chemistry | (3) 1H-10 |
| (183) | *Azotobacter vinelandii* | Optimizing conditions for transformation. | Keiji Yano | Tokyo University, Dept. of Agricultural Chemistry | (3) 1H-9 |

# Glossary

Aerobic: Living or acting only in the presence of oxygen.

Amino acids: The building blocks of proteins. There are 20 common amino acids.

Amino acid sequence: The linear order of amino acids in a protein. This sequence determines the properties of the protein.

Anaerobic: Living or acting in the absence of oxygen.

Antibiotic: A specific type of chemical substance that is administered to fight infections, usually bacterial infections, humans or animals. Many antibiotics are produced by using micro-organisms; others are produced synthetically.

Antibody: A protein (immunoglobulin) produced by humans or higher animals in response to exposure to a specific antigen and characterized by specific reactivity with its complementary antigen. (See also monoclonal antibodies.)

Antigen: A substance, usually a protein or carbohydrate which, when introduced in the body of a human or higher animal, stimulates the production of an antibody that will react specifically with it.

Applied research: Research to gain knowledge or understanding necessary for determining the means by which a recognized and specific need may be met (National Science Foundation definition). (See also generic applied research.)

Bacillus subtilis (B. subtilis): An aerobic bacterium used as a host in rDNA experiments.

Bacteria: Any of a large group of microscopic organisms having round, rodlike, spiral, or filamentous unicellular or noncellular bodies that are often aggregated into colonies, are enclosed by a cell wall or membrane, and lack fully differentiated nuclei. Bacteria may exist as free-living organisms in soil, water, organic matter, or as parasites in the live bodies of plants and animals.

Bacteriophage (or phage)/bacterial virus: A virus that multiplies in bacteria. Bacteriophage lambda is commonly used as a vector in rDNA experiments.

Basic research: Research to gain fuller knowledge or understanding of the fundamental aspects of phenomena and of observable facts without specific applications toward processes or products in mind (National Science Foundation definition).

Batch processing: A method of bioprocessing in which a bioreactor is loaded with raw materials and micro-organisms, and the process is run to completion, at which time products are removed. (Compare continous processing).

Biocatalyst: An enzyme that plays a fundamental role in living organisms or industrially by activating or accelerating a process.

Biochemical: Characterized by, produced by, or involving chemical reactions in living organisms; a product produced by chemical reactions in living organisms.

Biochip: An electronic device that uses biological molecules as the framework for molecules that act as semiconductors and functions as an integrated circuit.

Bioconversion: A chemical conversion using a biocatalyst.

Biodegradation: The breakdown of substances by microorganisms.

Biologics: Vaccines, therapeutic serums, toxoids, antitoxins, and analogous biological products used to induce immunity to infectious diseases or harmful substances of biological origin.

Biomass: All organic matter that grows by the photosynthetic conversion of solar energy.

Biopolymers: Naturally occurring macromolecules that include proteins, nucleic acids, and polysaccharides.

Bioprocess: Any process that uses complete living cells or their components (e.g., enzymes, chloroplasts) to effect desired physical or chemical changes.

Bioreactor: Vessel in which a bioprocess takes place.

Biosensor: An electronic device that uses biological molecules to detect specific compounds.

Biosynthesis: Production, by synthesis or degradation, of a chemical compound by a living organism.

Biotechnology: Commercial techniques that use living organisms, or substances from those organisms, to make or modify a product, and including techniques used for the improvement of the characteristics of economically important plants and animals and for the development of micro-organisms to act on the environment. In this report, biotechnology is used to mean "new" biotechnology, which only includes the use of novel biological techniques - specifically, recombinant DNA techniques, cell fusion techniques, especially for the production of monoclonal antibodies, and new bioprocesses for commercial production.

Catalysis: A modification, especially an increase, in the rate of a chemical reaction induced by a material (e.g., enzyme) that is chemically unchanged at the end of the reaction.

Catalyst: A substance that induces catalysis; an agent that enables a chemical reaction to proceed under milder conditions (e.g., at a lower temperature) than otherwise possible. Biological catalysts are enzymes; some nonbiological catalysts include metallic complexes.

Cell: The smallest structural unit of living matter capable of functioning independently; a microscopic mass of protoplasm surrounded by a semi-permeable membrane, usually including one or more nuclei and various nonliving products, capable alone, or interacting with other cells, or performing all the fundamental functions of life.

Cell culture: The in vitro growth of cells isolated from multicellular organisms. These cells are usually of one type.

Cell differentiation: The process whereby descendants of a common parental cell achieve and maintain specialization of structure and function.

Cell fusion: Formation of a single hybrid cell with nuclei and cytoplasm from different cells.

Cell line: Cells that acquire the ability to multiply indefinitely in vitro.

Chakrabarty decision: Diamond v. Chakrabarty, U.S. Department of Commerce, PTA, sec. 2105, 1980; landmark case in which U.S. Supreme Court majority held that the inventor of a new micro-organism, whose invention otherwise met the legal requirements for obtaining a patent, could not be denied a patent solely because the invention was alive.

Chromosomes: The rodlike structures of a cell's nucleus that store and transmit genetic information; the physical structure that contain genes. Chromosomes are composed mostly of DNA and protein and contain most of the cell's DNA. Each species has a characteristic number of chromosomes.

Clone: A group of genetically identical cells or organisms produced asexually from a common ancestor.

Cloning: The amplification of segments of DNA, usually genes.

Coding sequence: The region of a gene (DNA) that encodes the amino acid sequence of a protein.

Cofactors: Additional molecules needed for enzymatic function.

Commodity chemicals: Chemicals produced in large volumes that sell for less than $1 per pound. (Compare specialty chemicals.)

Complementary DNA (cDNA): DNA that is complementary to messenger RNA; used for cloning or as a probe in DNA hybridization studies.

Continuous processing: Method of bioprocessing in which raw materials are supplied and products are removed continuously, at volumetrically equal rates. (Compare batch processing.)

Cosmid: A DNA cloning vector consisting of plasmid and phage sequences.

Corporate venture capital: Capital provided by major corporations exclusively for high-risk investments.

Culture medium: Any nutrient system for the artificial cultivation of bacteria or other cells; usually a complex mixture of organic and inorganic materials.

Debt financing: The use of outside or borrowed capital to finance business activities.

Deoxyribonucleic acid (DNA): A linear polymer, made up of deoxyribonucleotide repeating units, that is the carrier of genetic information; present in chromosomes and chromosomal material of cell organelles such as mitochondria and chloroplasts, and also present in some viruses. The genetic material found in all living organisms. Every inherited characteristic has its origin somewhere in the code of each individual's DNA.

Diagnostic products: Products that recognize molecules associated with disease or other biologic conditions and are used to diagnose these conditions.

Directed Mutagenesis: Producing a mutation by altering specific nucleotides of the isolated gene.

DNA: Deoxyribonucleic acid.

DNA base pair: A pair of DNA nucleotide bases. Nucleotide bases pair across the double helix in a very specific way: adenine can only pair with thymine; cytosine can only pair with guanine.

DNA probe: A sequence of DNA that is used to detect the presence of a particular nucleotide sequence.

DNA sequence: The order of nucleotide bases in the DNA helix; the DNA sequence is essential to the storage of genetic information.

DNA synthesis: The synthesis of DNA in the laboratory by the sequential addition of nucleotide bases.

Downstream processing: After bioconversion, the purification and separation of the product.

Endorphins: Opiate-like, naturally occurring peptides with a variety of analgesic effects throughout the endocrine and nervous system.

Enkephalins: Small, opiate-like peptides with analgesic effects in the brain.

Enzyme: Any of a group of catalytic proteins that are produced by living cells and that mediate and promote the chemical processes of life without themselves being altered or destroyed.

Equity capital: Capital proceeds arising from the sale of company stock.

Equity investment: An investment made in a company in exchange for a part ownership in that company.

Escherichia coli (E. coli): A species of bacteria that inhabits the intestinal tract of most vertebrates. Some strains are pathogenic to humans and animals. Many nonpathogenic strains are used experimentally as hosts for rDNA.

Eukaryote: A cell or organism with membrane-bound, structurally discrete nuclei and well-developed cell organelles. Eukaryotes include all organisms except viruses, bacteria, and blue-green algae. (Compare prokaryote.)

Export controls: Laws that restrict technology transfer and trade for reasons of national security, foreign policy, or economic policy.

Feedstocks: Raw materials used for the production of chemicals.

Fermentation: An anaerobic bioprocess. Fermentation is used in various industrial processes for the manufacture of products such as alcohols, acids, and cheese by the action of yeasts, molds, and bacteria.

Food additive (or food ingredient): A substance that becomes a component of food or affects the characteristics of food and, as such, is regulated by the U.S. Food and Drug Administration.

Free-living organism: An organism that does not depend on other organisms for survival.

Gene: The basic unit of heredity; an ordered sequence of nucleotide bases, comprising a segment of DNA. A gene contains the sequence of DNA that encodes one polypeptide chain (via RNA).

Gene amplification: In biotechnology, an increase in gene number for a certain protein so that the protein is produced at elevated levels.

Gene expression: The mechanism whereby the genetic directions in any particular cell are decoded and processed into the final functioning product, usually a protein. See also transcription and translation.

Generic applied research: Research along the continuum between the two poles of basic and applied. This research may be characterized as follows: 1) it is not committed to open-ended expansion of knowledge as university basic research typically is but is less specific (more sidely applicable or "generic") than the typical industrial product or process development effort; 2) it has more well-defined objectives than basic research but is long term relative to product and proccess development; and 3) it is high risk, in the sense that the stated objectives may fail and the resources committed may be lost for practical purposes.

Gene transfer: The use of genetic or physical manipulation to introduce foreign genes into host cells to achieve desired characteristics in progeny.

Genome: The genetic endowment of an organism or individual.

Genus: A taxonomic category that includes groups of closely related species.

Germ cell: The male and female reproductive cells; egg and sperm.

Germplasm: The total genetic variability available to a species.

Glycosylation: The attachment of sugar groups to a molecule, such as a protein.

Growth hormone (GH): A group of peptides involved in regulating growth in higher animals.

Hormone: A chemical messenger found in the circulation of higher organisms that transmits regulatory messages to cells.

Host: A cell whose metabolism is used for growth and reproduction of a virus, plasmid, or other form of foreign DNA.

Host-vector system: Compatible combinations of host (e.g., bacterium) and vector (e.g., plasmid) that allow stable introduction of foreign DNA into cells.

Human chorionic gonadotropin (HCG): A hormone produced by human placenta, indicating pregnancy; widespread target of MAb developers to diagnose pregnancy at an early stage.

Human insulin (hI): Hormone that stimulates cell growth via glucose uptake by cells. Insulin deficiency leads to diabetes.

Human serum albumin (HSA): Abundant protein in human blood; as a product, used in highest quantities in medicine, primarily in burn, trauma, and shock patients.

Hybrid: The offspring genetically dissimilar parents (e.g., a new variety of plant or animal that results from cross-breeding two different existing varieties, a cell derived from two different cultured cell lines that have fused).

Hybridization: The act or process of producing hybrids.

Hybridoma: Product of fusion between myeloma cell (which divides continuously in culture and is "immortal") and lymphocyte (antibody-producing cell); the resulting cell grows in culture and produces monoclonal antibodies.

Hybridoma technology: See monoclonal antibody technology.

Immobilized enzyme or cell techniques: Techniques used for the fixation of enzymes or cells onto solid supports. Immobilized cells and enzymes are used in continuous bioprocessing.

Immune response: The reaction of an organism to invasion by a foreign substance. Immune responses are often complex, and may involve the production of antibodies from special cells (lymphocytes), as well as the removal of the foreign substance by other cells.

Immunoassay: The use of antibodies to identify and quantify substances. The binding of antibodies to antigen, the substance being measured, is often followed by tracers such as radioisotopes.

Interferons (Ifns): A class of glycoproteins (proteins with sugar groups attached at specific locations) important in immune function and thought to inhibit viral infections.

In vitro: Literally, in glass; pertaining to a biological reaction taking place in an artificial apparatus; sometimes used to include the growth of cells from multicellular organisms under cell culture conditions. In vitro diagnostic products are products used to diagnose disease outside of the body after a sample has been taken from the body.

In vivo: Literally, in life; pertaining to a biological reaction taking place in a living cell or organism. In vivo products are products used within the body.

Joint venture: Form of association of separate business entities which falls short of a formal merger but unites certain agreed on resources of each entity for a limited purpose; in practice most joint ventures are partnerships.

Leaching: The removal of soluble compound such as an ore from a solid mixture by washing or percolating.

Lignin: A major component of wood.

Lignocellulose: The composition of woody biomass, including lignin and cellulose.

Linker: A small fragment of synthetic DNA that has a restriction site useful for gene cloning, which is used for joining DNA strands together.

Liposome transfer: The process of enclosing biological compounds inside a lipid membrane and allowing the complex to be taken up by a cell.

Lymphocytes: Specialized white blood cells involved in the immune response; B lymphocytes produce antibodies.

Lymphokines: Proteins that mediate interactions among lymphocytes and are vital to proper immune function.

Messenger RNA (mRNA): RNA that serves as the template for protein synthesis; it carries the transcribed genetic code from the DNA to the protein synthesizing complex to direct protein synthesis.

Metabolism: The physical and chemical processes by which foodstuffs are synthesized into complex elements, complex substances are transformed into simple ones, and energy is made available for use by an organism.

Metallothioneins: Proteins, found in higher organisms, that have a high affinity for heavy metals.

Methanogens: Bacteria that produce methane as a metabolic product.

Micro-organisms: Microscopic living entities; micro-organisms can be viruses, prokaryotes (e.g., bacteria), eukaryotes (e.g., fungi).

Microencapsulation: The process of surrounding cells with a permeable membrane.

Mixed culture: Culture containing two or more types of micro-organisms.

Monoclonal antibodies (MAbs): Homogeneous antibodies derived from a single clone of cells; MAbs recognize only one chemical structure. MAbs are useful in a variety of industrial and medical capacities since they are easily produced in large quantities and have remarkable specificity.

Monoclonal antibody technology: The use of hybridomas that produce monoclonal antibodies for a variety of purposes. Hybridomas are maintained in cell culture or, on a larger scale, as tumors (ascites) in mice.

Mutagenesis: The induction of mutation in the genetic material of an organism; researchers may use physical or chemical means to cause mutations that improve the production of capabilities of organisms.

Mutagen: An agent that causes mutation.

Mutant: An organism with one or more DNA mutations, making its genetic function or structure different from that of a corresponding wild-type organism.

Mutation: A permanent change in a DNA sequence.

Myeloma: Antibody-producing tumor cells.

Myeloma cell line: Myeloma cells established in culture.

Neurotransmitters: Small molecules found at nerve junctions that transmit signals across those junctions.

New biotechnology firm (NBF): A company formed after 1976 whose sole function is research, development, and production using biotechnological means.

NIH Guidelines: Guidelines established by U.S. National Institutes of Health to regulate the safety of NIH-funded research involving recombinant DNA.

Nitrogen fixation: The conversion of atmospheric nitrogen gas to a chemically combined form, ammonia ($NH_3$) which is essential to growth. Only a limited number of micro-organisms can fix nitrogen.

Nodule: The anatomical part of a plant root in which nitrogen-fixing bacteria are maintained in a symbiotic relationship with the plant.

Nodulins: Proteins, possibly enzymes, present in nodules; function unknown.

Nucleic acids: Macromolecules composed of sequences of nucleotide bases. There are two kinds of nucleic acids: DNA, which contains the sugar deoxyribose, and RNA, which contains the sugar ribose.

Nucleotide base: A structural unit of nucleic acid. The bases present in DNA are adenine, cytosine, guanine, and thymine. In RNA, uracil substitutes for thymine.

Nucleus: A relatively large spherical body inside a cell that contains the chromosomes.

Oligonucleotides: Short segments of DNA or RNA.

Oligonucleotide directed mutagenesis: Directed mutagenesis by introducing a synthetic oligonucleotide into the isolated gene.

Organelle: A specialized part of a cell that conducts certain functions. Examples are nuclei, chloroplasts, and mitochrondria, which contain most of the genetic material, conduct photosynthesis, and provide energy, respectively.

Organic compounds: Molecules that contain carbon.

Pathogen: A desease-producing agent, usually restricted to a living agent such as a bacterium or virus.

Peptide: linear polymer of amino acids. A polymer of numerous amino acids is called a **polypeptide**. Polypeptides may be grouped by function, such as "neuroactive" polypeptides.

pH: A measure of the acidity or basicity of a solution on a scale of 0 (acidic) to 14 (basic). For example, lemon juice has a pH of 2.2 (acidic), water has a pH of 7.0 (neutral), and a solution of baking soda has a pH of 8.5 (basic).

Pharmceuticals: Products intended for use in humans, as well as in vitro applications to humans, including drugs, vaccines, diagnostics, and biological response modifiers.

Photosynthesis: The reaction carried out by plants where carbon dioxide from the atmosphere is fixed into sugars in the presence of sunlight; the transformation of solar energy into biological energy.

Plant Patent Act of 1930 (35 U.S.C ƒ5161-164): Confers exclusive license on developer of new and distinct asexually produced varieties other than tuber-propagated plants for 17 years.

Plant Variety Protection Act of 1970 (7 U.S.C. ƒ2321): Provides patent-like protection to new plants reproduced sexually.

Plasma: The liquid (noncellular) fraction of blood. In vertebrates, it contains may important proteins (e.g., fibrinogen, responsible for clotting).

Plasmid: An extrachromosomal, self-replicating, circular segment of DNA; plasmids (and some viruses) are used as "vectors" for cloning DNA in bacterial "host" cells.

Polymer: A linear or branched molecule of repeating subunits.

Polypeptide: A long peptide, which consists of amino acids.

Polysaccharide: A polymer of sugars.

Probe: See **DNA probe.**

Proinsulin: A precursor protein of insulin.

Prokaryote: A cell or organism lacking membrane-bound, structurally discreet nuclei and organelles. Prokaryotes include bacteria and the blue-green algae. (Compare **eukaryote.**)

Promoter: A DNA sequence in front of a gene that controls the initiation of "transcription" (see below).

Protease: Protein digesting enzyme.

Protein: A polypeptide consisting of amino acids. In their biologically active states, proteins function as catalysts in metalbolism and, to some extent, as structural elements of cells and tissues.

Protein engineering: Specific structural alteration of a protein by altering the gene for that protein.

Protoplast fusion: The joining of two cells in the laboratory to achieve desired results, such as increased viability of antibiotic-producing cells.

R&D limited partnership: A risk capital source and tax sheltered mechanism for funding the R&D of new products. It raises the potential rate of return to investors without adding extra cost to the corporation.

Recombinant DNA (rDNA): The hybrid DNA produced by joining pieces of DNA from different organisms together in vitro.

Recombinant DNA technology: The use of recombinant DNA for a specific purpose, such as the formation of a product or the study of a gene.

Recombination: Formation of a new association of genes or DNA sequences from different parental origins.

Regulatory sequence: A DNA sequence involved in regulating the expression of a gene.

Replication: The synthesis of new DNA from existing DNA and the formation of new cells by cell division.

Resistance gene: Gene that provides resistance to an environmental stress such as an antibiotic or other chemical compound

Restriction enzymes: Bacterial enzymes that cut DNA at specific DNA sequences.

RNA: Ribonucleic acid. (See also **messenger RNA.**)

Scale-up: The transition of a process from an experimental scale to an industrial scale.

Selection: A laboratory process by which cells or organisms are chosen for specific characteristics.

Semiconductor: A material such as silicon or germanium with electrical conductivities intermediate between good conductors such as copper wire and insulators such as glass.

Semiconductor device: An electronic device that uses a semiconductor to limit or direct the flow of electrons. Examples are transistors, diodes, and integrated circuits.

Semiconductor industry: Companies that manufacture semiconductor devices. As used in this report, the description of the semiconductor industry is that deriving from the period between 1947 (discovery of the transistor) to the early 1960's.

Small Business Investment Corporations (SBICs): Private companies licensed by the Samll Business Association (SBA) and owned by stockholders who have made investments in exchange for equity. SBICs are required by SBA to invest or loan money exclusively to U.S. small businesses.

Specialty chemicals: Chemicals, usually produced in small volumes, that sell for more than $1 per kg (50¢ per pound). (Compare commodity chemicals.)

Species: A taxonomic subdivision of a genus. A groupd of closely related, morphologically similar individuals which actually or potentially interbreed.

Startup financing: Financing usually supplied by venture capitalist to fund the early R&D, production, sale of a new company's products.

Steriod: A group of organic compounds, some of which act as hormones to stiumulate cell growth in higher animals and humans.

Strain: A group of organisms of the same species having distinctive characteristics but not usually considered a separate breed or variety. A geneticlaly homogenous population of organisms at a subspecies level that can be differentiated by a biochemical, pathogenic, or other taxonomic feature.

Substrate: A substance acted upon, for example, by an enzyme.

Technology Transfer: The movement of technical information and/or materials, used for producing a product or process, from one sector to another; most often refers to flow of information between public and private sectors or between countries.

Therapeutics: Pharmaceutical products used in the treatment of disease.

Thermophilic: Heat loving. Usually refers to micro-organisms that are capable of surviving at elevated temperatures; this capability may make them more compatible with industrial biotechnology schemes.

Totipotency: The capacity of a higher organism cell to differentiate into an entire organism. A totipotent cell contains all the genetic information necessary for complete development.

Toxin: A substance, produced in some cases by disease-causing micro-organisms, which is toxic to other living organisms.

Transcription: The synthesis of messenger RNA on a DNA template; the resulting RNA sequence is complementary to the DNA sequence. This is the first step in gene expression. (See also **translation.**)

Transformation: The introduction of new genetic information into a cell using naked DNA.

Transistor: An active component of an electrical circuit consisting of semiconductor material to which at least three electrical contacts are made so that it acts as an amplifier, detector, or switch.

Translation: The process in which the genetic code contained in the nucleotide base sequence of messenger RNA directs the synthesis of a specific order of amino acids to produce a protein. This is the second step in gene expression. (See also transcription.)

Transposable element: Segment of DNA which moves from one location to another among or within chromosomes in possibly a predetermined fashion, causing genetic change; may be useful as a vector for manipulating DNA.

Vaccine: A suspension of attenuated or killed bacteria or viruses, or portions thereof, injected to produce active immunity. (See also **subunit vaccine.**)

Vector: DNA molecule used to introduce foreign DNA into host cells. Vectors include plasmids, bacterio-phages (virus), and other forms of DNA. A vector must be capable of replicating autonomously and must have cloning sites for the introduction of foreign DNA.

Venture capital (venture capital funds): Money that is invested in companies with which a high level of risk is associated.

Wild-type: The most frequently encountered phenotype in natural breeding populations.

Yeast: A fungus of the family Saccharomycetacea that is used especially in the making of alcoholic liquors and as leavening in baking. Yeast are also commonly used in bioprocesses and in cloning.

# Appendix: Literature Support to Biotechnology Panel*

In addition to technical assessments, an important objective of the JTECH Program is the identification of relevant and timely Japanese S&T source material for the technical areas addressed by the panels. Thus, panel members are requested to evaluate, from their technical perspective, the usefulness of the various types of source materials provided. Panel members are also encouraged to give recommendations on Japanese source material in their technical areas which they feel should be acquired and dissiminated to the U.S. technical community on a continuing basis.

In this appendix, we summarize:

- the JTECH approach in providing literature/translation support to the panels, and a description of the general types of material provided,
- open-source Japanese literature available in the biotechnology area, and
- literature provided to the biotechnology panel by the JTECH staff.

A.1 Approach

The following approach has been followed in providing panel members with pertinent source material:

1. Initially, at the panel "kickoff" meeting, the JTECH staff provides general source material for panel members' review. This material is mainly in English and usually of an overview nature. Also presented and discussed is background information on the various types of Japanese publications, technical society and working group meetings, organizations, etc. which are

---

* This appendix written by the JTECH Program Staff; the staff member responsible for literature and translation support is Dr. Y. Kim.

potential sources of information for particular technical topics.

2. Each panel member then ascertains particular source material (including papers in Japanese) needed for their technical topic.

3. The JTECH staff then collects, evaluates, and translates those requests by the panel members.

(Steps 2 and 3 are iterative, with several interactions taking place during the panel assessments together with trips to Japan by Y. Kim to collect information.)

4. Toward the end of the panel life, the panel members, together with the JTECH staff, assess the source materials found to be most useful for the technical area. Also, panel members are encouraged to offer recommendations related to more effective methods of acquisition/translation/dissemination for the U.S. technical community.

Under the JTECH literature support approach, two factors are emphasized:

A. Timely, Pertinent Source Material. In addition to providing panelists with readily available "open" material (such as excerpts from Japanese technical journals and periodicals), the emphasis of the approach is in providing recent "semi-open" sources of Japanese research accomplishments and plans (such as results of Japanese ad hoc seminars and working groups). (These different categories of source material are discussed in the next section.)

B. Selectivity. To be effective under practical time and budget constraints, it is important that some selectivity be exercised in the source material provided. In providing the panelists with translated material, the following selective approach is used: (1) The titles of papers from a Japanese document are translated and distributed to panelists; (2) for seemingly interesting titles, a panelist requests that certain abstracts be translated (this step is usually done by telephone); and, (3) based on the abstracts, the full translation of the most relevant papers is performed. (In applying this procedure for the Biotechnology Panel, only about 10% of the total Japanese source material collected required full translation, as indicated later in Section A.5.)

## A.2 Categories of Source Material

In the JTECH approach for providing literature support to panelists, we consider the Japanese S&T source material in terms of three categories. These categories are defined below, and the material provided to the biotechnology panel in each category is listed in the next section.

### A. General Material (mostly in English)

These are materials initially provided to assist panel members in identifying specific source material needed for their particular area of expertise. This material is typically shallow in technical depth, but provides a general program background. Included are popular overview articles, previous technology assessment studies, and translations of Japanese government publications on overall programs, research objectives, participating organizations, funding levels, etc.

### B. Open Material

This category consists of "openly available," regularly published technical publications. A list of the technical journals pertinent to biotechnology is given in the next Section.

### C. Semi-Open or Closely-Held Material

The emphasis under the JTECH program is in providing panelists with this category of source material (described below), which is of a more timely benefit, as illustrated in Figure A-1.

Japanese academic societies hold biannual or annual meetings, and participants are provided with the abstracts of the meeting. Many manuscripts of these abstracts are hand-written; they are not reproduced in regular journal publications. These abstracts contain current and useful information, but they have to be accessed through privately-held material. In this sense, the meeting abstracts should be regarded as semi-open.

In the Japanese high tech community, frequently ad hoc seminars and conferences are organized to pool and disseminate the state-of-the-art research information. Such meetings are attended usually by the invitees only, and the conference proceedings are printed in Japanese with most manuscripts being handwritten to allow entry of the latest information. Some of the more formal technical information contained in these proceedings eventually work their way through the appropriate professional journals published in English, but with a typical delay of six to eighteen months.

Japanese industries participate in the national R&D projects under a "Research Association" type arrangement. Technical progress made in these projects is disseminated first in the technical committees composed of member companies, and refined (edited) technical papers reach public domain at a much later time. In highly competitive R&D, timely access to raw data is important. (The ongoing national projects in the biotechnology area are: Senescence Control Index, Bioreactor, Artificial Organs, Intelligent Machines, Biologically Active Substances, & New Microorganism Utilizing Technology.)

Most Japanese high-technology industries maintain in-house R&D centers that are staffed with leading technical expertise. Many companies publish periodical reports that contain some useful technical information. Of course, commercially sensitive technical information is closely-held as proprietary.

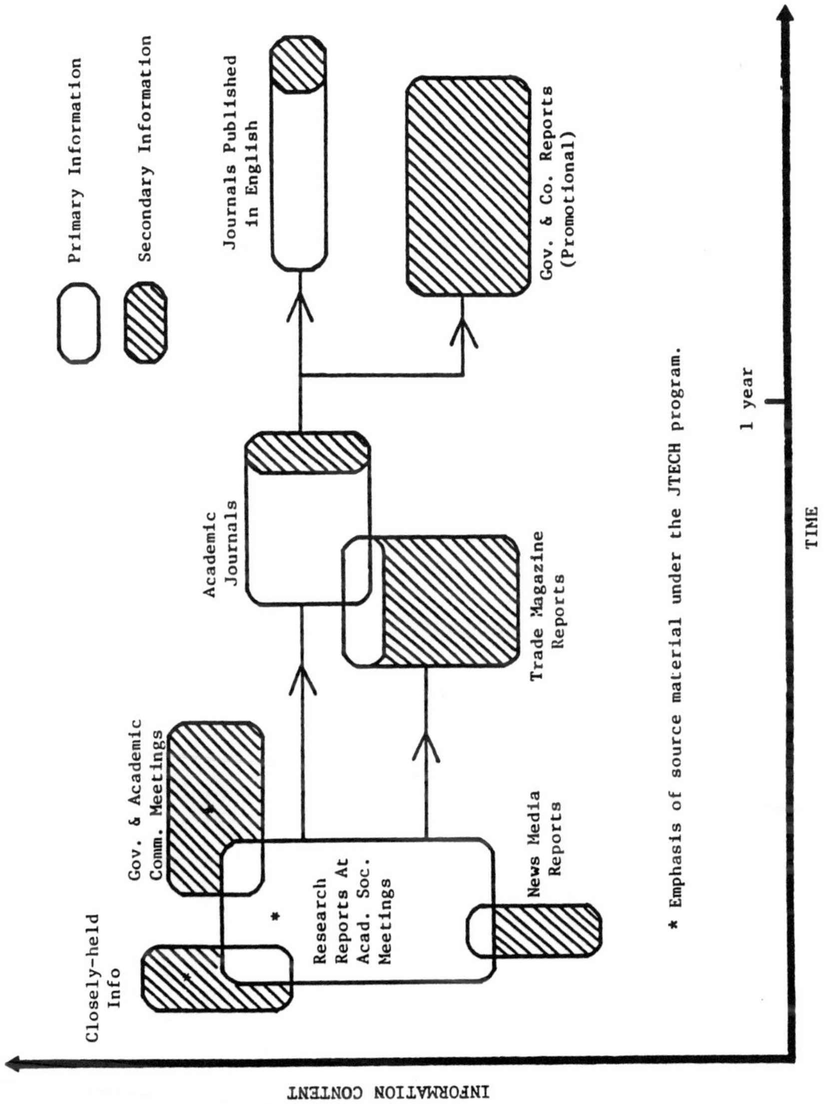


Figure A-1. Flow of Technical Information from Japan.

## A.3 Japanese Literature in Biotechnology

### A. Reference Sources (General material, mostly in English)

In Japan, the governmental promotion of life science began in 1971 at the recommendation of the Council for Science and Technology (CST), the nation's highest science and technology policy-making body. However, the national focus on modern biotechnology based on recombinant NA began to emerge only in the early 1980's. The biotechnology boom in the United States was already underway with over one hundred start-up ventures. As has been the case with many high technologies, Japan is now in a serious "catch-up" mode in biotechnology. The Japanese activities in biotechnology during the past decade and plans in the coming year can be found in many sources. However, we find the following references to be most advantageous:

(1) Trends in Emerging Technologies: Japan's Industrial Technology Innovation Toward the 21st Century.

This is a 900 page book published in October 1983 under the auspices of MITI and STA. While it covers many different "next generation industrial technologies," over 100 pages in the third chapter are devoted to biotechnology. Governmental policies on and implementation of rDNA related R&D are described in detail, including the activities of the related biotechnologies.

(2) Genetic Engineering Industry Handbook, 1981-1984 (Idenshi Sangyo Nenkan, 1983-1984) is published by Science Forum, a venture publishing company located in Tokyo. This book contains a wide variety of comprehensive information on Japanese activities in biotechnology.

(3) Biotechnology in Japan: The Reference Source. Japan Pacific Associates, located in Palo Alto, CA., translated and edited the handbook described in (B).

Of the general source material presented at the initial panel meeting (#1 - #10), #2—Directory of Japanese Researchers in Biotechnology, is taken from reference (C) above.

B. Japanese Technical Journals and Periodicals

Journals pertinent to Biotechnology are:

1) Tanpakushitsu Kakusan Koso, (Protein, Nucleid Acid, Enzymes).

2) Kagaku to Seibutsu, (Chemistry and Biology).

3) Nihon Nogei Kagaku Kai Shi, (Japan Agricultural Chemistry Association Magazine).

4) Nihon Saikin Gakukai Shi, (Japan Bacteriology Association Magazine).

5) Hakko Kogaku Kai Shi, (Fermentation Industry Association Magazine).

6) Hakko to Kogyo, (Fermentation and Industry).

7) Seikagaku, (Biochemistry).

8) Seibutsu Kagaku, (Biological Sciences).

9) Seibutsu Kagaku News, (Biochemistry News).

10) Gendai Kagaku, (Modern Chemistry).

11) Soshiki Baiyo, (Tissue Culture).

12) Iden Gaku Zatsu Shi, (Japanese Journal of Genetics).

5. Japanese Science & Technology Literature Available in the United States

The Library of Congress has an impressive holding of Japanese S&T literature, including almost all of the open-source material mentioned above. Most periodicals appear to arrive with a four to six month delay. The cataloging of Japanese S&T literature is done mostly in terms of romanization of Japanese titles. Therefore, it is rather difficult to make use of these materials unless the researcher is familiar with the Japanese language.

A.4 Literature Provided to JTECH Biotechnology Panel

A. General


1. James C. Abegglen and Akio Etori, "Japanese Technology Today," Scientific American, 1984.

2. A. Alun, "Japanese Biotechnology: Search Begins for Superbugs," Nature, Volume 310, July 19, 1984.

3. Commercial Biotechnology: An International Analysis. OTA Report, Office of Technology Assessment, January 1984.

4. Directory of Japanese Researchers in Biotechnology, pp. 281-302.

5. "Science in Japan," Nature: International Weekly Journal of Science. Volume 305, Number 5933, September 29-October 5, 1983.

B. Open


1. M. Ikehara, "New Manufacturing Process for Human Growth Hormone by Means of Gene Manipulation," Technocrat, Volume 16, number 11. November 1983. pp. 22-26.

2. M. Ohmasa, "Possibility of Cell Fusion," Technocrat. Volume 17, number 4. April 1984. pp. 24-26.

3. M. Okanishi, "Development of Host-Vector Systems in Actinomycetes," Technocrat, Volume 17, number 3. March 1984. pp. 10-24.

4. "Biotechnology in Japan News Service," Japan Pacific Associates.

5. "Development of Bioreactor for Blood Analysis," Technocrat. Volume 16, number 6. June 1983. pp. 27-29.

6. "Development of Prefabricated Gels for DNA Sequencing, Computer Software for Structure Analysis of DNA Sequences," Science & Technology in Japan, July-September 1984. p. 31.

7. "Discovery of Externally Secretory Gene," Technocrat, Volume 17, number 2. February 1984. p. 43.

8. "Japan Bioindustry Letter," BIDEC (Bio-industry Development Center). Volume 1, Number 3. June 1984.

9. "Life Science Research at Physical and Chemical Research Institute," Science News. n.d.

10. "Mitsui Pharmaceuticals Commencement of Research on Singular High HCG Monoclonal Antibodies (MCA), RIA & Screen Diagnosis," Nikkei Biotechnology, May 21, 1984. Nikkei McGraw Hill Co.

11. "Nihon University & Fuji Rebio in Joint Development of NonA. Non B Hepatitis Monoclonal Antibodies (MCA) Diagnostic Kit," Nikkei Biotechnology, May 21, 1984. Nikkei McGraw Hill Co.

12. Report on Inquiry No. 10 "On Basic Plan for R&D of Pioneering and Basic Technologies in Life Sciences," Science & Technology in Japan, July-September 1984. pp. 8-10.

13. "Research & Development of Life Sciences in Japan," Science and Technology in Japan, Japan Ministry Agency for Science & Technology. April/June 1983. pp. 8-16, 40-41.

14. Sample pages of selected Japanese biotechnology periodicals

a) Japanese Journal of Genetics
b) Chemistry & Biology
c) Protein, Nucleic Acid, & Enzyme
d) Modern Chemistry
e) Nippon Nogeikagaku Kaishi (J. of Agricultural Chemical Society of Japan)
f) Biological Science (Tokyo)
g) Hakkokogaku Kaishi (Society of Fermentation Technology)
h) Journal of Biochemistry

15. "Science & Technology Forum '84," Science & Technology in Japan, July-September 1984. pp. 20-21.

16. "Setup for Promoting the Development of Life Science," Science & Technology in Japan, July-September 1984. pp. 31-32.

17. "Tokyo University Applied Microbiology Symposium: Discovery of New Physiological Active Substance From Genetic Research on Polyproteins," Nikkei Biotechnology, May 21, 1984. Nikkei McGraw Hill Co.

18. "Trend of Chemical Sensors in Japan." Technocrat, Volume 17, number 1. January 1984. pp. 10-19.

C. Semi-Open or Closely-Held

1. Herman W. Lewis, "Biotechnology in Japan," National Science Foundation. June 18, 1984.

2. Gary Saxonhouse, "Biotechnology in Japan." n.p., n.d.

3. Formation Mechanism of Multi-Cell Systems, Special Research Project, Funded by the Ministry of Education, Japan. Progress Reports for FY 1983.

4. Japan Agricultural Chemical Society, 1984 Annual Meeting. April 1-4, 1984. Translations of session titles

5. Japan Biochemical Society, 1983 Annual Meeting, September 29-October 1983. Translations of session titles

6. Japan Genetic Society, 55th Annual Conference. October 8-10, 1983. Translations of session titles

7. Japan Molecular Biology Society, 6th Annual Convention, August 22 - 25, 1983, Session Schedule.

8. Recombinant DNA as Manifested in Physiological Mechanisms, sponsored by the Ministry of Education in Japan. March 1984. Translations of abstracts

9. "Study of Federal Biotechnology Issues,"--Report to Division of Policy Research and Analysis National Science Foundation, prepared by Arthur D. Little, Inc. July 12, 1984.